Explanatory Particularism in Scientific Practice

Explanatory Particularism in Scientific Practice

MELINDA BONNIE FAGAN

OXFORD
UNIVERSITY PRESS

OXFORD
UNIVERSITY PRESS

Great Clarendon Street, Oxford, OX2 6DP,
United Kingdom

Oxford University Press is a department of the University of Oxford.
It furthers the University's objective of excellence in research, scholarship,
and education by publishing worldwide. Oxford is a registered trade mark of
Oxford University Press in the UK and in certain other countries

Published in the United States of America by Oxford University Press
198 Madison Avenue, New York, NY 10016, United States of America

British Library Cataloguing in Publication Data
Data available

Library of Congress Control Number: 2024950516

ISBN 9780198897453

DOI: 10.1093/9780191998874.001.0001

Printed and bound by
CPI Group (UK) Ltd, Croydon, CR0 4YY

FSC www.fsc.org
MIX
Paper | Supporting responsible forestry
FSC® C013604

The manufacturer's authorised representative in the EU for product safety is
Oxford University Press España S.A. of El Parque Empresarial San Fernando de Henares, Avenida de Castilla, 2 – 28830 Madrid (www.oup.es/en or
product.safety@oup.com). OUP España S.A. also acts as importer into Spain
of products made by the manufacturer.

Acknowledgments

Ideas in this book have developed over more than a decade, and influences are too numerous to record here. With that caveat, this book has benefited greatly from comments and criticism from Hanne Andersen, William Bechtel, Erin Beeghly, Dick Burian, Daniel Burnston, Julia Bursten, Jordi Cat, Sara Chant, Carl Craver, Adrian Currie, Haixin Dang, Marina DiMarco, Steve Downes, Elihu Gerson, Tom Gieryn, Richard Grandy, Sara Green, Matt Haber, Joyce Havstad, Frank Hindriks, Oleg Igoshin, Phyllis Illari, Johannes Jaeger, Katie Kendig, Kareem Khalifa, Soazig Le Bihan, Jim Lennox, Elisabeth Lloyd, Helen Longino, Miles MacLeod, Michela Massimi, Elijah Millgram, Roberta Millstein, Monica Morrison, Nancy Nersessian, Thi Nguyen, Michael O'Rourke, Anne Peterson, Katie Plaisance, Angela Potochnik, Aleta Quinn, Collin Rice, Sarah Richardson, Federica Russo, Fred Schmitt, Miriam Solomon, Jim Tabery, Miriam Thalos, Rose Trappes, Mercedes Marie Ward, and Andrea Woody. Many thanks to all; all errors are of course my own. Chapter 7 owes much to discussions with scientists Robert Fujinami, Hans Haecker, Vicente Planelles, Zac Stephens, Dean Tantin, and Arabella Young. Many thanks to them for their time and insights. Of special value were my experiences teaching material related to this book to students at the University of Utah. My deepest thanks to students in my Fall 2019 Philosophy of Science Seminar and Spring 2023 Undergraduate Senior Seminar.

Early versions of my particularist account were presented to various audiences, who provided valuable feedback, both positive and negative. My thanks to Fall 2013 Fellows of the Pittsburgh Center for Philosophy of Science, colloquium audiences in the Philosophy Departments of the University of Utah, University of California San Diego, University of Colorado Boulder, University of Washington Seattle, History and Philosophy of Science Departments at the University of Indiana Bloomington and the University of Pittsburgh. Thanks also to workshop participants at the Centre of Excellence in the Philosophy of the Social Sciences (University of Helsinki), Copenhagen University, Michigan State University's Center for Interdisciplinarity, Wissenschaftskolleg zu Berlin, Colorado State University, the University of Idaho Philosophy Club, and Stanford University, and audiences at sessions of the 2013 Eastern APA (Winter 2013), Philosophy of Science Association meetings (2014, 2018, and 2022), Intermountain West Graduate Student Conference (2015), Society for the Philosophy of Science in Practice (2015 and 2016), European Philosophy of Science Association meeting (2015 and 2023), History of Philosophy of Science meeting (2016), Integrated History and Philosophy of Science (&HPS6), the 8th Pittsburgh Fellows Quadrennial

Conference (2016), Intermountain Philosophy Conference (2019), International Conference on Engaging Ethics and Epistemology in Science (2022), and the American Association for the Advancement of Science (Pacific Division 2023).

Funding for this project was provided by a Faculty Innovation Fellowship from the Rice University Division of Humanities, a Scholar's Award from the National Science Foundation (Award No. 1354515), the Sterling M. McMurrin Endowment Fund, and multiple research awards from the University of Utah College of Humanities.

Contents

Figures and Tables

Figures

Tables

The tendency of the sciences has long been an increasing proclivity of separation and dismemberment. . . . The mathematician turns away from the chemist; the chemist from the naturalist; the mathematician, left to himself, divides himself into a pure mathematician and a mixed mathematician, who soon part company; the chemist is perhaps a chemist of electro-chemistry; if so, he leaves common chemical analysis to others; between the mathematician and the chemist is to be interpolated a 'physicien' (we have no English name for him), who studies heat, moisture, and the like. And thus science, even mere physical science, loses all traces of unity.

William Whewell, 1834

Introduction

This is a book about scientific explanation. There have been many such, as explanation is a major topic in philosophy of science. In that field, one proposing a new theory of explanation generally first clears the ground of alternatives. This book does not follow that format. Other philosophical theories of explanation (causal, mechanistic, topological, counterfactual, mathematical, eikonic, minimal model, and more) aren't conflicting alternatives to the view proposed and developed in subsequent chapters. Rather than clearing the theoretical ground, I aim to chart a new path through it. This new path is marked by three guiding ideas: explanation as collaborative activity, diversity as an epistemic resource, and engaging scientific practice. This trio of commitments, developed and articulated in the following chapters, repurposes rather than refutes other philosophical accounts of explanation. After introducing these themes, the rest of this chapter previews my overall argument and main thesis, as well as some key results.

Three themes

The three guiding themes together define this book's starting-point. All three are grounded on scientific practice, and (provisionally) justified on that basis. They are basic commitments of my particularist approach to studying scientific explanation. Although I do not defend these commitments explicitly, my hope is that they are vindicated by results. In this sense, the book is one long argument for my particularist approach to studying scientific explanation. I now set out these three guiding ideas. First: explanation is a form of *collaborative activity*; a special case of social action. This motivates my use of social action theory (Chapter 3). But the theme of collaboration runs deeper. One goal of this book—the most modest—is to explore the idea that explanation is collaborative; to study scientific explanation through that lens. Explanation so conceived is *pro-social*. Features of inquiry that encourage collaboration are conducive to explanation; features necessary for collaborative activity are necessary for explanation. This opens up a normative dimension of explanation heretofore unexamined by philosophers of science. The idea also has descriptive implications. It's a commonplace that science is a social activity. But few works in philosophy of science make this aspect central. "Science," on my view, is a sprawling, dynamic array of lines of research programs, clustered into groups of various grains, scales, and demographic sizes.

Explanatory Particularism in Scientific Practice. Melinda Bonnie Fagan, Oxford University Press.
 DOI: 10.1093/9780191998874.003.0001

These groups include epistemic communities, each aiming at explanation (among other goals).[1] Constructing an explanation, I propose, is a collaborative social activity grounded in a particular social context. Explanations bear the stamps of their contexts, each emerging as an epistemic achievement against a particular background. This background often includes deeply-held commitments, central to a scientific community's practice and group identity. To put the point another way: a major part of becoming a scientist is taking onboard a certain view of explanation. Different scientific communities (e.g., welfare economics, developmental psychology, tumor immunology) have very different views about what a good explanation looks like and how it should be constructed.[2]

This points us to the second theme: *diversity as epistemic resource*. In social epistemology, the diversity at issue is generally of individuals: diverse individuals produce better scientific knowledge.[3] In this book, the theme plays somewhat differently. I consider diversity primarily as an attribute of scientific communities. Individual epistemic agents figure as participants in collaborative activities. The aspects of diversity that matter, for my purposes, are features of scientific communities that contribute to producing an explanation. But these cover a lot; methodological features, broadly conceived (see Chapter 1). Scientific communities vary widely with respect to these features, and so explanatory products are also very diverse. A second aim of this book is to show how this form of diversity can be an epistemic resource; to show epistemic benefits of diverse scientific explanations.[4] Commitment to 'understanding through difference' also shapes my overall argument. Diverse theories and viewpoints can be an epistemic resource in philosophy, too. The various philosophical accounts of explanation already on offer are, on my approach, resources for engaging collaborative explanation in scientific practice. One needn't obliterate the alternatives before proposing a new view. Instead, a new approach throws new light on earlier theories. In this book, the hoped-for new light concerns how philosophical theories engage scientific practice.

This is my third theme: *engaging scientific practice*. Commitment to this idea is widely-endorsed in philosophy of science today. But studies of scientific explanation endorse it rather selectively (see Chapter 1). The dialectical goal of this book is to provide a more inclusive, open-ended way for philosophers to study explanation

[1] Scientific groups are complexly organized; partly-overlapping and cross-cutting rather than a tidy hierarchy of inclusion. Chapter 1 presents my concept of an epistemic community in more detail.

[2] Sometimes these matters are debated within a single epistemic community—such debates are often intense, and can lead to a schism; two scientific communities where before there was one.

[3] Different theories in social epistemology offer different interpretation of what "better" consists in, the knowledge-production process, which aspects or modes of diversity are epistemically significant, and limits to beneficial diversity (if any).

[4] Throughout this work, I'll sometimes refer to explanation as an activity, sometimes as a product of that activity. It should be clear in context which sense is meant. Because I see this process–product relation as close and non-dichotomous, this equivocation should be innocuous.

in scientific practice. To do so, I take diverse explanatory practices in scientific communities as the starting-point. Philosophical theories of explanation come into play *after* empirical study of one or more particular scientific communities. Their role is heuristic; philosophical theories are resources for characterizing particular communities' explanatory practices and products in context. Taking this approach requires giving up *explanatory monism*: the view that there is one substantive philosophical account of *all* scientific explanation, which it is philosophers' task to discover and articulate. Rejecting monist aspirations doesn't entail rejecting philosophical theories framed by that assumption. It does repurpose those philosophical theories—into a diverse set of resources for studying scientific explanations in practice. In this role, those theories directly engage scientific practice, potentially playing impactful roles in significant scientific challenges and debates today (see Part III).

The name "explanatory particularism" highlights rejection of explanatory monism. Of course, "explanatory pluralism" is a more obvious moniker. But that name is already taken. In philosophy of science today, "explanatory pluralism" refers to theories that admit two or three distinct kinds of scientific explanation, instead of just one. Explanatory particularism is a form of pluralism about explanation, but is more radical than other views going by that name. It's more radical in several respects: requiring deeper engagement with scientific practice; departing further from assumptions about explanation in widely-held in philosophy of science today; and using diverse philosophical resources to analyze explanation in scientific practice. So a different moniker is appropriate—hence, *explanatory particularism.*

Argument preview

A different path need not wipe out others. The main goal of this book is to present explanatory particularism and its implications—a positive project. This overall argument reflexively enacts the guiding idea of diversity as a resource, not a problem to be overcome. The argument unfolds in three parts, which organize the book. Part I introduces the main ideas of explanatory particularism and develops them in accordance with the three guiding themes. The result is a philosophical framework that adapts insights about models and modeling (from philosophy of science) and collaborative activity (from philosophy of social action). Part II applies this framework to an important class of cases: interdisciplinary research (IDR) contexts. This yields further results, including a non-exhaustive set of norms for combining diverse scientific explanations. This collaborative aspect of (some) explanations articulates pro-social ideals for understanding in scientific practice. Part III applies these results to two detailed case studies, showing how pro-social norms for combining different explanatory models can yield insights for scientists and philosophers. The concluding chapter indicates further cases for particularist study.

Why focus on interdisciplinarity? This aspect of scientific practice hasn't been studied in terms of explanation (at least, not by philosophers). But that approach yields some valuable insights. Many IDR projects involve combining explanatory resources from different scientific communities. Notoriously, these efforts often fail. Different views of explanation are often implicated in these failures. Explanatory particularism clarifies this obstacle to interdisciplinarity. Briefly: for any particular scientific community, explanations are significant products. Members of those communities tend to be very invested in the values, methods, and concepts associated with explanation in their field. The flipside of this is that researchers committed to one community's view of explanation often find the explanations of other communities wanting. The latter don't look like 'real explanations,' but 'mere descriptions' or suchlike. It's common for scientists to draw lines between 'us and them' in a way that asserts 'our' explanatory prowess over 'theirs.' Philosophy of science has tended to exacerbate this tendency by valorizing particular styles of explanation over others. Explanatory particularism offers a new approach, showing how and under what circumstances this explanatory obstacle to interdisciplinarity can be overcome.

I'll briefly preview some other main results. One is a general framework for analyzing IDR.[5] This reveals a new category, which I term "low-effort interdisciplinarity." As the name suggests, this mode of research is less demanding than convergence or transdisciplinary research. Another general framework defines a taxonomy of relations between distinct explanatory models. Combining the two yields minimal conditions for connecting explanatory models constructed in different communities. The result of this special case of collaborative social activity is an interdisciplinary explanation (IDE). Further analysis of this aspect of explanation clarifies epistemic benefits of IDE, which dovetail with social benefits. On the more practical side, conditions for IDE are further operationalized into concrete guidelines for low-effort interdisciplinarity. These guidelines can be used to facilitate IDE projects, as well as indicating when such efforts are likely to fail. Further results concern understanding. I show how particularist explanations can provide understanding beyond the boundaries of particular scientific communities, distinguish several related aspects of understanding advanced by IDE, and, building on these results, propose a new, pro-social ideal of unity of science. On this view, science in practice produces a patchwork of explanations, advancing understanding via a profusion of variously-connected models. Through a delicate web of specific, context-dependent comparisons and contrasts, a form

[5] General claims are incompatible with strict particularism. But my version of particularism is not thoroughgoing. Part I adapts other philosophical resources to articulate a general framework that applies to particular cases of scientific explanation.

of scientific objectivity emerges that particularists can endorse. The resultant image of science is pro-social and collaborative, flattening traditional epistemic hierarchies and affording many different possible relations between explanations within and across scientific communities. In this way, explanatory particularism invites connections between scientific and other, non-scientific communities. Science so conceived is not demarcated from other epistemic efforts, but woven into the expansive tapestry of social life.

Plan of the book

The book's argument builds up to this image of science in stages. Chapter 1 introduces the basic ideas of explanatory particularism, then positions the view against its dialectical background of philosophical studies of scientific explanation. The latter comprise a tradition going back at least seven decades. A full survey of this tradition would take multiple volumes; Chapter 1 does not attempt it. Its more modest goal is to show the enduring influence of explanatory monism on philosophical studies of scientific explanation. The monist approach is then criticized, motivating explanatory particularism. This does not refute other philosophical theories of explanation currently on offer, but repurposes them. The next stages of my argument are more positive. Chapters 2 and 3 develop the basic ideas of explanatory particularism using philosophical resources not directly connected to studies of explanation. These adaptations of other areas of philosophy are reflexively consistent with the book's guiding themes, and with collaborative values articulated in later chapters. There are other ways one could develop the basic ideas of explanatory particularism; the framework articulated in Part I is not the only possible one.

The resource tapped in Chapter 2 is another area of philosophy of science: studies of scientific models and modeling. The main thesis of this chapter is that particularist explanations are a kind of model. The first part of the chapter clarifies the meaning of that thesis, building on the work of Giere, Morgan, Morrison, and other philosophers who examine models in scientific practice. Adapting and combining their insights, I argue that particularist explanations should be conceptualized as models, in a loose, inclusive sense of the term. As with explanation, to endorse this meaning of "model" is not to refute all other philosophical accounts of models and modeling. Because my goal is not to offer a theory of scientific models and modeling, I remain noncommittal about most debates in that area. Regarding models' functions/uses/roles in science generally, I claim only that explanation is one of the uses of models in science—and what that amounts to varies across scientific communities. The next part of Chapter 2 illustrates that thesis about explanatory models with an example from scientific practice. This case study (embryo models in stem cell research) shows how intricate and complex model–model relations can be, even within a single scientific community. The final part of the chapter proposes a general framework for analyzing model–model relations in terms of stage of construction, degree of connection, and type of relation (direct/indirect, similar/different). Later chapters put this framework to use.

Chapter 3 uses resources from another area of philosophy: social action theory. The latter is a distinct sub-area, featuring ideas and methods of argument unlike those of philosophy of science today. Its theories are suitable for my purpose here, however, because a guiding theme of the book is that explanation is (a product of) a form of collaborative activity. I adapt philosophical resources from social action theory, reflexively instantiating the kind of connection between different epistemic communities that comes to the fore in Part II. I first briefly survey the philosophical literature on social action and collective intention, to arrive at a thin consensus account of the latter. Next, I further articulate that consensus account, applying it to the case of epistemic agents working in scientific communities to collaboratively explain some target of interest. I illustrate the core ideas with another extended example—this one in the style of social action theory; i.e., first-person reflection on the character of one's own or another's everyday experience. (This is a form of case study not often employed in philosophy of science, and for that reason might be jarring to some readers. One philosopher's anecdote can be another's main evidence base! This anticipates some key results of Part II.) My general social action framework, applied to the case of scientific explanation, yields a set of norms for constructing such an explanation: non-exclusivity, incompleteness (humility), publicity, and respect for others as full and equal partners in explanatory activity. These norms constrain the attitudes of participants constructing an explanatory model. In this way, social action theory offers a normative framework for analyzing particularist explanation. The centerpiece of this framework is a three-layered interactive structure of collective intention, which in turn implies norms for collaborative activity. These constraints are grounded in a practical stability requirement for collective intention—and, for explanation in particular, on shared understanding as a final end.

Modeling and social action frameworks are integrated at the end of Part I, completing my general particularist approach to studying explanation. However, full articulation of this particularist account is beyond the abilities of any single scholar. The full, substantive theory requires broad, inclusive study of a wide variety of particular explanations in scientific practice, past and present. There is already a rich literature of case studies from history and philosophy of science (and allied fields) that goes some way toward that end. But of course there is more to be done. Parts II and III of this book make what I hope are fruitful efforts in that larger, community-scale project. Part II applies Part I's framework to an important class of cases: interdisciplinary (IDR) contexts. In this application, Chapter 4 is largely preparatory. This chapter reviews the scholarly literature on IDR (broadly and as pertaining to explanation) and positions my explanatory particularist framework alongside that scholarly background. That survey and comparison yield several insights. First, the core ideas of explanatory particularism comport well with IDR scholarship. Second, the general taxonomy of model–model relations distinguishes several categories of IDR. One of these is novel: relatively "low-effort" interdisciplinarity. Finally, as previewed above, diverse views of explanation across the sciences pose a significant obstacle to IDR. This doubles as a challenge for explanatory particularism: How is interdisciplinary explanation (IDE) possible? How can particularist explanations get past the boundaries of their specialized communities of origin? Chapter 4 begins to answer that challenge; the next two chapters complete the task.

Chapter 5 applies the social action framework to the activity of connecting ("bridging") disparate explanatory models constructed in different scientific communities. For this collaborative activity to succeed, connecting efforts from each specialization must cohere with one another (in a sense to be explicated). This implies a set of minimal requirements for connecting two or more disparate explanatory models ("bridging models")—constraints on the bridge itself, as well as on participants' attitudes. Importantly, these constraints are very weak; e.g., no element of an IDE need conform to standards of all involved specializations at once—not even the bridge. There's no general requirement for an overall IDE to function univocally. Connections between explanatory models are as diverse, or more so, as what they connect. This type of IDR is, accordingly, "low-effort" relative to more resource-intensive transdisciplinary IDE efforts. Again, an extended example illustrates: the array of life science models concerning hemoglobin—a more complex, realistic IDE project than simple pairwise cases. The last section of the chapter begins to articulate the epistemic and social significance of bridging models.

Chapter 6 continues that task, exploring implications of the particularist account for understanding. One key result is a mode of understanding characteristic of IDE projects, which I term "relational understanding." This mode or aspect of understanding, I argue, extends to other scientific contexts as well. However, relational understanding is not a general, monist theory of scientific understanding—this result doesn't supersede my overall particularist account. Rather, it is its culmination. Social, relational understanding is an aspect of scientific understanding revealed by applying my particularist framework (Part I) to the question of interdisciplinary explanation (IDE) (Part II); a new theoretical approach applied to a heretofore neglected class of cases (at least as far as philosophical studies of explanation are concerned). Making IDR, its challenges, and aspects of understanding implicated in meeting those challenges, central to philosophical study of explanation, is itself an important achievement of explanatory particularism. The preview above lists some other key results: ideals of unity of science, objectivity, a pro-social image of science and its epistemic authority, and more. These emerge at the culmination of my particularist account, the conclusion of Part II.

Part III, the final part of the book, consists of case studies. Two detailed examples apply the IDE framework to one successful (Chapter 7) and one unsuccessful (Chapter 8) interdisciplinary context. Each illustrates key ideas from earlier chapters, as well as affording further insights about those particular cases. Chapter 7 examines efficient cooperation between immunology and epidemiology, deep differences in the two fields' modeling and explanatory practices notwithstanding. Chapter 8 diagnoses one failed theoretical agenda in systems biology. These detailed case studies also showcase the role of other philosophical theories of explanation, repurposed upon relaxing monist aspirations (see Chapter 1). As noted, those other theories aren't conflicting alternatives to explanatory particularism. Instead, they play a vital heuristic role in studying explanations in scientific practice—and, for unsuccessful IDR projects especially, valuable tools for progress. Chapter 9 concludes by summarizing the main results and identifying a compendium of cases for future work.

A final methodological note. As we all work within some disciplinary perspective, a key challenge throughout this book is how to examine science (and philosophy) while remaining reflexively consistent—that is, without imposing one field's technical flourishes on another. Building up and adapting philosophical resources in Part I is my response to that challenge. Accordingly, the book's argument and organization are unlike traditional philosophical studies of scientific explanation. The aim of this book is not conversion, but understanding. I'm not trying to convert readers to explanatory particularism. I don't expect scholars committed to one philosophical tradition to suddenly take up this one instead. I do hope to show such readers what it would be like to do so—to see a bit into the other side. That outcome exhibits an aspect of understanding across scientific fields, or between scientific fields and other sectors of society, that results from applying my account to those cases. This book is written for and out of the epistemic context of philosophy of science today, responding both to that field's traditions and to the state of science today, conceived in terms of loosely connected, patchy, dynamic epistemic communities.

PART I
INTRODUCING EXPLANATORY PARTICULARISM

1
Explanatory Particularism: Basics and Background

Scientific explanation has long been a core topic in philosophy of science. Many different theories have been proposed, spurring debates about laws of nature, causality, relations between mathematics and the world, and other issues. This book offers a new philosophical approach to studying scientific explanation, which opens up new lines of inquiry and debate. The basic idea is to take scientific practice as primary—including social organization and associated diversity of explanatory styles across the sciences. All parties to philosophical debates about explanation recognize that explanations differ across the sciences. But this diversity is often treated as an obstacle to be overcome; observed complexity that it is a philosopher's task to clarify into one or a few explanatory forms. What if we take explanatory diversity across the sciences at face value? Or, further, as a positive epistemic resource? That's the approach taken in this book; what I term *explanatory particularism.*

As a theory of explanation, explanatory particularism does not lend itself to economical expression. The full account, indeed, is beyond the scope of any monograph. The goal of this book is to sketch the overall contours of such an account—a full particularist account of explanation across all scientific communities—and indicate how this sketch could be filled out by a community of philosophical researchers. This chapter begins that task, introducing the basic ideas and background commitments of explanatory particularism (Section 1.1), showing the contrast with established philosophical tradition (Section 1.2), and motivating the particularist approach relative to that tradition (Section 1.3). The last part of the chapter sets the stage for the rest of Part I (Chapters 2 and 3).

1.1 The very idea

The term "particularism" applied to explanation suggests that a single, general, substantive account of explanation is not to be had. The connotation holds true: I propose a single general schema for studying explanation, substantively filled out in myriad different ways. Why not call this "pluralism"? Pluralists about explanation also reject the idea of a single, general, substantive account of explanation. But explanatory particularism goes further than current versions of explanatory

Explanatory Particularism in Scientific Practice. Melinda Bonnie Fagan, Oxford University Press.
 DOI: 10.1093/9780191998874.003.0002

pluralism.[1] Explanatory pluralists today argue for two or three distinct types of explanation, or a tractable set of explanation-defining properties that variously cluster. My proposal is different. Explanatory particularism conceptualizes explanations as products of particular contexts—of epistemic communities. More precisely:

> (EP) A scientific explanation is a product of a particular scientific community that is significant, specialized, and advances understanding for that community.

EP is the basic thesis of explanatory particularism—general though not very substantive. There is some content, however. EP's central concepts and associated ideas form the background for all the chapters to follow. So the first task is to make this background clear. I'll work through each of the main ideas in EP.

Particular scientific community: For explanatory particularism, the social organization of science is fundamental. A particular scientific community is a social group; a sociological entity. Science's social organization is very complex, involving groups at a variety of scales or grains:[2] research teams, laboratories, larger collaborations, multi-institution assemblages, sub-fields, traditional disciplines, and more. At the coarsest grain, we can conceive the totality of scientific inquiry (at a given time) as 'the scientific community.' At the other extreme are small groups of individual scientists: research partners and teams. All scientific groups are of course composed of individual scientists. The latter often work in small research teams that co-publish results and interpretations thereof. At a coarser (but not maximally coarse) grain, the scientific community is subdivided into traditional disciplines: biology, chemistry, physics, etc. The particular scientific communities referenced in EP are of intermediate grain: more inclusive than co-authoring research teams yet narrower than traditional disciplines. I'll refer to these intermediate-grain communities as "scientific fields" or "specializations" (using the two terms interchangeably). A specialization includes a number of working scientific groups or teams (both sizes and numbers of which vary widely), and may cut across traditional disciplines. The list of such specializations is long and continually expanding: molecular biology, integrative systems biology, synthetic biology, immunology, structural biology. . . Although there aren't rigorous individuation criteria for scientific communities at this grain (so far as I am aware), in practice we can identify them pretty well. Much of the social infrastructure of science—societies, journals, regular meetings, pedagogy—serves and delineates scientific communities at this intermediate grain. Importantly, one needn't assume

[1] See Section 1.2.

[2] The term "level" is increasingly questioned by philosophers of science, and carries a load of accreted conceptual baggage from over-use. I avoid it where possible.

anything about explanation to identify a particular scientific community. One just needs to look at the social epistemic practices and infrastructure in some area of science. To be sure, the social organization of science is dynamic; particular scientific communities appear, die out, split, merge, and more. In general, scientific communities present a moving target for philosophical analysis. But that is just to say that particular scientific communities are historical entities. History of science, like sociology of science, is a natural partner for philosophical studies committed to EP.

Significant products: Scientific communities are epistemic communities. Epistemic communities produce knowledge (among other things). So any scientific field that persists awhile yields some epistemic products. Among these epistemic products, some stand out as *significant*: major achievements that are recognized, highlighted by the community, often communicated to a wider audience. Explanations, according to EP, are a class of such achievements. It follows that explanation is a central task and aim of science—this is another background assumption of explanatory particularism, alongside the community structure of science as a dynamic array of specializations. Of course, explanation is not science's only task, and not every scientific activity aims at explanation. However, plausibly, any scientific community includes explanation among its aims. *Significance* is a further requirement: an EP-explanation is a major achievement for some particular scientific community. That doesn't preclude explanations being credited as achievements of individual scientists who are members of those communities. But for EP-explanation, allocation of credit is a side-issue. Significance in this context has a social aspect going beyond an individual's epistemic achievement. To be sure, items of scientific knowledge (including but not limited to explanations) are proposed by individual scientists or research teams. Their proposals are then subjected to criticism by a wider community of peers.[3] The process that produces an explanation may involve content, as well as criticism, from many different members of a scientific community. That is, an EP-explanation may emerge from the combined efforts of many research teams or other groups within that community. Alternatively, its content may be due to a single scientist or research team (although this is probably rare in science today). But even in these cases, an EP-explanation emerges as a *significant* achievement from a process of criticism within that scientific community. In this sense, a scientific explanation is produced by and is an achievement of that community, howsoever credit for the achievement is allocated within that community or beyond its boundaries. Importantly, EP is not committed to any philosophical account of achievement or significance. Just as with specializations, we identify

[3] This characterization of science owes much to Longino's pioneering work on the social epistemology of science (1990, 2002). EP does not presuppose Longino's account of objectivity or empirical knowledge, nor the form of pluralism she endorses, however (see Fagan, under review).

significant scientific achievements by looking directly at scientific practice. In practice, scientists tell us—and the world—about the achievements that matter for their community. Scientific explanations come to philosophers' attention precisely because they are a big deal to scientists in some particular community. How else would we learn about those explanations, if not from scientists?

Advancing understanding: Not all of a scientific community's significant epistemic achievements are EP-explanations. Those that are, advance understanding for community members. Such an advance is another general requirement for EP-explanation. But the requirement is very thin; the concept of understanding here is open and unrestricted.[4] What understanding and an advance thereof amounts to varies widely across scientific communities. Rather than pre-judge the issue, explanatory particularists look to scientists to identify and communicate the achievements that advance understanding in their fields. So I agree with philosophers who closely link explanation and understanding. But I am a particularist about both. On the particularist approach, an explanation and its associated form of understanding are identified and characterized by directly studying a particular scientific community. That is, if one asks a representative sample of scientists in a community 'Does achievement A advance our understanding?' and they mostly answer 'yes,' then A is an explanation produced by and for that community. The preceding question immediately prompts at least two further queries: 'Understanding of what?' and 'How so?' We want to know the target of explanation, and the features of A that afford advance in understanding thereof. So EP directs philosophers of science to study explanation in this way: examine these significant epistemic products in their local contexts of construction and use. Explanatory particularism is an approach to studying explanation, as much as a thesis about explanation.[5] Importantly, 'understanding' is not thereby permanently relegated to an obscure black box. Instead, the concept is filled in piecemeal by studies of particular scientific fields and their explanatory achievements. The target of explanation, similarly, is whatever members of the particular scientific community at issue aim to understand. An EP-explanation, in the relevant community, grounds statements of the form: 'Now we better understand X,' where X is some topic of inquiry. EP places no restriction on X. The reason for leaving targets and features of explanation open, in general, is straightforward. Targets and features of explanation (as conceptualized here) are very diverse in scientific practice.

[4] Chapter 6 presents a view of understanding associated with one important application of EP, and positions that view in relation to other philosophical accounts of understanding on offer.

[5] As a thesis about explanation, EP is strikingly thin. That is deliberate. EP is not itself the full particularist account, but a starting-point for detailed, fine-grained study of explanation across the entire range of scientific communities. Explanatory particularism is an approach to explanation open to the full range of achievements of understanding recognized as significant across the sciences.

Three preliminary examples

A few brief examples will illustrate this explanatory diversity. Later chapters delve more fully into particular cases. These brief treatments are preliminary, not full particularist studies of explanation.

Example 1: the Higgs mechanism

In particle physics, the Higgs mechanism is a mathematical innovation playing a key explanatory role in the Standard Model of fundamental particles and forces.[6] It provides a unified treatment of weak nuclear and electromagnetic interactions among the various physical particles of the Standard Model, within a theoretical framework organized around symmetries in particles represented as quantum fields. In the Standard Model's theoretical framework, particles are represented as quantum fields having mass, charge, magnetic moment, energy levels, and spin. The Standard Model is designed to allow exact solutions to be calculated and compared to experimental results involving those properties. Symmetries (and approximate symmetries) offer informative and calculation-friendly patterns among the menagerie of experimentally-identified particles. The goal is a (mathematically) unified treatment of all fundamental particles and their properties. Gauge theories have mathematical properties that fit well with, and are taken to explain, these fundamental properties and patterns. But there are also interactions among particles, and to represent weak nuclear and electromagnetic interactions in an empirically adequate way, the Standard Model needs symmetry *breaking*. The Higgs mechanism is an achievement that satisfies these two important but seemingly incompatible constraints on fundamental physical theory: a unified treatment based on symmetries, accommodating interactions among particles through symmetry breaking. It is a refined gauge theory that explains weak nuclear and electromagnetic interactions among fundamental particles via spontaneous symmetry breaking, in a manner consistent with empirical data. This is a significant accomplishment: an elegant and highly distinctive integration of physical theory, unificatory aims, modeling constraints, and empirical data.

Example 2: the double helix

The structure of deoxyribonucleic acid (DNA) is a seminal explanatory achievement of molecular biology.[7] The double helix model represents the three-dimensional spatial arrangement of chemical groups making up the DNA macromolecule: phosphates, sugars, and four kinds of nucleotide base.

[6] This example is based on Kibble (2015) and King (2018).
[7] From Watson and Crick (1953) and myriad later sources.

Phosphate-sugar-base 'units' are arranged in two linear chains, coiled around one axis to form a helix, with hydrogen bonds between pairs of bases linking the chains. Only certain combinations of base-pairs can fit into this chemically stable structure: adenine with thymine, guanine with cytosine. Specific base pairing entails that one DNA chain can serve as an unambiguous template for the other. So the structure suggests a functional explanation of how DNA transmits hereditary information and (with additional molecular machinery) encodes the amino acid sequence of proteins. The latter is an explanatory factor in an extremely wide range of biological phenotypes. Unlike the Higgs example, the double helix model is not based on mathematical theory and does not aim to unify all biological phenomena. It was constructed to accommodate a wide range of experimental results in accordance with accepted principles of biochemical structure. Rather than fundamental symmetries, the double helix is interpreted causally or functionally: breaking, binding, unwinding, transport, synthesis, etc.

Example 3: density functional theory

Density functional theory (DFT) plays a significant explanatory role in nanoscience.[8] Its name notwithstanding, DFT is not a theory in the traditional sense, but a method borrowed from computational chemistry, using simulation to characterize the behavior of (modeled) nanoscale materials. Our best physical theories don't work at the nanoscale; one can't use the Schrödinger equation (for example) to predict the behavior of nanoscale materials. Instead, DFT is used to construct a simplified model of such materials, compressing the properties of many interacting electrons into a single lump quantity, "electron density." That quantity is determined by many simulation "experiments" for a given model. The hope is to discover through these simulations stable design rules for reliably building nanoscale technologies. Lenhard (2006) argues that nanoscience aims at "understanding by control" rather than "theory-based explanations."

Even in this tiny, narrow sample, the explanations differ widely.[9] Their linguistic aspect alone ranges from mathematical formalism to thick description of molecular interactions, and design principles for technological applications. They are also metaphysically diverse. The Higgs mechanism is essentially abstract, while the structure of DNA can be realized abstractly or concretely. DFT makes use of abstract mathematical theory but requires instantiation in a computer. Their respective targets also differ: properties and interactions of fundamental particles,

[8] This example is based on Lenhard (2006).

[9] Many other examples could be provided. Relevant philosophical case studies include Goodwin (2013), on different explanatory interests of organic and quantum chemistry; Franklin (2018), on the different explanatory standards employed by philosophers accounting from renormalization group explanation of universality; see also Woodward (2003, 231–233).

inheritance across generations of living things, behavior of novel nanomaterials. And the contrasts go further. Each example invokes different concepts and locates explanatory power differently. The form, content, and mode of understanding (what we might call the style of explanation)[10] varies for each. All three are important achievements in their respective fields. Yet what they accomplish is very different. The Higgs mechanism is a mathematical construct that balances unificatory aims, modeling constraints, and empirical accuracy. The double helix model was constructed to accommodate a wide range of experimental results and bits of biochemical theory, enabling a flood of causal and structure/function inferences. Simulation models of nanomaterials are constructed by mingling laws of quantum physics with simulation methods, demonstrating understanding through reliable engineering design. In EP's terms, each shows up as an achievement by successfully accommodating constraints and exhibiting values of its particular community context.

These examples also illustrate another important feature of explanations in scientific practice: they are constructed via complex processes. Not only are the epistemic achievements of particular communities diverse—the products—but the processes leading to each, taken separately, involve a variety of resources. Some of those resources put their stamp on the product, informing the features in virtue of which it is deemed significant and advancing understanding. In the above examples, criteria for a significant explanatory achievement are informed not only by each field's theoretical background (physics, biochemistry) but also its technological context and conceptual and normative epistemic background. Mathematical advances in gauge theory, proficiency in x-ray crystallography, and innovations in both nanomaterials and computer simulation are central to each example, respectively. More generally, an EP-explanation successfully accommodates many different contextual factors: epistemic norms and values, methodological constraints, current state of technology, and more. Each of the three examples above (the Higgs mechanism, the double helix, DFT for nanoscience) was constructed in the context of a distinct set of goals, methods, technologies, and norms. Each advances understanding some object of inquiry: fundamental particles, macromolecules implicated in fundamental processes of life, the behavior of novel nanomaterials. And—it bears repeating—these examples deal with an extremely narrow range of specializations, with relatively close historical connections and subject matter: particle physics, (early) molecular biology, and nanoscience. The full range of scientific explanations in practice is much, much more diverse, encompassing social, behavioral, and historical sciences.

[10] The term "style" is usually associated with art, occasionally extended to scientific reasoning (e.g., Crombie 1994). A generic meaning applies straightforwardly to particularist explanation: a style "designates recognizably shared traits of created forms that transcend the individual producer" (Bredekamp et al. 2015, 18).

Specialization

This brings us to the last key concept of EP. Unlike the requirements of significance and advancing understanding, *specialization* isn't a feature that philosophers can pick up directly from scientists in their communities. Specialization is an important aspect of scientific practice, but not one scientists themselves are likely to communicate directly. An explanation produced by a particular scientific field is *specialized*, meaning that it's constructed in the context of a distinct set of goals, methods, and norms, reflected in that explanation's form, content, and role. Building on the point made above, EP-explanations are constructed so as to satisfy various contextual constraints of particular scientific communities, in a way that exhibits a community's epistemic values associated with understanding. By balancing various contextual features, an EP-explanation appears as a significant achievement in a particular scientific community. In this way, EP-explanations are localized to their epistemic communities of origin, answerable to the goals, methods, and standards of those contexts. They are specialized products, constructed by and for communities of experts.

The three examples above illustrate this specialization. For readers unfamiliar with the relevant scientific communities, those passages are uninformative technical jargon. Even these brief treatments don't convey understanding to just anyone. Each of the three achievements is 'fitted' for members of a particular epistemic community, who share a wide variety of substantive background assumptions about explanations and methods for constructing them. In this sense, to appreciate an explanation as a scientific achievement, one needs to be an expert. However, the idea of *specialization* is more apropos for my purposes than the dichotomy of experts vs. non-experts. No one today is an expert in 'science,' full-stop. Maybe that was possible long ago.[11] But scientific inquiry today is epistemically fragmented, siloed into myriad highly specialized communities. The latter are often very technical. It takes years of training and practical experience to become a full-fledged member. Much of this training and education bears on explanation. And much of it occurs outside the classroom, as fledgling scientists gain practical experience with scientific work. Any such experience is in one field or another, unavoidably. There is no general training in "how to do science."[12] In training to become a member of a particular field, one learns about the kind of understanding sought concerning objects of inquiry, the features of epistemic

[11] A case could be made for William Whewell (whose quote opens the book) as the last human individual with a reasonable claim to 'total scientific knowledge' (thanks to Elijah Millgram for this idea).

[12] There are general science courses, but those are only the beginning of specialized scientific training.

achievements that are valued, the problems considered significant, the kinds of answers that advance understanding in the desired mode, and the methods used to gain such understanding. Learning how to contribute to explanations in the style of a particular scientific community is a major part of the training involved in becoming a member of that community.[13]

At this point, I've clarified the basic ideas of explanatory particularism. My argument for endorsing these ideas is in Section 1.3. As a preview, however, it's worth noting that EP is grounded on several well-established features of scientific practice, weaving these strands into a coherent package. Scientific communities are very diverse, differing in the objects they seek to understand, their preferred methods and technologies, and the conceptual and normative epistemic background against which explanatory achievements are recognized. Different communities impose different norms and standards on construction and evaluation of explanations. Those standards provide specific constraints on what a successful explanation should be like the features it should exhibit, as such, and the ways it represents or otherwise relates to the target of explanation. The upshot is that although the goal of explanation is broadly shared across the sciences, it is conceptualized and accomplished in diverse and locally specific ways. An explanation produced by a particular scientific field is specialized in that it reflects that field's distinctive goals, values, methods, and presuppositions. Of course, explanations are not the only products of scientific communities that are specialized in this way. But they are among the most important. Scientific explanations are highly valued by scientists—EP builds this into the very idea of a scientific explanation.[14] Their importance derives in part from the epistemic role of advancing understanding. An EP-explanation plays this role in virtue of being specialized; i.e., exhibiting a scientific community's epistemic and methodological values, striking a harmonious balance between the various knowledge-relevant virtues prized by that particular field, alongside other contextual features. Because of their significance and epistemic role, explanations are specialized, local, and context-dependent. An EP-explanation is a major achievement of understanding that ties together diverse conceptual and technological resources in a particular social context, in a way that satisfies the goals, methods, and standards of that context. This summary ties together the basic ideas of explanatory particularism, epitomized in (EP). The rest of the book develops this view and explores its implications.

[13] These insights are anticipated in Woody (2015), discussed in Section 1.2.

[14] A number of philosophical accounts of explanation ignore this feature, citing examples of remarkable triviality. This indicates a gap between (some) philosophers' concerns and scientists' explanatory practices (see Section 1.3).

What's new?

Explanatory particularism has obvious implications for philosophical study of explanation. As noted, it offers an approach to such study, prioritizing scientific practice and social aspects of science. This social epistemic approach aims to reveal ways explanations in different areas of science differ from one another, as well as similarities. Explanatory particularism does *not* preclude there being shared aspects of explanation across multiple specialized scientific fields, alongside their differences. Rather, it precludes limiting explanation to just those shared aspects, elevating these to universal criteria for explanation. The combined emphases on scientific practice, diversity, and social aspects of science set explanatory particularism apart from other philosophical approaches to explanation. Section 1.2 limns these contrasts in detail. Here, I briefly codify EP's distinctive themes:

Prioritizing scientific practice: The explanatory particularist approach directs philosophical studies of explanation to engage closely with scientific practice. Specifically, philosophers should approach this topic by attending first to scientific practice rather than presupposing any substantive philosophical account of explanation, as sketched previously. The substantive account for EP emerges piecemeal, from a range of such studies, community by community.

Diversity: All scientific fields aim at knowledge and understanding, but they do so in highly diverse ways. Explanatory particularism takes this diversity as a starting point and resource, rather than as a challenge to be overcome by analytic ingenuity.

Social aspects of science: The notion of *a particular scientific community* is a sociological concept. In this sense, EP is predicated on the view that science is social. It is widely recognized that science is a social activity, and that its social organization has epistemic significance. However, other philosophical accounts of explanation do not take these facts as central or fundamental.[15]

It's important not to overstate EP's novelty. Many philosophers of science emphasize the above themes in their work. Most broadly, philosophy of science in practice (PSP) is increasingly prevalent in our field. PSP's guiding idea is that philosophy of science should be grounded on, and responsive to, the work of actual scientists. For example, the mission statement of the Society for Philosophy of Science in Practice includes methodological commitments to "productive interaction between philosophical reasoning and a study of actual scientific practices,

[15] One exception: Woody's (2015) functional account of explanation. Woody's account, though distinct from EP and its development in later chapters, has important commonalities with my view.

past and present" and to "elucidat[ing] what kinds of activities are required in generating knowledge."[16] EP fits well with these commitments, conceptualizing explanations as localized to particular contexts of construction and use: specialized scientific communities whose epistemic values (and other features) are reflected in the explanations they produce. Initially a minor strand of the discipline, PSP has grown in prominence since the early 2000s, with many top journals and conferences encouraging this approach. Under the aegis of PSP, and related projects in history and philosophy of science, quite a few scholars are already using something like the particularist approach proposed here.[17] Broadly speaking, my novel contribution (so far) is just the label "explanatory particularism" and the assertion that all these projects are contributions to philosophical study of explanation.

That assertion, however, sets explanatory particularism at odds with longstanding philosophical tradition. Nearly all philosophical accounts of explanation, past and present, aim to capture interesting features shared by many (or all) explanations; to articulate and explicate essential criteria shared by all (good) scientific explanations. However, such an approach is bound to miss any philosophically interesting features of explanations *not* shared across the sciences. Explanatory particularism reverses this traditional orientation, highlighting specialized features of explanations for particular communities that are likely to vary widely across scientific fields. In this sense, explanatory particularism is firmly non-traditional. To see the contrast, we need to discern the monist tradition in philosophical studies of explanation.[18]

1.2 Seven decades of explanatory monism

Explanation has long been a central topic in philosophy of science. Long-running inquiry and debate have produced numerous philosophical accounts of the nature, form, and value of explanation: asymptotic, mathematical, 3M-mechanistic, dynamic mechanistic, equilibrium, general counterfactual dependence, topological, minimal model, kairetic, functional, game theoretic/pragmatic, causal pathway, eikonic, and many more.[19] Across these myriad proposals there is little

[16] See <https://philosophy-science-practice.org/about/mission-statement/>.

[17] See, for example, <http://integratedhps.org/>. Much recent philosophical work on scientific explanation takes a PSP approach, offering fine-grained studies of explanation as well as methods and norms essential to construction of explanatory models in various scientific fields; e.g., neuroscience (Craver 2007), molecular biology (Darden 2006), chronobiology (Bechtel 2011), theoretical physics (Batterman 2002, Bokulich 2008), and chemistry (Woody 2015). Rice's (2015, 2018) studies of idealized model explanations also include particularist ideas.

[18] This treatment is appropriately reflexive: the context in which explanatory particularism arises is important for understanding the view itself. Philosophical inquiry, like all fields, has a context, in which history of debate plays a large and often unacknowledged role.

[19] See, respectively: Batterman (2002), Lange (2016), Kaplan and Craver (2011), Bechtel (2011), Potochnik (2017), Reutlinger (2016), Huneman (2010), Rice (2015), Strevens (2008), Woody (2015), Mantzavinos (2016), Ross (2021), Bokulich (2018)—a partial list.

consensus, even on what the project is. Philosophical studies of scientific explanation variously refer to their results as "models," "theories," "conceptions," "views," or "modes." For simplicity, and to reduce the load of philosophical baggage carried at the outset, I will use "account" as a generic term for any philosophical proposal that aims to analyze, explicate, or otherwise characterize scientific explanation.

Although explanatory particularism breaks with tradition in philosophical studies of explanation, it does not thereby rebut or reject all its predecessors. My project is reformative, not eliminative. The goal is to redirect philosophical debate about scientific explanation, not jettison earlier insights. Because explanatory particularism is not directly opposed to any earlier account, my argument for EP does not take the form of rebutting its predecessors one-by-one. Rather, I oppose a background assumption shared by many (but not all) prior philosophical accounts of explanation. It's that very assumption that pits those accounts against one another as incompatible alternatives: *explanatory monism*. While explanatory particularism does not exclude any specific characterization of *some* scientific explanations, it does exclude the presumption that there is one substantive and illuminating philosophical account of *all* scientific explanation. In rejecting monism of this kind, explanatory particularism departs from established philosophical tradition. This section shows how deeply explanatory monism is entrenched in that tradition.[20] Not only is monism presumed by several influential philosophical accounts, but its assumption also frames debate around accounts of explanation that are not inherently monistic.

Deductive-nomological explanation

Although far from the first philosophical account of explanation, Carl Hempel's deductive-nomological (D-N) account has the distinction of being "widely accepted [in the 1960s and early 1970s] by scientifically minded philosophers; indeed, it qualified handily as the received view" (Salmon 1989, 4).[21] So it's a handy starting-point for sketching the philosophical tradition that persists to the

[20] Readers familiar with this background may prefer to skip to Section 1.3. A thorough survey of philosophical accounts of explanation is beyond the scope of this book. The classic survey is Salmon (1989). Although now rather dated, no other monograph has succeeded it. The extent of the philosophical literature on explanation today defies thorough review in a single tractable work. Detailed, more focused surveys include Curd et al. (2013) Chapter 6, Reutlinger and Saatsi (2018), and Woodward and Ross (2021).

[21] Hempel referred to his view as a "model" of explanation, but, as noted in the main text, I'll use the more generic "D-N account"—or just "D-N" for brevity. The *locus classicus* for D-N is Hempel and Oppenheim's paper, 'Studies in the logic of explanation,' which appeared in *Philosophy of Science* (1948). Hempel subsequently developed the account, with expanded treatments appearing in *Aspects of Scientific Explanation* (1965) and *The Philosophy of Natural Science* (1966). The key ideas are consistent across these different presentations (although there are some new developments; see Salmon 1989, Curd et al. 2013).

present day. Hempel initially defined an explanation as a deductive argument that derives a description of the phenomenon explained from premises including at least one law of nature (Figure 1.1). This is the founding member of a family of philosophical accounts centered on the idea that explanation is *of* phenomena (events, patterns, processes, etc.) *by* "covering laws." On the covering-law account, all explanation is by logical derivation from laws of nature. D-N explanation exemplifies the commitments of mid-twentieth-century logical empiricism. The covering-law account is general and exhibits internal coherence and logical clarity. Subsumption is logical, such that scientific knowledge is systematized to form an explanatory hierarchy of more and less general laws, from which (when supplemented with antecedent conditions) all possible events can be derived, either deductively or with high probability via strong inductive argument. The associated vision of science is a linguistic arrangement of ordinary, publicly-accessible phenomena into orderly regular patterns, codified as laws. The laws form a derivational system, the more restricted subsumed by more general laws of nature.[22]

Antecedent conditions (realized before or at time of E)	C1, C2, C3 . . .Ck
Laws describing uniformities of nature (all F are G)	L1, L2, L3 . . .Ln
--	----------
Explanandum (sentence describing phenomenon to be explained)	E

Figure 1.1 Schema of deductive-nomological (D-N) explanation. In Hempel and Oppenheim's initial presentation, a D-N explanation is a valid argument that logically derives a description of the phenomenon to be explained (explanandum) from a set of true premises (explanans) that (i) essentially includes at least one law of nature and (ii) has testable, empirical content.
Source: adapted from Hempel (1966).

Explanatory monism is a central presumption of D-N and other covering-law accounts. Logical empiricists viewed logic as a universally applicable tool for analyzing statements. The concepts of logical subsumption and laws of nature introduce order and organization onto this all-encompassing logical field. D-N states conditions under which a collection of statements qualifies as a scientific explanation. There is no place for the thought that any explanation could fall outside D-N's logico-linguistic scope. Hempel and Oppenheim (1948) explicitly present D-N as a set of "general principles" for explanation that apply not only to physics, but also to biology, psychology, sociology, economics, and linguistics (140–146). Notably, they argue that explanations in human and biological sciences that appear to invoke some form of teleology (e.g., explanations of intentional human behavior,

[22] Hempel did not satisfactorily work out the logic of this derivational system of laws, but the idea is a vital part of D-N, as well as unificationism (see next subsection).

evolutionary adaptations, trajectories of biological development) either conform to D-N or should be rejected as scientifically unacceptable. D-N's intended scope includes all "branches of science," not just the causal explanations in physics that furnish Hempel's own motivating examples (135–136).[23] Monist aspirations are also behind Hempel's later extensions of D-N to encompass explanations of statistical regularities and high-probability events (1965). Deductive-statistical (D-S) and inductive-statistical (I-S) explanation bring the latter, respectively, into the systematic logical hierarchy of laws and predictable observations. Both are framed in terms of familiar logical distinctions between deductive and inductive inference, universal and statistical laws. That is, D-S and I-S are grounded on the same categories and concepts as D-N, so all three naturally bundle together to comprise the classic covering-law account of explanation.[24]

Significantly, later extensions of D-N are *not* grounded on examples of explanation in scientific practice—unlike the original 1948 arguments. Subsequent debate about the covering-law account has tended to follow Hempel in this regard. To this day, philosophical studies of explanation tend to proceed in terms of concepts and categories internal to philosophy of science. In the 1960s and 1970s, when Hempel's account was the primary focus of debate, the main topics were logical derivation and subsumption, laws of nature, and inter-theoretic reduction. Although topical foci have changed over the past half-century, the internalist bent of philosophical discussion of explanation is robust. Even D-N's downfall attests to its monist commitments. Hempel's account famously allows for many counterexamples, featuring hexed salt, syphilitics, men taking birth-control pills, and other curiosities. Such counterexamples are widely taken to be sufficient reason to reject the covering-law account of explanation. This dialectical situation presumes explanatory monism—not just across the sciences, but in everyday life. Hempel's covering-law account was proposed, received, and eventually rejected, as an attempt to analyze all explanations without exception. Though the covering-law account today has few if any defenders, its core ideas remain influential. This includes the monist presumption that there is a single general substantive account of all scientific explanation, which it is the philosopher's task to articulate.[25] With his influential account, Hempel set the monist agenda for philosophical studies of explanation that persists today.

[23] This version of explanatory monism also amounts to a form of reflexivity for philosophy of science: D-N (and its later developments) offer general principles that cover the full range of explanations, much as scientific laws cover the phenomena they explain. This reflexivity, carried forward in the philosophical debate about explanation, leads to slippage between philosophers' own standards for a philosophical account of explanation and the standards those accounts propose for scientific explanation.

[24] See Salmon (1989, 9).

[25] See Cartwright et al. (2020) for a recent (qualified) defense of covering-law explanation.

Explanatory unification

A different interpretation of the monist presumption spurred another classic account: explanatory unification, aka unificationism. The unificationist account begins with Michael Friedman's paper, 'Explanation and Scientific Understanding' (1974). Friedman argues that a unifying explanation contributes to understanding by "reducing the total number of independent phenomena that we have to accept as ultimate or given" (15). Philip Kitcher's successor account (1981, 1989) addressed technical problems in Friedman's analysis, replacing laws with the notion of "argument patterns." But Kitcher preserved many of Friedman's ideas about explanatory power and understanding. Kitcher's account, now considered the classic statement on explanatory unification, defined explanation of a phenomenon as derivation of a description of that phenomenon under a general argument pattern belonging to a core set ("the explanatory store") of such patterns that together provide the best systematization of the entirety of scientific knowledge at a particular time (1989, 430). A scientific community's explanatory store is "the set of arguments that achieves the best tradeoff between minimizing the number of premises used and maximizing the number of conclusions obtained," reducing the number of facts that must be accepted as primitive in achieving understanding of the world (Kitcher 1989, 431). So, on this view, explanatory power stems from coverage of a wide range of phenomena by one or a few laws. Explanation proceeds through logical subsumption, as in D-N, but explanatory power stems from the breadth of applicability of an argument pattern rather than in the logical connection constituting an argument itself. Unifying explanation so defined is the integument of logically systematized scientific knowledge.

The unificationist account is committed to explanatory monism in at least two significant respects. First, as an alternative and successor to the covering-law account, explanatory unificationism inherits its predecessor's monist aspirations. Second, explanatory unificationism is distinctively committed to *the oneness of science*.[26] That is, the unificationist account analyzes scientific explanation from a global perspective, in terms of the totality of scientific knowledge. Friedman's analysis refers to the total number of logically independent lawlike sentences accepted at a given time by the scientific community. Explanation is relative to the entirety of scientific knowledge at a given time (i.e., the set of all laws and facts accepted). Kitcher's analysis retains this global perspective. Although he does not explicitly define "the scientific community," Kitcher seems to envisage it as unstructured and

[26] 'Unity of science' is the more familiar term—but that is better reserved for the upshot of explanatory unification (or another systematizing relation for scientific knowledge). What I am calling 'the oneness of science' is a prior assumption, associated with taking what might be called a global (or monistic) perspective on science.

undifferentiated (except by explanatory unification). All explanatory argument patterns accepted within a scientific community at a particular time are lumped together in the explanatory store. The oneness of science is further inscribed in the idea of a single system of explanatory arguments. From a global vantage-point, explanation consists in the systematization of scientific knowledge via a small number of brute laws (Friedman) or argument patterns (Kitcher). Unifying explanation amounts to systematization, or logical ordering, of the totality of knowledge had by *the* scientific community at a given time.[27] This is a strikingly monist vision of the scientific community and its explanatory practice.

Although explanatory unificationism has few defenders today as an account of scientific explanation, several of its themes remain prevalent in philosophy of science. One is the association of explanation with simplicity; i.e., to explain is to simplify one's description of the world—to impose clarifying simplicity on a complex welter of facts. This notion fits well with the monist presumption, encouraging philosophers of science to adopt an abstract, top-down view of the sciences. Relatedly, the unificationist account is explicitly hierarchical: a wide, diverse array of facts subsumed by a small set of patterns or principles. Kitcher's technical analysis of argument patterns allows for more elaborate stratification, but the idea of explanatory hierarchy remains central. In their influential introductory commentary, Curd and colleagues note that explanatory unification is distinctive in "its 'top-down' focus on the most general unifying theories in science as providing the best, and perhaps the only, acceptable explanations" (2013, 783). To put the point another way, explanatory unification is not a neutral way of looking at science, even though "the scientific community" is treated as an undifferentiated monolith. For a unificationist, explanatory power is not evenly distributed across the sciences, but concentrated in high-level general principles that have explanatory priority over other epistemic items. This hierarchical, logically stratified image of science is implicit in the covering-law account as well (as Kitcher notes). D-N is hierarchical in that "explanation of a phenomenon . . . consists in its subsumption *under* laws or *under* a theory" (Hempel and Oppenheim 1948, 152; italics mine). The unificationist account inherits and elaborates on this idea, positing science as stratified by explanatory relations. Monism, explanatory hierarchy, and the association of explanation with simplicity together form a powerful conceptual package, a persistently influential background that continues to frame philosophical studies of scientific explanation.

[27] This is not too far from Hempel's image of an orderly system of laws, with various facts derived from laws of nature and specific regularities derived more general principles. Unificationists, however, define explanation as the systematization of scientific knowledge, not as the logical derivation that could produce such systematization. One could fairly say, though (as Kitcher does; 1981, 508), that the unificationist account is a further development of ideas already present in D-N.

Pragmatic explanation

Around the same time as unificationist proposals, Bas van Fraassen offered a very different account, premised on the idea that explanation is a communicative act (1977, 1980).[28] On this pragmatic approach, an explanation answers a "why-question" in terms of a range of alternative answers (e.g., "Why P rather than R, S, T . . .?"). The set of alternatives for any particular question is determined by the context in which it is asked. A general "logic of why-questions" formalizes and extends this basic contrastive idea.[29] For a question of the form "Why P?" the logical components are P, contrast class X, and a "request specification" or relevance relation R. So a pragmatic explanation has the form "P rather than X because A," where A is a true proposition that stands in relation R to <P, X>, P is true, and all members of X are false.[30] In science, A is a theory, P is some observed fact or pattern, and X is determined by the theory's domain and the questioner's beliefs and interests. The value taken by variable R is also context-dependent; a function of the questioner's scientific background and the state of state of scientific knowledge at the time. The pragmatic account is intended as fully general, covering all possible cases of explanation. In this respect, it adheres to the monist tradition. But unlike other monist accounts, the pragmatic view arguably achieves the goal of accounting for explanatory practice (i.e., requesting, giving, and evaluating explanations) in all its diversity.[31] This is because the pragmatic account places few a priori restrictions on explanation; only its contrastive character is presupposed. The features in virtue of which an explanation is good or satisfactory are largely determined by the context of the inciting question.

The pragmatic account defines explanation as context-dependent and interest-relative, and arguably accommodates the full range of explanatory diversity in scientific practice. These features are common ground with explanatory particularism. Indeed, one might consider the latter a successor to the pragmatic account; a program for filling out various instantiations of van Fraassen's A, R, and X.[32] However, there are several important differences between my view and van Fraassen's. First, as noted, the pragmatic account is a general analysis of explanation, specified by the logic of why-questions and its contrastive relations. Explanatory particularism does not propose a general analysis of all explanation in terms of the logic of questions and answers, nor that explanation is essentially contrastive. Rather, my general proposal is that an explanation is a specialized,

[28] See also Achinstein (1983, 1985).

[29] The scholarly term for this branch of logic is "erotetics," so van Fraassen's analysis is called "the erotetic account."

[30] See van Fraassen (1980, 144).

[31] See Lloyd and Anderson (1993).

[32] Elisabeth Lloyd, for one, has endorsed this view (personal communication). Many thanks to her for clarifying commonalities between van Fraassen's classic account and explanatory particularism.

significant product of a particular scientific community. Second, the pragmatic account limits explanation to the linguistic realm. In this respect, explanatory particularism is more permissive—an explanation can be anything that a specialized scientific community recognizes as an achievement in understanding, linguistic or not. Third, the pragmatic account has no counterpart to EP's significance requirement. According to EP, not every satisfactory answer to a question that arises in a scientific community qualifies as an explanation. Rather, an EP-explanation is a major achievement that advances understanding in such a community. Fourth, EP's general characterization is not all there is to explanatory particularism. Details of a specialized scientific explanations matter, and so articulating the full particularist account of explanation is a project for a community of philosophers studying science in practice. van Fraassen (1977, 1980) is uninterested in exploring the diversity of explanations across the sciences.[33] His general analysis allows for many such details but does not include them. What matters for the pragmatic account is that any such case fits with the general formula; nothing is ruled out a priori. The pragmatic account allows for, but says no more about, diversity of explanations in scientific practice.[34] For an explanatory particularist, in contrast, what matters are the details of any specialized explanation; the features in virtue of which it counts as an achievement. Explanatory particularism takes the diversity of specialized scientific explanations as the main focus of philosophical study. As an approach to studying explanation, this is very unlike van Fraassen's pragmatic view, although there are important commonalities.

Causal accounts ascendant

The three classic accounts discussed so far are explicitly monistic. D-N and unificationism aspire to articulate the nature of and conditions for explanation across all sciences. All three are staples of introductory textbooks, part of the training of several generations of philosophers of science. Of the three, the

[33] van Fraassen (1977, 1980) takes explanation to be outside the domain of scientific knowledge. The latter, for van Fraassen, consists of logical and probabilistic connections between theories and observations. In this he accords with D-N and the logical empiricist tradition (albeit developed in a unique, aesthetically elegant way). Very roughly, he sees theories "latching onto" to the world via a delicate skein of exact correspondence between mathematical patterns. Scientific theories as such are properly judged in terms of this correspondence, grounded in confirmation by observations. Truth, empirical adequacy, or acceptability, are (arguably) central scientific virtues. But explanatory power is external to science, driven by and relative to our interests and purposes in asking "why X?" van Fraassen (2008) offers a somewhat different view of science, but this work has not, to my knowledge, influenced philosophical discussion of scientific explanation to date.

[34] More recent pragmatic accounts retain the traditional goal of a single, universal account of explanation, but focus on its communicative role. For example, Mantzavinos (2016) proposes to capture the diversity of scientific explanation with a new unified framework: the concept of explanatory games, and a small set of fixed kinds of rules that define and govern those games.

covering-law account casts the longest shadow. Even as serious objections and counterexamples piled up, D-N passed from "the received view" to a contrastive spur toward alternatives. Long after most philosophers ceased to defend it, the covering-law account was the touchstone and starting point for philosophical study of explanation. But not anymore. Today, philosophical debate about explanation is oriented around causal accounts, chiefly Woodward's interventionism (2003). The shift was long in coming; causal accounts of explanation have been in the ascendant arguably since the late 1970s.[35] Their starting point is the idea that explanations are not arguments, but instead pick out causal relations in the world. The locution "pick out" is deliberately vague. Different theories of causality entail different causal accounts of explanation (e.g., Salmon 1984, Humphreys 1989, Dowe 2000, Woodward 2003). One philosophical theory of causality can further diffract into different accounts of explanation if the latter include different conditions specifying the qualities of causal relations that figure in good or acceptable scientific explanations: robustness, specificity, invariance, depth, and the like (see, e.g., Hitchcock and Woodward 2003; Strevens 2008; Woodward 2010). Just as the covering-law account directs philosophical debate about explanation to laws of nature, causal accounts merge philosophical studies of explanation with analysis of causality.

The ascent of causal accounts of explanation began with Wesley Salmon's causal–mechanical (C-M) account of explanation (1978). Salmon analyzed explanation in terms of a substantive metaphysics of causality: a C-M explanation correctly identifies the causal processes and interactions that produce one or more effects of interest in the world. Reflection on contrasts with D-N led to Salmon to distinguish "ontic" and "epistemic" views of explanation (1978, 1989). The ontic view localizes explanation to real causal relations in the world, the epistemic to human reasoning and inference. Ontically speaking, one feature of the world explains another in virtue of some real, objective relation (usually causal) that holds between them. This starkly contrasts with the D-N account of explanations as arguments, the unificationist account of explanation as argument systematization, and the pragmatic account of explanation as the context-dependent communicative act of answering a why question.[36] Although C-M has faded from prominence, causal

[35] See, e.g., Saatsi and Reutlinger's introduction to their recent collection *Explanation Beyond Causation* (2018).

[36] Craver (2014) offers a valuable elaboration on the ontic/epistemic distinction, distinguishing four "modes in which people (including scientists) talk about explanation": communicative, ontic, textual, and cognitive (30). The communicative mode refers to one person's explaining something to another; explanation as a verb. This communicative activity is between epistemic agents, involving some form of knowledge-transmission from speaker to hearer. The ontic mode, in contrast, locates explanation in the real, "objective" world, independent of our beliefs and epistemic situation. The textual mode of explanation locates explanatory power in an "explanatory text"; i.e., a scientific theory or model, which is said to explain some phenomenon of interest. Lastly, explanation may be conceptualized as a cognitive activity—specifically, the activity of understanding, which Craver takes to be a kind of psychological process with epistemic significance.

accounts of explanation remain dominant in philosophy of science and the ontic–epistemic distinction is alive and well. Influential successors to Salmon's "causal turn" include the new mechanist account and interventionism.

The new mechanist approach to explanation was inaugurated by "Thinking about Mechanisms," co-authored by Peter Machamer, Lindley Darden, and Carl Craver (2000).[37] A mechanistic explanation, as the name suggests, describes a mechanism. A mechanism, according to new mechanists, consists of multiple components that engage in causal relations or activities spatially and temporally organized so as to produce, underlie, or maintain some phenomenon of interest, such as protein synthesis or long-term potentiation.[38] So mechanistic explanation is a special case of causal explanation. Such an explanation describes the relevant components, causal relations, and spatial–temporal organization so as to show how they produce, underlie, or maintain the phenomenon that is thereby explained. Mechanistic explanation is causal, although exactly how these causal aspects of mechanisms should be analyzed is the subject of considerable debate.[39] One analysis often deployed in this regard is James Woodward's interventionism (2003). The latter is the most widely-accepted philosophical account of explanation today, defining an explanation as revealing causal dependency relations that are analyzed as a distinctive kind of counterfactual.

According to interventionism, a cause is some item represented by a variable, changes in the value of which make a difference to the value of another variable, under a set of specified conditions that serve to isolate the relation at issue in the manner of a controlled experiment. An intervention is, roughly, an idealized hypothetical experiment that isolates a causal relationship between variables X and Y, if any exists, by breaking all potentially confounding causal ties that could produce correlated changes in the values of X and Y. More precisely, "X is a type-level cause of Y in some background condition Bi . . . if and only if there is a possible intervention on X in Bi such that if such an intervention were to occur, the value of Y or the probability distribution of Y would change" (Woodward 2013, 45). The last clause in this quote is the schema for an interventionist counterfactual, the class of statement Woodward uses to analyze the meaning and truth conditions of causal claims. Causal relations are defined as invariant (i.e., the same functional form maps values of X onto values of Y) under some (not all)

[37] Earlier works, notably Bechtel and Richardson (1993), also contributed importantly to the new mechanist account. But the 2000 paper more explicitly aims at an account of explanation (in molecular biology and neuroscience). Craver (2007) is the most developed account of the latter.

[38] This terminology follows Machamer et al. (2000) and Craver (2007); see also Glennan (1996, 52); Bechtel and Abrahamsen (2005, 423), Craver and Darden (2013). The disjunction "produce, underlie, or maintain" is from Kaplan and Craver (2011). Note that this definition is narrower than that proposed by Machamer and colleagues (2000), being restricted to what Craver (following Salmon) refers to as "constitutive mechanisms."

[39] For a range of recent views, see Glennan (1996, 2002, 2017) Woodward (2002, 2013), Bogen (2005), Waters (2007), Ross (2015).

possible interventions. An interventionist explanation aims to answer "what-if-things-had-been-different?" questions about Y, thereby enabling prediction and control of Y's value across a range of circumstances. Causal relations of explanatory interest to us often exhibit properties like robustness, specificity, or depth, which Woodward and other interventionists analyze in terms of their counterfactual theory.

While none of these influential causal accounts is inherently monistic, debate about them is heavily influenced by explanatory monism. All three were introduced as *one important kind of explanation*—which evidently allows for others. Indeed, Salmon (1984, 1989) explicitly denied that C-M covers all cases of scientific explanation. His final word on the subject was a dualistic integration of alternative accounts, combining "bottom-up" causal and "top-down" covering-law views of explanation. Salmon conceptualized the complete philosophical account of explanation as a collaborative effort. Similarly, Machamer, Darden, and Craver (2000) identified mechanistic explanation as the predominant kind of explanation in "many fields of science," leaving the scope of their account "an open question" (1–2). Their initial claims were limited to molecular biology and neuroscience. Woodward's interventionist account was introduced as "a comprehensive account of causal explanation that applies to a wide variety of causal and explanatory claims in different areas of science and in ordinary life" (2003, 4)—a wide swathe of cases to be sure, but a far cry from all explanation. However, in subsequent philosophical discussion, the modesty of these initial causal proposals fell away. Over years of philosophical debate, causal accounts tend to 'harden' from initial pluralism to stark monism. Others have noted this trend. For example, in the introduction to their 2018 edited volume on non-causal explanation, Alexander Reutlinger and Juha Saatsi (2018) note that "Salmon's pluralist outlook and his portrayal of the history of the debate (articulated in his *Four Decades*) were largely lost in subsequent philosophical work" (1). Philosophical debate proceeded as though Salmon's C-M, and other causal theories, were monistic. Although few if any new mechanists argue that *all* scientific explanation is mechanistic, the arc of debate has steadily extended its range. Perhaps most stridently, Craver (2007) and Kaplan and Craver (2011) argue that all explanation of the brain is (or should be) mechanistic, effectively presuming explanatory monism in a limited domain. Woodward's interventionism is now the dominant philosophical account of explanation, the presumption that it covers all cases of genuine explanation often taken as a default. Against this background, Woodward's own recent suggestion (2019) that there may be multiple kinds of explanation (e.g., law-based in physics, mechanistic in biology) appears as a significant, even original, concession. Causal accounts of explanation are defended or criticized as *the* account of explanation across all sciences (or, in the case of mechanism, a broad class of sciences), even when their innovators initially deny this monist presumption. 'Creeping monism' of this sort attests to the entrenchment and ongoing influence of monistic assumptions about explanation in philosophy of science. Recent debates reveal monist commitments in other ways, too.

Recent debates

As noted, interventionism has replaced the covering-law account as hegemonic focus of discussion. Current debate is largely framed around interventionism's central concepts and their application to scientific examples. The latter are plentiful; interventionism accords well with many examples of explanation from scientific practice. This has generated a sizeable literature extending interventionism beyond its original grounding in quantitative social science, notably to biology (e.g., Craver 2007, Woodward 2010).[40] These vindications of Woodward's account are tempered by study of its limits: examples of scientific explanation that do not fit interventionist requirements (e.g., Batterman 2010, Huneman 2018). Two decades of confirming and disconfirming studies has led to lively ongoing debate over how to theorize the full range of admitted scientific examples. No one today (to my knowledge) argues that there are no causal explanations in science. The question is whether there is also non-causal explanation, or whether a suitably sophisticated causal account can subsume all genuine cases of explanation across all the sciences.

So one strand of current debate about explanation can be summed up as "causal or non-causal?" But the overall dialectical situation is more complex. In a recent overview, Alexander Reutlinger and Juha Saatsi (2018) distinguish three main positions: causal monism, general monism, and pluralism. Causal monists hold that all explanation is causal explanation (e.g., Strevens 2008, Skow 2015, Potochnik 2017). These accounts are accompanied by careful explication of the meaning of "causal," together with other conditions and constraints that aim to capture the role and significance of explanation across all the sciences. General monists allow for causal and non-causal explanation in the sciences, proposing a more abstract and inclusive analysis of all scientific explanation. The prevailing approach is to generalize Woodward's counterfactual theory, analyzing explanation in terms of "generic counterfactual dependence" rather than causal dependence only (e.g., Bokulich 2011, Reutlinger 2018). According to this "liberal consensus," to borrow Kuorikoski's (2021) phrase, an explanation reveals counterfactual dependence relations among phenomena of interest, with those relations further subdivided into several types. All explanations display some version of a counterfactual dependence relation between the explanandum and explanans. Different dependency relations correspond to different types of explanation: causal, mathematical, structural, perhaps more. There is significant debate within the liberal

[40] There is also a substantial theoretical literature exploring interventionism's conceptual and metaphysical implications, although this is less relevant for comparison with EP.

consensus. For example, Povich (2018) argues that this general account requires an "ontic conception of scientific explanation," while Bokulich (2018) defends a model-based "eikonic conception." Most general monists agree, however, that all explanations reveal dependency relations of some kind or other among phenomena of interest, such that "explanatory power consists is being able to answer what-if-things-had-been-different questions" (Povich 2018, 119).[41] So we can distinguish two main flavors of monism—causal and general—both importantly centered on Woodward's interventionism.

The third main position is explanatory pluralism. In contrast to monists, these theorists propose distinct analyses for causal and non-causal explanations. They agree with general monists that science includes both causal and non-causal explanation, but disagree that a single general account can subsume all varieties of both. Non-causal explanations are positively explicated in various ways: as mathematical, structural, topological, unifying, and more (e.g., Huneman 2010, Batterman and Rice 2014, Rice 2015, Lange 2016, Reutlinger and Saatsi 2018). So there are a variety of pluralist positions on offer, depending on which kinds of non-causal explanation are admitted. A recent special issue of *Perspectives on Science* (Kostić 2019) aims to explore and systematize these varieties into a unifying philosophical framework, using examples from physics and theoretical evolutionary biology. A reader might well wonder: isn't explanatory particularism just another flavor of explanatory pluralism? Why introduce it as a new approach, complete with new name? Both explanatory pluralism and explanatory particularism reject monism, demanding more than one substantive philosophical account to cover all cases of scientific explanation. So they have that in common. But there are important differences too. Most saliently here, explanatory particularism relates differently to the monist tradition of philosophical studies of explanation.

Pluralists about explanation are largely working within the monist tradition. They take as their point of departure the dominant account of explanation on offer today: causal explanation (i.e., interventionism or some elaboration thereof). To show the limits of this causal account, they look to examples from theoretical physics and evolutionary biology. These are the specializations traditionally given the most attention and accorded the highest prestige by philosophers of science. Epistemic kudos philosophers bestow on these fields make it hard to dismiss these examples as unscientific, second-rate, or unworthy of philosophical consideration. Pluralists argue that these high-status examples do not fit the criteria for causal or generic counterfactual explanation. On the positive side, they offer alternative accounts of what does the explanatory work in their examples. Their alternatives tend to have firm roots in mid-twentieth-century philosophy of science: unification, universality, mathematical structure. The goal of these pluralist arguments

[41] But see Khalifa et al. (2021) for an inferential alternative.

mirrors that of explanatory monism; a substantive account of non-causal explanation as *one kind of thing* (perhaps two, at most). For example, Reutlinger and Saatsi define explanatory pluralism as the thesis that there are different types of explanation covered by "two (or more) distinct theories of explanation" (2018, 5) while Pincock defines it as the idea that "explanations come in several types" (2018, 39). Pluralism of this sort is just one step away from monism. The goal is to characterize non-causal explanation positively and substantively, as distinctively mathematical, structural, topological, etc.—a counterpart to causal explanation. Overall, scientific explanations are to be analyzed as causal, mathematical (and/or structural, etc.), or a combination of the two.

In considering this version of pluralism, my mind always leaps to a scene from *The Blues Brothers*, where the bar-owner tells the titular duo "We play both kinds, country *and* Western!" Of course, that's an intuitive response, not an argument. It's not even a good analogy! 'Causal vs. non-causal' is logically exclusive, and the differences claimed to hold between causal and structural explanation (for example) are clear and significant.[42] But the intuition isn't groundless either. Having only two or at most three ways of thinking about scientific explanation is prima facie very limited, given the richness and diversity of efforts at understanding in scientific practice. The main contrast between explanatory pluralism and explanatory particularism is that the latter leaves the number of distinct kinds of scientific explanation unlimited. Indeed, classifying all scientific explanations into one or a few kinds is entirely beside the point for a particularist. The goal instead is to examine major achievements that advance understanding in a wide variety of particular scientific communities, past and present, yielding an array of 'thick' accounts of those explanations and the contextual factors that shape their construction and use. This does not exclude the result that some features of particular explanations and related community factors are widely shared across many scientific fields, or even all. But that result would emerge (if it does) from comparative study of a wide range of particularist accounts—a later stage of investigation than we presently occupy. Where explanatory pluralism admits two (possibly three) kinds of explanation rather than one, explanatory particularism places no limits on the variety of scientific explanations.

Explanatory particularism does not work within the monist tradition of studying explanation. It departs from that tradition by straight-up abandoning the search for a substantive account of explanation for all the sciences. The goal is not to 'cover' all scientific explanation with two such accounts—causal and non-causal—but instead to examine the full range of diversity of explanation in scientific practice; how they are constructed in particular contexts, in accordance with

[42] A better analogous typology of music would be both country and pop. It's not funny, but it does highlight the limitations of dualism in this sphere.

the values and constraints of those communities, and how those constructs advance understanding according to members of those communities. Philosophical commitments—including substantive theories about explanation—come in later, in analyzing the explanatory practices of a particular scientific community. This allows for the possibility that one scientific explanation could exhibit features (virtues) associated with multiple different philosophical accounts of explanation: supporting generalizations, revealing patterns of causal dependence, revealing basic principles or core mechanisms, and so on. EP also allows for philosophical study of how different scientific explanations relate to one another—an issue that has received startlingly little attention in philosophy of science, except for debates about reductionism.

Engaging diversity: new accounts

Very recently, some philosophers of science have sought to engage the diversity of scientific explanations in practice. These "diversity studies" of explanation (my term) share important themes with EP; their proponents are fellow-travelers with explanatory particularists. Yet none (to my knowledge) go as far as EP in departing from philosophical tradition. Some of these new accounts retain the monist assumption, arguing that the variety of scientific aims leads to different instantiations of one kind of explanation across all scientific practice. For example, Potochnik (2017) argues that all explanations capture causal patterns; i.e., convey accurate information about causal dependence and its scope. However, which pattern is explanatory for a given research project depends on the phenomenon of interest, research methods, and the project's specific goal. Although a version of causal monism, Potochnik's view admits a wide variety of causal explanations at the grain of research programs, researchers' interests, and cognitive aims. Her pluralist account also allows for different ways that distinct explanations might relate to one another: as competitors, independent autonomous explanations, or causally integrated. Potochnik and de Oliveira (2020) apply this account to cognitive science, identifying several different explanatory styles at work in that field: "different, independent explanations . . . provide cross-cutting accounts, sometimes even of the same phenomenon, that capture different causal patterns" (1314). Similarly, Pincock (2018) endorses pluralism about explanatory relevance relations within a monist approach, while Hochstein (2017) distinguishes several goals for scientific explanation, arguing that satisfactory scientific explanation (at least of complex phenomena) requires multiple models to adequately describe the "physical mechanisms that generate and sustain the target phenomenon" (1106). These recent accounts analyze explanation across the sciences in terms of 'unity in diversity,' reconciling explanatory monism with diverse explanations in scientific practice.

Other recent accounts endorse pluralism about explanation, but reframe the debate so as to challenge the binary of causal vs. non-causal explanation. Kostić (2019) proposes that mathematical explanations are non-causal to different degrees, "reflected in the level of explanatoriness, unification, modal strength, and explanatory depth" (4). The more non-causal a mathematical explanation, he proposes, the better it works in combination with a causal explanation to satisfy this set of explanatory desiderata. Ross (2021) challenges the distinction between causal and topological explanation, arguing that borderline or combined cases appear in scientific practice also, citing examples from systems biology, biochemistry, and ecology. Mixed cases such as causal topological explanation, she argues, afford greater engagement with diverse explanatory practices in science than standard causal and non-causal accounts alone. Ross' querying of the causal vs. non-causal distinction is a noteworthy step away from the traditional philosophical framework inherited from monism. Rice and Rohwer (2021) go even further in engaging diversity of explanations in scientific practice, proposing that explanation should be reconceived as a cluster concept: "there are multiple subsets of features that are sufficient for providing an explanation, but no single feature is necessary for all explanations" (1025). Surveying the philosophical literature, they recognize at least seven different overlapping subsets of features defining a kind of explanation: covering-law, causal, mechanical, unificationist, statistical, mathematical, and equilibrium explanations.

These recent proposals indicate growing appreciation of the diversity of explanations in scientific practice and the limits of traditional philosophical frameworks in engaging that diversity. Emphasizing the diversity of scientific aims and research programs, as well as challenging the traditional causal vs. non-causal binary, these new accounts share much with explanatory particularism. However, the latter departs more thoroughly from the tradition of philosophical debate about scientific explanation. Consider: Rice and Rohwer survey the philosophical literature to identify the overlapping subsets noted above, showing that none of the individual proposed features is necessary. They use that same literature to define the different subsets. This ties their cluster-pluralism to the established philosophical framework for debating explanation, albeit in a more open-ended and less restrictive way. But in using the philosophical tradition to define the categories of possible scientific explanations, Rice and Rohwer retain that tradition's limitations. The philosophical literature does not, and has not for seventy years, examined scientific practice in a representative or evenly distributed way. Nearly all the cases considered in the literature that gives rise to the criteria Rice and Rohwer consider are from physics or theoretical biology. This is a very, very incomplete and skewed sample of scientific explanations. The subsets that Rice and Rohwer recognize still reflect philosophical biases and assumptions, rather than scientific practice itself. Similarly, Ross and others interested in explanatory diversity are working within the established framework that has shaped recent debate. Explanatory particularism dispenses with that framework, such that the terms of philosophical debate do not define all the possible options for scientific explanation. In this way, explanatory particularism is committed to more thoroughgoing engagement with scientific practice, while breaking more decisively with the tradition of philosophical debate about explanation. One could think of EP as a further distance along the path started by these recent accounts emphasizing diversity in explanations across the sciences.

One other recent account of explanation shares more with EP, notably its social aspect. Woody's functional account (2015) takes the role of explanation in scientific communities as primary, directing philosophers to study explanation via close examination of particular practices. Her proposal, based on detailed studies of chemistry, has significant parallels with EP. Shifting focus from individual explanations to explanatory practices in particular fields, Woody argues that "the practice of explanation serves to maintain coherence and cohesion within and across scientific communities" (86). The ideal gas law, for example, constitutes "an ideal of intelligibility for chemistry." On Woody's account, a scientific community is held together by shared standards for intelligibility, embodied and inculcated by explanations. These standards, and associated explanations, can be normatively evaluated in light of community aims. Those aims are diverse, both within and among scientific communities. EP could be fairly considered a further development of Woody's view. There are some salient differences though. Woody emphasizes the pedagogical role of "known explanations" and the role of explanatory discourse in establishing and maintaining group identity. In contrast, my version of explanatory particularism focuses more on cutting-edge research, explanatory diversity, and relations between different scientific communities/explanations. Yet the parallels are significant. Woody's and my accounts are both versions of explanatory particularism. Many other scholars (notably in history and philosophy of science) take a particularist approach to studying science, which includes explanation. So it's important not to overstate the novelty of my proposal here. EP expresses ideas and themes that many philosophers of science are already working with, but which have not (yet) had much impact on current debates about scientific explanation. Recent efforts to engage diversity of scientific explanation from within the monist tradition share some of those themes, but do not go as far.

To sum up: for more than seven decades, explanatory monism has been a philosophical attractor; the default assumption for philosophical studies of scientific explanation. Departures from explanatory monism are so far very modest and highly contested. Most pluralists about scientific explanation tend to be committed to only a few types (causal and structural, say). Conceived monistically, the different philosophical accounts of explanation relate to one another as incompatible alternatives, each seeking to rebut the others or to identify an area of science in which their account holds sole sway. Only very recently has explanatory diversity entered the discussion. But it's important to recognize the significance of this move, and the similarities between EP and recent "diversity accounts" of scientific explanation. Explanatory particularism is not entirely an outlier in the current debate. It is, instead, a more radical fellow-traveler of other recent accounts. The goal of this section has been to show the background and dialectical context of EP, positioning my account by contrast to others on offer. In the process, I've argued that explanatory monism is a deeply-entrenched background assumption that has shaped philosophical studies of and debates about explanation for many decades. I haven't, however, argued against explanatory monism and its associated approach to studying scientific explanation. The next section takes on that task.

1.3 Why explanatory particularism?

So far, I've introduced the basic ideas of explanatory particularism, shown how it departs from the established monist tradition, and positioned it in relation to other recent accounts with shared emphasis on the diversity of explanations in scientific practice. The next question is: why? Why be an explanatory particularist, given the profusion of other accounts already available? What motivates explanatory particularism, in our current dialectical context?

Let's first be clear what the reason is *not*. I'm not arguing that all accounts of explanation currently on offer are hopelessly flawed and therefore must be rejected. I don't think any are hopelessly flawed.[43] But I do think that no single such account, even one that admits diverse scientific aims, can adequately characterize the full variety of explanations—achievements of understanding—across all scientific communities. My target is the monist tradition, not individual accounts within it. The various philosophical ideas about explanation that have accumulated over the past seventy years are a hugely valuable resource. But that resource is best deployed as a heuristic toolkit for studying scientific explanations in practice, not a conceptual taxonomy forcing competition between incompatible alternatives.[44] Accordingly, my motivating argument for EP in this section is about philosophical method and how it engages scientific practice. My aim is to transform the philosophical debate, not to introduce a new challenger into a *Highlander*-esque melee. So the argument that follows is against explanatory monism and the strictures its entrenchment places on philosophical debates about explanation. What I'll aim to show is that explanatory particularism is a better-grounded approach to studying scientific explanation than explanatory monism, given several uncontroversial

[43] This tolerance of diversity in accounts of explanation is a reflexive application of ideas about social understanding that emerge in later chapters (esp. Chapter 6). Other philosophical accounts of explanation aren't refuted, but reconceptualized in the particularist framework.

[44] In the same spirit, see Ishida (2020) on realism and instrumentalism.

facts about science and philosophical method. I'll start with a simple (although not decisive) argument along these lines.

A pessimistic pluralist induction

The simplest argument against explanatory monism is that it's been tried for a long time and hasn't worked. This is not to say that seventy years of studying explanation has been time wasted. It is to say that over this interval no single substantive account of explanation has been proposed that plausibly covers all and only cases of scientific explanation, and not for lack of trying. Philosophical accounts of explanation committed to the monist presumption have, so far, been unsuccessful in satisfying it. So it is time to abandon that assumption. Mantzavinos (2016) articulates the situation clearly:

> a cursory glance at the prevailing scientific practices shows that the unitary models of explanation on offer have only a limited range of application. It is simply a matter of fact, thus, that their resources cannot capture anything but the explanatory activities of *some* areas of (theoretical) science. The claim that each of them raises, that is, that it is supposed to accommodate *all and every* scientific activity is not tenable. (8–9; italics in original)[45]

The history of philosophical studies of scientific explanation supports a *pessimistic induction* for explanatory monism. In the past, a number of universal, monistic accounts of scientific explanation have been proposed. For all these, counterexamples have been found that contradict their universality. (The only arguable exceptions to this generalization, pragmatic accounts, succeed by making all but the erotetic character of explanation context-dependent.) Based on this track-record, one can reasonably infer that counterexamples will be found for any substantive universal account of scientific explanation. One can conclude that no single substantive account of scientific explanation can adequately capture all explanations across all of science. Therefore, monism about explanation should be rejected. To accomplish the philosophical goal of accounting for explanatory activities in "all and every" field of science, multiple accounts of explanation are needed.

Explanatory particularism does not ineluctably follow from this pessimistic induction, of course. Explanatory pluralists can agree as well. But if monism is rejected, then the appeal of limited versions of pluralism is also considerably

[45] Mantzavinos also concurs with EP that "the goal of a philosophical account of explanation should not be to capture *the* explanatory relation, but rather to capture the many ways in which explanations are provided in the different domains of science" (2016, 19). However, the way he implements this suggestion is distinct from explanatory particularism (another fellow-traveler).

lessened. As we've seen, the framing of recent debate ("causal or non-causal?" "two or three kinds of explanation?") is largely inherited from monist commitments. Absent monism, why should the causal/non-causal distinction be fundamental for studies of scientific explanation? The traditional list of theoretical virtues (accuracy, consistency, broad scope, simplicity, and fruitfulness) affords more diversity, and it is widely agreed that these different virtues can be interpreted, ordered, and weighed in different ways depending on the case—or, as EP has it, the community context. If we agree that there is more than one kind of scientific explanation, why restrict the number to two or three? That restriction is a relic of monism, filtered through the prism of current philosophical debate. We should let features of scientific practice, not antecedent philosophical commitments, define the terms of debate about scientific explanation. Explanatory particularism does exactly that.

Philosophy of science, in practice

The pessimistic induction is not a decisive argument against explanatory monism. But it sets the stage for a stronger argument, centered on how philosophical methods engage scientific practice. In brief: the practice of philosophy of science, together with diversity of explanations in scientific practice, undercuts explanatory monism. The problem for monism arises from three uncontroversial facts:

- Science today is a vast and heterogeneous human enterprise.
- A single scholar can be deeply acquainted with no more than a few fields of science today.
- Philosophers of science tend to work as lone individuals or in very small groups.

It is a truism that no single person today can be acquainted with all scientific knowledge; at least, not in any depth or with any sophistication. Recognizing this limitation is an important insight of contemporary epistemology. No single scholar, no matter how brilliant, can engage expertly with more than a few fields of science today.[46] Yet to engage with scientific explanations so as to be able to articulate key features of their construction, evaluation and use requires becoming deeply conversant with the practices of particular fields. Consider the three examples of scientific explanation from Section 1.1: the Higgs mechanism, the double helix, DFT for nanomaterials. The descriptions provided there,

[46] Thagard makes a similar point (1997, 258).

jargon-loaded though they seem, are very sketchy and simplified in comparison with their full scientific presentations. To fully appreciate, gain understanding from, and be in a position to analyze even one of these examples would require considerable study. Just as in science, one must specialize to study the sciences philosophically. Even a philosophical duo or small working group can engage at most with a few narrow slices of explanatory practice across the sciences. And here features of philosophical method become important. Philosophers of science, like scholars in the humanities more generally, tend to work as lone individuals or in very small groups. Although there have been few empirical studies of this to date, the pattern is well-known. Philosophical accounts of explanation, like other accomplishments in the field, are credited to one or two individuals.

It follows, then, that every philosophical account of scientific explanation has a parochial starting point, grounded on exemplars of explanation from the scientific fields with which its author (or pair of authors) is best acquainted. While a philosopher might be deeply conversant with a particular tool widely used across many different scientific fields (say, statistical tests or microscopy), they cannot be expected to understand how that tool is deployed in each particular field. The "founder effect" of pioneering philosophers of science taking physics as their exemplar is widely recognized. But less discussed is the inevitability of some version of this founder effect, given the individualistic organization of philosophical inquiry. Scientific practices are specialized and diverse, with explanations among the most distinctive products of an epistemic community. No individual philosopher can develop this expertise for more than a few specialized fields. Different philosophical accounts of explanation conform to and illuminate practices of different scientific communities, each seen through a lens of philosophical commitments. In twentieth-century philosophy of science, examples of scientific practice were almost exclusively confined to theoretical physics. The first philosophical accounts of explanation accordingly emphasized features of explanation in that field: general laws covering a wide range of phenomena, rigorous derivation, logical structure. Salmon's early account of statistical explanation (1971) appealed to microstatistical physical processes and built on Reichenbach's pioneering theory of causal inference, which itself emerged from the latter's investigation of the implications of relativity theory. Salmon's C-M account (1984) retained the focus on physics, emphasizing fundamental causal processes and interactions, with examples almost entirely taken from fields of physical science.[47] More recent

[47] Salmon's account of statistical explanation is, admittedly, less tied to physics than his later C-M view. But the general point holds; Salmon's statistical account is firmly rooted in statistical methodology—itself a distinct scientific specialization, albeit one with links to many other scientific fields (but not all—and more importantly, not without transformation in accordance with the purposes and standards of those contexts).

accounts of explanation are grounded on other scientific fields. For example, the mechanistic view of explanation takes as its starting point molecular biology and neuroscience (Machamer et al. 2000), while Woodward's interventionism (2003) is based on econometrics.

Traditional philosophical studies of explanation are thus encumbered by tension between methodological presumption and actual practice. On the one hand, the entrenched tradition in philosophy of science is to propose a single substantive account of explanation for all the sciences; to analyze the nature and requirements of explanation. On the other, no single philosopher or duo can be well-acquainted with the full range of diversity in explanations offered across the sciences—even limiting the analysis to the present day.[48] Explanations in theoretical physics do not resemble those of molecular biology, while the field of systems biology is in part defined by rejection of the latter's style of explanation. The rise of philosophy of science in practice, mandating close attention to how scientists actually arrive at and use explanations, makes the diversity all the more salient. Yet, if the traditional philosophical project is to succeed, diversity in explanatory practices across the sciences must be epiphenomenal, masking a shared conceptual core that can be philosophically analyzed and explicated. The most popular current candidate for this conceptual core is the counterfactual dependence relation defined in or generalized from Woodward's interventionism (2003). But any monistic account of scientific explanation will be in the same position vis-à-vis scientific practice. In this way, philosophical tradition primes us to look past—to overlook—diversity in scientific practice. Each case of scientific explanation is to be analyzed and shown to exhibit the shared conceptual core; the proposed essence of scientific explanation. The explanation as it appears in scientific practice—with all the specific characters distinctive of the field in which it is constructed and used—is to be analyzed away. Limited pluralist proposals don't overcome this Procrustean tendency. The idea that there are at most few types of scientific explanation is a philosophical construct, an artifact of long-running debate about scientific explanation in which the main alternatives figure as covering-laws or causes. Instead of being framed by the terms of internecine philosophical debate, studies of scientific explanation should be grounded on explanations in actual scientific practice.

The preceding considerations do not show that no monist account of scientific explanation can possibly succeed. There might be a shared conceptual core for all achievements of understanding in all scientific fields, past and present. EP does not exclude that possibility. But supposing there is such a shared core, philosophers of science are not currently in a position to characterize it. That is because

[48] Bringing history of science into play greatly increases the diversity at issue.

philosophical studies of scientific explanation have not examined most scientific fields. Instead, those studies are and have been highly selective, concentrating on slim stock of examples from a narrow range of sciences. In the mid-twentieth century, such extrapolation was supported by hierarchical views of the organization and unification of scientific knowledge (e.g., Oppenheim and Putnam 1958). But reductive hierarchies of this sort are hard to defend today. Yet philosophical debate about explanation remains focused on a very narrow range of examples. This reveals a second, subtler problem: explanatory monism encourages dialectical cul-de-sac when applied to cases. The cul-de-sac arises when philosophers with different views of explanation consider and reconsider the same examples, offering different accounts of what is doing "the real explanatory work" in the case at issue. The cases discussed are overwhelmingly from theoretical physics, evolutionary biology, or everyday life. The scientific examples that come under scrutiny in these debates are from scientific fields that have already been extensively studied by philosophers, who grant their explanations high epistemic status. Debate then focuses on a few key examples from these high-prestige fields (e.g., renormalization group explanations in physics). The dialectic proceeds as follows: one philosopher offers a fine-grained account of what is doing the explanatory work in their chosen example: mathematical reasoning, mathematical structures, or the like. Another responds by accounting for the same example in terms of causal dependence, another variety of counterfactual dependence, or a different non-causal feature.

So far, so good. The problem is that these debates cannot be settled by studying scientific practice. How can we settle what is doing the real explanatory work, without begging the question by presupposing one or another philosophical account of explanation is correct? These are internal debates that turn on philosophical ideas about laws, causality, mathematical reasoning, and more. I am not suggesting that these questions are unimportant. The problem is that engagement with scientific practice is hijacked by these debates internal to philosophy. There is no impetus to examine explanatory practices in fields less-studied by philosophers: enormous swathes of life, behavioral, social, and historical sciences. That is the cul-de-sac: debate settles on a few selected examples, the setting for conflicting philosophical theories and commitments. In this way, philosophical debate about explanation loses contact with much of scientific practice. Monism about explanation, even for pluralists who accept its terms of debate, effectively limits philosophers' view of science, encouraging a kind of myopia centered on traditionally well-studied fields and examples. To be clear: this is not a problem for those interested in clarifying the nature of (say) mathematical structures and their role in reasoning. But it is a significant limitation on engaging explanation in scientific practice. The established pattern of debate between explanatory monists and pluralists encourages losing touch with (or never seeing at all) much of that practice. The upshot is to insulate philosophers from the full range of explanatory activities across the sciences, while a few examples receive disproportionate attention.

Example: the operon model

Philosophy of biology exhibits a peculiar variant of this pattern, in studies of the operon model of gene regulation (Jacob and Monod 1961). The operon model played a seminal role in early molecular biology and continues to figure in molecular and systems-level explanations of gene regulation. But its philosophical treatment is puzzling. Nearly every major philosophical account of explanation—certainly all that have been applied to molecular biology—uses the operon as a supporting example (see Table 1.1). All agree on the basic biological model, and that this model

Table 1.1 Philosophical accounts of the operon model as explanation

Name	Date	Topic	Conclusion
Schaffner	1969	D-N explanation	yes
Hull	1972	theory reduction	no
Schaffner	1974	theory reduction	mixed
Darden & Maull	1977	theory reduction interfield theories causal explanation	no yes yes
Kimbrough	1979	theory reduction	no
Kitcher	1984	theory reduction unifying explanation	no yes
Richardson	1996	abstract networks D-N explanation	yes yes?
Beatty	1997	biological laws theory pluralism causal mechanisms	no yes yes
Woodward	2002	interventionism	yes
Levy & Bechtel	2013	abstract networks detailed mechanisms	yes no
Bich et al.	2015	principles of bio-organization	yes
Fagan	2016	interventionism collaborative explanation	no yes

explains regulation of gene expression.[49] But what the latter claim amounts to runs the full gamut of philosophical views on biological explanation. The operon model has been interpreted as an instance of D-N explanation (Schaffner 1969), a step toward reduction of biological phenomena to physical and chemical laws (Schaffner 1974, Kimbrough 1979), an interfield theory linking genetics and biochemistry (Darden and Maull 1977), an example of anti-reductionism for Mendelian and molecular genetics (Kitcher 1984), an illustration of highly abstract models of gene regulatory networks (Richardson 1997), one of a heterogeneous collection of theories or mechanisms required to "cover the domain" in the absence of biological laws (Beatty 1997), an example of mechanistic explanation supporting interventionism over causal process theories (Woodward 2002), an example of an abstract model that contradicts the basic mechanistic view (Levy and Bechtel 2013), and an example satisfying general principles of biological organization (Kauffman 1976, Bich et al. 2016). Schaffner's 2016 book briefly alludes to the operon model as an example of partial reduction hinging on explanatory generalizations—bringing the example full-circle (though not back to D-N). My own 2016 paper compares Woodward's interventionist treatment of the operon model to Jacob and Monod's original 1961 version, arguing that contrasts between the interventionist version and the original model provide space for another account. There are probably treatments I have missed, and likely others coming in the future. What to make of this embarrassment of philosophical accounts of the operon? There are two traditional answers.

"Anything goes"

One is "anything goes"—the operon model is a mirror in which philosophers see whatever view of scientific explanation they prefer: fundamental laws, description of causal mechanisms, abstract general principles, counterfactual dependencies, whatever. This is not, of course, an appealing response for philosophers, but it may seem intuitive given the dialectical situation around the operon. But this relativist response does not hold up, upon closer inspection. Before 1990 (roughly), the philosophical discussion about the operon model centered on critical assessment of D-N and associated ideas on theory-reduction. Schaffner defended (and modified) the traditional view, while others in the nascent field of philosophy of biology (including Hull, Kitcher, Darden and Maull) used the operon model to argue against reduction of Mendelian by molecular genetics. For over two decades, there was sustained philosophical debate about the operon model, concluding with decisive rejection of D-N for this and similar biological cases. The trajectory of

[49] More details of this case in Fagan (2016a).

philosophical debate shows that not just *anything* goes, for a philosophical account of explanation to fit the operon model. That model does not indiscriminately support every philosophical account of explanation.

"There can be only one"

However, the operon model does fit multiple such accounts—if not *anything*, many things go.[50] And this too may seem unacceptable. The traditional philosophical response to the question 'How can a single scientific case support multiple views of explanation?' is '*It doesn't*.' Different philosophical accounts of the operon are mutually exclusive alternatives, each aiming to capture the essence of the operon model as a scientific explanation in a winner-takes-all competition. However, this response is unsatisfactory, because there are no clear grounds for deciding the contest. To see this, we need to look closely at philosophical treatments of the operon model after about 1990. There is no sustained philosophical debate about the operon from this time. Instead, philosophers of biology writing about the operon embed their views in the details of particular research programs. For example, Richardson looks to theoretician Stuart Kauffman (who wrote about the operon model in the 1960s and thereafter); Levy and Bechtel to Uri Alon's 'motif' approach to systems biology; Woodward adapts Judea Pearl's concept of intervention for econometrics. Darden and Maull use influential textbook treatments of molecular biology as their starting point; similarly for Kitcher.

Given the grounding of each in scientific practice, it is not clear how or why any of these accounts could be decisively rejected. The D-N account *was* rejected, as discussed above. But the same cannot be said of the other accounts. Supporters of each make a good case for their favored account of explanation, grounded in scientific practice. The operon model is often presented as a basic mechanism of gene regulation, grounded in experimental successes of molecular biology. Its relations include specific causal processes and a wide variety of difference-making relations, which afford fine-grained prediction and control over systems to which it applies. Scientists also construe it as an abstract circuit, a regulatory genetic motif, and a multilevel systems model.[51] Its scientific significance is largely due to its being offered as a general mechanism for gene expression; i.e., a small set of principles covering an enormously wide range of cases, which comports with covering-law explanation and explanatory unification. If all are grounded in scientific practice,

[50] For brevity, I will hereafter omit the qualifier "except for D-N"; the different philosophical accounts of the operon discussed in what follows should be understood to refer to those not eliminated through philosophical debate.

[51] Jacob himself characterizes the operon model as a "network of specific, genetically determined circuits" (Jacob 1965, 158).

then how can we select one among the different philosophical accounts of the operon, without begging the question? If a philosopher is already committed to a particular account of explanation, then the traditional response ("there can be only one") will seem obvious—the operon is an explanation of that sort, and all the others are wrong. But if we have already settled on the answer, then why study the operon model—or any scientific examples—in the first place? If the result is a foregone conclusion, then nothing is gained by support from case studies. And if something is gained from the case—if the operon example does offer support for one's favored account—then other accounts enjoy comparable support. The question of how to discriminate among these various accounts of the operon arises as before. On what grounds could a philosopher argue, without assuming the answer in advance, that the operon model explains via fundamental biological principles (say) and *not* by describing a detailed causal mechanism or revealing counterfactual dependence relations? This leads us to the dialectical cul-de-sac discussed above. That debate could continue ad infinitum, sustained by philosophers' differing intuitions and reflective commitments about what explanation really is. Meanwhile, most explanatory achievements of "functional biology" after 1970 remain unnoticed by philosophers of science.

Explanatory particularism leads us out of this cul-de-sac, offering a more fruitful approach for engaging explanations in scientific practice. Rather than pitting different philosophical accounts against one another, we should attend to *all* the operon model's diverse explanatory uses. Sometimes it is used like a general covering-law or principle of gene regulation. Sometimes it is cited as a simple causal mechanism; a basic genetic regulatory circuit. Sometimes these two uses coincide.[52] Sometimes the model is elaborated with more causal and spatiotemporal details, in ways that accord with Darden and Craver's view of mechanistic explanation. Sometimes manipulation and control are emphasized; here the interventionist account comes to the fore. Once the operon model ceases to be the occasion of philosophical turf war, we can turn to more fruitful questions. In what contexts are these various uses significant, and how do they advance understanding in those contexts? How are different explanatory uses of the operon model related to one another (if they are)? Some progress toward an answer emerges from the reflections here. Each philosophical account focuses on one aspect of the operon model, identifying it as *the* source of explanatory power. Darden (with Machamer and Craver) defines mechanistic explanation as a description of a causal process in terms of "bottom-out" activities and entities, with no gaps between start-up and termination conditions. Woodward singles out relations exploitable for purposes of manipulation and control (causal dependencies) as those that are important

[52] See Cartwright et al. (2020) for a rather different argument for the claim that (some) mechanistic and covering-law explanations coincide.

for explanation. Richardson follows Kauffman's articulation of "statistical mechanics for living systems," aiming to identify general principles (or laws) for biological networks. Levy and Bechtel defend the importance of abstract schematic explanations derived from mathematical models and simulation, while Moreno and colleagues seek a unified theory of biological organization and regulation, based on a small set of principles.[53] Each of these accounts can be defended in terms of norms and standards of explanation in *particular fields or research programs*: molecular biology, neuroscience, econometrics, systems biology, and so on. The different philosophical accounts of the operon model are grounded on ideas and practices in particular scientific fields. None displaces or subsumes the others—just like the scientific communities that value this explanatory achievement, each in its own fashion.

Given this situation, the most straightforward and well-motivated starting point for philosophical studies of explanation is the 'home context' of the explanation at issue. Science is increasingly interdisciplinary, specialized, and tied to technological change. Going forward, then, significant scientific explanations will increasingly require contributions from multiple fields and disciplines. In such interdisciplinary contexts, a model supporting different styles of explanation is a valuable resource: a versatile point of contact between fields with diverse explanatory aims and commitments. The variety of explanatory uses of the operon may therefore be an explanatory virtue in its own right (see Part II). For now, there are three EP-takeaways from the operon example. First, the diverse philosophical case studies of the operon, rather than supporting competing theories of scientific explanation, provide a lens through which to discern a variety of explanatory achievements in scientific practice. Second, philosophers of science should emphasize the ways different "local" theories of explanation can overlap or support one another in significant explanatory achievements. Third, philosophers should look beyond familiar cases to study explanations in more scientific communities, including those traditionally neglected: nearly all laboratory sciences, engineering, social and behavioral sciences, communities operating outside academia, and many more. Insights about scientific explanation emerge from considering a variety of in-depth examples of explanations in particular scientific communities, relating these to one another and to philosophical ideas about explanation. Fruits of the long monist tradition are reconceptualized as a resource; conceptual tools for analyzing EP-explanations, their changes, and relations to one another.

[53] To be fair, Levy and Bechtel's approach is pluralist—they claim only that some biological explanations are abstract, succeeding because of (not despite) absence of mechanistic details.

Beyond explanatory monism

I have argued that there are good reasons for philosophers of science to abandon monist commitments that have directed and constrained the study of scientific explanation for (at least) seventy years. Explanatory monism is not well-motivated from the standpoint of scientific or philosophical practice. It has been tried for roughly a human lifetime and has, so far, not succeeded (pessimistic pluralist induction). Philosophers' lone-wolf methods are at odds with the heterogeneity and specialization of science, so any substantive philosophical account of explanation is unavoidably parochial (philosophy of science, in practice). Compounding this limitation, monist commitments lead philosophical studies of explanation into dialectical cul-de-sacs, with no impetus to look beyond a few over-studied cases. Today—in our own philosophical context—explanatory monism is an unreasonable aspiration, as attested by recent studies on the diversity of scientific explanations. Explanatory particularism offers a way forward via engagement with a wider range of scientific communities and their explanatory practices; a way out of the cul-de-sac. For scholars of explanatory diversity, EP is but a short step further along their path. There is nothing to lose by endorsing it but the limitations of monist assumptions and entrenched epistemic hierarchies. Philosophers committed to explanatory monism might understandably find explanatory particularism unappealing. My hope is that people in this group recognize explanatory particularism as a legitimate approach to studying explanation, even if they do not take it themselves. (Part II argues that such an outcome is epistemically valuable in its own right.)

What about more limited forms of explanatory pluralism, admitting two or three types of scientific explanation? I've argued that the restriction to a few distinct types of explanation is a relic of the monist framework. To press a point made above: what grounds the typologies of "causal vs. non-causal" or "causal or mathematical" in studying scientific explanation? If the ground is purely philosophical, why impose those categories on the sciences? Absent claims of epistemic imperialism from philosophers (which scientists laugh off or ignore), there must be some ground for these categories in scientific practice. But if so, then explanatory practices across the sciences should partition into 'territories' characterized by different types of explanation. That is, we should see major categories or branches of science (natural and social, theoretical and experimental) map onto the types of explanation. But the different types of explanation identified by philosophers do not map onto branches of science in any clear-cut way. Instead, their territories overlap, leading to persistent debate over exactly what does the explanatory work in particular cases: the dialectical cul-de-sac.

One pluralist response to this situation is to impose increasingly fine-grained distinctions on explanations in scientific practice, to partition explanatory power in any particular case to, say, mathematical and causal aspects. This is done by distinguishing explananda: specific questions answered by different kinds of scientific explanation (e.g., Bokulich 2018; Chirimuuta 2018, 166–169).[54] But imposing these distinctions drives a further wedge between philosophy and scientific practice. An achievement like the operon model is dissected—atomized—into distinct kinds of answers to specific questions. The locus of scientific interest and explanatory value from a particular community's point of view is fragmented. This obscures the explanatory achievement in its own context, making it hard to see why the operon model (say) has impacted biology, in various ways, for more than fifty years. Explanatory particularism does not demand that we throw away philosophical accounts honed in the monist tradition. But it does demand that we reconceive those accounts' relation to scientific practice. Instead of partitioning away scientific explanatory achievements to resolve philosophical competition over cases, we should look to those achievements on their own terms. Hard-won philosophical accounts can then be used to clarify those achievements.[55]

This section has argued that, unlike explanatory monism and limited forms of explanatory pluralism, EP is solidly grounded on unexceptionable facts about science and philosophy. I want to conclude by addressing a further factor that may block EP's acceptance, despite the foregoing arguments. That is the hierarchical image of explanation inculcated in philosophical education. Few would dispute that there are many scientific explanations, which vary in their apparent features over time and across disciplines. A monist aims to show that all these are instances of the same single thing—to articulate a general account that captures the unchanging essence of scientific explanation. Hempel further posited a logical system of scientific explanations, organized by laws ordered by generality. Unificationism defines explanation in this image: all our scientific knowledge arranged in a pristine logical lattice, the grasping of which amounts to understanding. The idea of explanatory unification captures, not what philosophers find in the sciences, but what they believe scientific explanation to be like on the basis of their specialized training. Many philosophers of science associate scientific explanation with a logically ordered system of laws covering facts, reducing the bewildering diversity of observations to clear logical order, explanatory power flowing from a small "core set" of simple fundamental laws or principles. This image of scientific explanation is, for many philosophers of science, almost synonymous with "explanation" and "understanding." It is part of the background framework of our epistemic community—just as ideas about explanation and understanding are for scientists

[54] Interventionism builds in a version of this fragmentation of explanatory achievements, conceptualizing explanation in terms of answering a range of 'what-if-things-had-been-different-questions' involving substitutions in the values of variables representing features of a causal system or background conditions.

[55] See Part III for examples.

in their own particular fields.[56] The difference is that the philosophers' image is attributed to science, not applied to their own communities. The full range of explanatory practices across the sciences is obscured by this basic piece of philosophical training. Explanatory particularism replaces the traditional hierarchical image of explanation with the complex, variegated social organization of science.

1.4 Summary and next steps

This chapter has introduced and motivated the core ideas of explanatory particularism, positioning that approach vis-à-vis the established monist tradition of philosophical studies of explanation. Section 1.1 presented thesis EP, unpacking its basic ideas and background assumptions. Section 1.2 contextualized EP by describing the monist tradition from which it departs. Section 1.3 argued that EP is well-motivated as philosophy of scientific practice, and today, better-motivated in this regard than monism about explanation. Some of that argument is negative, criticizing the monist tradition. Those critiques are not refutations, however. Explanatory particularism is a *departure* from tradition, not erasure of it. This 'non-refuting' argument strategy is reflexively appropriate for my account. A core theme of this book is that diversity (in explanations that advance understanding) is a resource, not an obstacle to be overcome. This extends to philosophy as well as science. My commitment to pluralism is expressed in the book's argument and organization of chapters, as well as their content. As part of a more diverse collection of approaches, monism too can be valuable—indeed, it has yielded over seventy years of insights about scientific explanation. I do not advocate tossing those aside and starting over, if indeed that were possible. Rather, I propose that we now take a different grip on those insights about explanation and carry them forward on another path. The same reflexive commitment is behind the lengthy discussion of other accounts of explanation. Explanatory particularism has its own community context; relations to neighboring views and accounts. Comparisons and contrasts among these both clarify and contextualize my own view.

The label "particularism" is apt, implying that a single, general, and substantive account of explanation is not to be had.[57] Monist aspirations to such an account have greatly influenced philosophical studies of scientific explanation, from the heyday of logical empiricism onward. Explanatory particularism thus departs from entrenched tradition, and for that reason may seem wrongheaded to many

[56] To be clear: EP does allow for explanations with this pattern. But it doesn't define explanation in accordance with that image.

[57] Although explanatory particularism is a form of pluralism, that term is not as apt, because "explanatory pluralism" today refers to views very unlike mine. Also, explanatory particularism is not to be confused with moral particularism, though there are some shared themes (thanks to Elijah Millgram raising this point).

philosophers of science. Yet, as I've argued, it's actually the monist tradition that is in tension with scientific and philosophical practices. Explanatory particularism is grounded on uncontroversial facts about explanation in scientific practice, and further motivated by features of philosophical method. Its core thesis (EP) directs philosophers studying explanation to engage closely with scientific practice in one or a few particular communities. Scientific explanations so conceived are significant products—achievements—of epistemic communities. They are *specialized*: suited to and reflecting the goals, standards, and methods of the community in which they are constructed and used. Scientific explanations are local achievements that advance understanding of topics of inquiry in a particular field. What an explanatory achievement consists in, the ways in which it is valuable, what understanding amounts to—all these vary widely across scientific communities. Most philosophically interesting features of scientific explanations—explanatory virtues, locus, and nature of explanatory power—are context-dependent. To gain understanding from an explanation, one must connect with the particular community in which that explanation is produced. Philosophical study of scientific explanation should begin with explanations in scientific practice, rather than with the categories and concepts that philosophers traditionally use to analyze explanation. That's the method associated with EP; a form of naturalistic philosophy.[58]

However, hewing closely to scientific practice might be too much of a good thing. If traditional analyses and typologies of explanation are eschewed, then what added value is provided by philosophy? Is the proposal just to replace philosophy with descriptive case studies—empirical science studies, the stuff of history and sociology of science? In that case, explanatory particularism is born redundant, so to speak. But there is another way. To be sure, a particularist's philosophical framework for analyzing explanations cannot itself prescribe substantive conditions for scientific explanation. To do otherwise would be to contradict the very idea of explanatory particularism—that a substantive account of scientific explanation should be gleaned by widespread, detailed study of particular scientific fields. I shall, instead, construct a philosophical framework oriented toward the *conditions for* constructing and using an explanation; i.e., a significant achievement of understanding for a scientific community.[59] Those conditions are at once epistemic and social. Articulating this framework is the task of the next two chapters.

To briefly preview: my social epistemic framework for analyzing scientific explanation is grounded on two philosophical resources. The first is from philosophy

[58] Thanks to Dan Burston for noting this connection with naturalism.

[59] Isn't this just another form of explanatory monism? To be drawn into that argument is to choose a terminological dispute over practical studies of scientific explanation. To those who consider EP just another form of monism, I would ask them to consider the substantive comparisons and contrasts between EP and other accounts, discussed in this chapter. If they concede that EP is different from other accounts, even if all are labeled "monism," that is enough for my purposes.

of science: conceiving epistemic products as models. The second is from social philosophy: analyses of collective action and intention. A different selection or combination of resources would yield a different version of explanatory particularism.[60] My two framing resources correspond to epistemic and social aspects of particularist explanations. The epistemic aspect is explicated in terms of models (Chapter 2). The social side is explicated in terms of collaborative activity (Chapter 3). This framework develops the basic ideas of EP to articulate a fuller version of explanatory particularism. That completes the task of Part I: setting out my explanatory particularist framework as a new approach to philosophical study of explanation. Part II applies that framework to interdisciplinary contexts.

[60] That's appropriate; I'm not claiming to exclusive rights to EP. As noted, many scholars already use the associated approach.

2

Explanatory Models

This chapter and the next develop a philosophical framework for particularist explanations. Of course, that framework shouldn't pre-judge scientific communities' explanatory achievements. Explanatory particularism asserts that those achievements are diverse, to be examined on a case-by-case basis. The framework instead specifies *conditions* for construction and use of particularist explanations. Those conditions are both epistemic and social; articulated in this and the next chapter, respectively. The goal of this chapter is two-fold. The first is to clarify and motivate the following thesis: as epistemic products, particularist explanations are fruitfully conceptualized as *models*. My argument for this thesis connects the basic ideas of EP from Chapter 1 with insights from philosophy of science on models and modeling. The second goal is to develop that thesis into a general modeling framework for studying particularist explanations. The latter sets particularist studies of scientific explanation apart from the work of historians and sociologists of science (see Chapter 1). It is, of course, not wholly particularistic. There's no inconsistency here; explanatory particularism does not imply thoroughgoing particularism about everything.[1] Nonetheless, although I don't argue for particularism about models in science, my thesis and resulting framework are compatible with that view. This too is rooted in the philosophy of science literature, which includes some influential particularist ideas.

This chapter proceeds as follows. Section 2.1 clarifies the sense of "model" at issue, indicating its grounds in the philosophy of science literature. Section 2.2 shows that this meaning dovetails with EP's basic ideas and associated approach. The result is a thesis interpreting EP in terms of models: EP_M. Section 2.3 illustrates this elaboration of EP with a detailed example: embryo models in stem cell research. This case highlights the variety and intricacy of model–model relations in a diverse scientific community. Section 2.4 builds on this result, proposing a general framework for characterizing such relations (applied in Part II). Insights from using this framework feed back to contribute to philosophical studies of models in scientific practice, which come to the fore in later chapters. So, this chapter, like the next, is largely groundwork for Parts II and III.

[1] See Mills (2005) for limits of thoroughgoing particularism.

Explanatory Particularism in Scientific Practice. Melinda Bonnie Fagan, Oxford University Press.
 DOI: 10.1093/9780191998874.003.0003

2.1 Meanings of "model"

In this section, I clarify a meaning of "model," grounded in the philosophical literature, which fits the main ideas of explanatory particularism. In science, the term "model" is used in many ways, and it's difficult to identify a precise general meaning. The meaning I propose will be familiar to most philosophers of science. But it's a rather loose concept, nothing like a rigorous definition or analysis of scientific models. Several philosophers of science have proposed definitions and classifications of models in science (e.g., Suárez 2004, Giere 2004, Weisberg 2013, Morrison 2015). Why not just use one of their analyses? The reason is that these philosophical accounts (and others) are motivated by concerns distinct from those of this book. Philosophical accounts of models and modeling aim to characterize the nature and role of (scientific) models, and to specify how models fulfill those roles. Major topics of debate include what relation models bear to their worldly targets, how models enable "surrogate reasoning" about those targets (if they do), and the significance of models misrepresenting their targets.[2] Philosophers who study models and modeling in scientific practice are rightly concerned with these questions. But my concern is to articulate a sense of "model" that doesn't pre-judge matters that EP leaves open—a meaning inclusive enough to accommodate all potential explanatory achievements across the sciences. As a by-product, this sense of "model" also leaves open many questions that general accounts aim to answer. Insofar as possible, I remain neutral as to the nature, types, and roles of models in science, as well as debates about scientific representation and model-based inference. Where my conception of models is not neutral is in its inclusiveness. The meaning of "model" for explanatory particularism is thinner and less restrictive than most definitions on offer in philosophy of science. For this reason, particularist accounts of models and modeling are natural partners of EP.[3] Insofar as the arguments of this and subsequent chapters succeed, they support model particularism. However, a principled argument for that view is beyond the scope of this book.

What does it mean to say that particularist explanations are models? We can start with a basic idea, adapted from Giere's agent-based analysis of scientific representation (2010):

[2] See Frigg and Nguyen (2018).

[3] Thanks to an anonymous reviewer for raising this point. Particularist accounts of models include the 'mediators' view (Morgan and Morrison 1999), deflationism (Odenbaugh 2021, Downes 1992, 2020), and Parker's adequacy-for-purpose account (2020).

(M) A model figures in the four-place relational schema < S, M, X, P >, where S is an agent, M is a model, X is M's target, and P is the purpose (or purposes) for which M is constructed and/or used.[4]

For reasons noted previously, (M) is not an account of models and modeling in scientific practice. It is, instead, a resource adapted from others' accounts to develop my account of particularist explanation. Schema (M) states that a model figures in a four-place relation between an agent, a model, a target, and purposes. What are those relations, exactly? (M) is adapted from Giere's agent-based account of scientific representation, which states that "[a]gents (1) intend; (2) to use model, M; (3) to represent a part of the world W; (4) for purposes, P. So agents specify which similarities are intended and for what purpose" (2010, 274).[5] In Giere's schema, then, S relates to M as users to a tool; M relates to W via similarity, in respects and to degrees determined by P; and P is had by S, cognitively and conatively. (M) is, roughly speaking, a liberalized or relaxed version of Giere's schema. It's relaxed in that it doesn't assume that models are representational, that their targets are parts of the world, or that they relate to those targets via similarity. (M) allows for all these situations, but does not require them. There are some requirements though; (M) is not vacuous. I'll discuss its content point by point.

[1] A model is directed toward some target (M-to-X).
This directed relation is essential. There are no models simpliciter; a model must be *of* something other than itself.[6] A model on its own, isolated, and lacking any target whatsoever, is a contradiction in terms. So the concept of a model is relational, and that relation is directed: from a model to its target. Many philosophers who study models and modeling take this directed relation to be representational. (M) does not assume this. The nature of model–target relations is left uncharacterized in general. In particular cases, of course, there will be something to be said about this relation. Motivation for this lack of specificity is partly an effort at philosophical neutrality, and partly an implication of explanatory particularism. As to the former: representationalism about models in science is a popular position, but there are structuralist, inferentialist, and fictionalist accounts on offer as well. In keeping with its ecumenical orientation, explanatory particularism strives to be

[4] Although Giere's work is important and central (esp. his 1988), it is not the sole basis for (M); many other philosophers have contributed to the notion of "model" clarified here. These include van Fraassen (1980), Lloyd (1988), Cartwright (1999), Teller (2001), and Godfrey-Smith (2006).

[5] This agent-based account updates Giere's classic account of models and scientific representation (1988), which identifies a model as an abstract entity that is defined by a set of statements and perfectly satisfies one or more laws (sets of equations). Van Fraassen's semantic account is also seminal for much recent work on models and modeling (1980), on which the view presented here draws. Thanks to Lisa Lloyd for raising this point.

[6] See Keller (2000).

neutral with respect to ongoing debates about models in scientific practice. So (M) allows for, but does not presume, representationalism about models in science. To be sure, many scientific models function by representing their targets. If those models are explanatory, then that representational role will emerge from empirical study of those models in their community context.

There are also particularist grounds for leaving the M-to-X relation open and unrestricted. The class of cases with which this book is concerned are those where 'M' is a significant achievement that advances understanding in some scientific community. There must be some target for that achievement; an answer to the question "understanding of what?" The directedness of the model–target relation coincides with the directedness of the explanation to what is explained (explanans to explanandum, to use standard philosophical terminology). But what that target is, and the nature of the model–target relation, must be left open to avoid pre judging features of a community's explanatory achievement. Explanatory particularism places no restrictions on the explanans–explanandum relation or on what can play the role of explanandum (i.e., target). Those specifications are relative to particular contexts; particular scientific communities.[7] Philosophers accustomed to more traditional approaches to studying explanation might ask at this point: don't explanatory models by definition include both explanandum and explanans? Causal models, for example, usually include variables for both causes and effects.[8] The EP framework insists on the distinction between model and target; between scientists' constructs and that which they seek to understand. A model is directed to its target, but distinct from that target. Moreover, in scientific practice, an explanatory model's target is often vague, fuzzily-bounded, ambiguous, or variable—hard to get a fix on. As with the rest of explanatory practice, explanatory targets change over time within a specialized field, just as major achievements of understanding do. The model–target distinction allows for such variability and dynamism, as well as keeping the notions of model and target distinct. Some explanatory models represent explananda and explanans; others don't. The EP-models framework allows for all such cases.

[2] Model use is intentional.
A model's scientific role is relative to and partly determined by its users' intentions and purposes. This amounts to rejecting the idea of a dyadic model–world relation. Instead, some agent S is involved in that relation too. That involvement is by

[7] Note that Giere's assumption of realism is also relaxed; a model's target need not be some part of the real world. Often that will be the case, but the requirement is not built into (M). The target is just whatever M is directed toward by S in light of P. (This is why (M) substitutes Giere's 'W' for 'X'.)

[8] Thanks to an anonymous reviewer for raising this question.

way of S's purpose or purposes P. Although some philosophers have defended dyadic accounts of models, the idea that agents are implicated in their use is widely accepted today. In Giere's schema, as noted, agents specify which similarities between models and targets are intended, and for what purpose. The *Stanford Encyclopedia of Philosophy* entry on scientific representation states that "scientific representation is an intentional concept, depending on factors such as a user's intentions, purposes and objectives, contextual standards of accuracy, and intended audiences" (Frigg and Nguyen 2018). Boesch (2019) gives a helpful list of recent accounts that "give prominence to the agency and use of scientists in offering an account of scientific representation" (2307). Many non-representational accounts of models highlight users' inferences and/or attitudes; intentionality is not restricted to representational accounts. On the whole, (M) follows the philosophy of science literature in presuming that model use is intentional and relative to agents' purposes.[9]

[3] Models may be abstract or concrete.
The relational, agent-based schema applies regardless of the ontological nature of variable M. The sense of "model" that I am concerned with here places no restrictions on the kind of thing that can play the role of M in the schema <S, M, X, P>. Most philosophers of science take abstract models as primary. However, any account of models in scientific practice has to recognize that some models are, and fulfill their epistemic roles by being, material objects (see Section 2.3). Others, by the same token, are systems of equations or another species of abstracta. As computer simulations play increasingly prominent roles in some sciences, the question of how to conceptualize these models (as abstract, concrete, or in some other category) becomes more urgent. Schema (M) is neutral on these debates.

[4] Models are intentionally constructed.
This expands on point [2]. The agents whose purposes determine aspects of the model–target relation include those who construct a model as well as its users. This point addresses a lacuna in Giere's agent-based schema. In (M), models relate to agents not only as tools to users, but also as constructs to builders. Purposes of both are included in P. Constructive activities are left open and unrestricted; idealization, approximation, and abstraction are important in some model-building projects, but not universal. The philosophical account that grounds this part of (M) is the "models as mediators" view pioneered by Mary Morgan and Margaret Morrison (1999). Their approach is not an incompatible alternative to Giere's but the other side of its coin, so to speak. Giere's account is focused on *uses* of models

[9] This aspect of EP-models is further articulated in Chapter 3.

by epistemic agents, while Morgan and Morrison's focuses on how models are *constructed*. Both aspects are important for explanatory particularism, and thus for (M).

[5] Model-construction and use are context-dependent, localized to epistemic communities.
This is a straightforward implication of explanatory particularism. Moreover, as with the previous point, it is central to the "models as mediators" view. That account is supported by detailed case studies from many different fields.[10] Defenders of the mediators view argue, case by case, that model-construction and use is responsive to epistemic background and context. This is very much in line with explanatory particularism. Morrison (2015) notably advocates studying models in their scientific context, clarifying a model's explanatory role in terms of the specific ways that model is constructed and used in a particular context. (Indeed, Morrison's approach to explanation, understanding, models, and representation is particularist, and there is considerable common ground between her "deflationary" account and my own.) More broadly, philosophical studies of models and modeling in scientific practice generally take context-dependence as a given.

[6] Models are constructed from heterogeneous resources.
This point follows directly from EP. Like other models, particularist explanations are constructed by scientists, typically via a complex process that involves various technical resources, inferences, and epistemic norms. This point is also central to the 'models as mediators' view. Models are "autonomous" in virtue of not being derived from theory or phenomena. Instead, diverse resources go into construction and use of a (mediating) model; these "make the model a unique kind of entity" (Morrison 2015, 22). That is, models, by virtue of their heterogeneous construction, are partially autonomous from both theory and aspects of the real world (see Morrison and Morgan 1999, McCoy and Massimi 2018). Morgan, Morrison, and their collaborators emphasize both the context-dependence and heterogeneity of model-construction; for example, "a process of integrating and molding a set of heterogeneous ingredients, such as metaphors, mathematical formulae, policy views, and theoretical assumptions" (Mattila 2005, 538). Explanatory particularism further emphasizes the heterogeneity and context-dependence of models in science—not only in their construction, but also in their form, content, and use.

There is one significant contrast between (M) and the models as mediators view, however. From emphasizing models' autonomy, it is but a short step to casting

[10] A selection of these appears in Morgan and Morrison's edited volume (1999); many others can be gleaned from later literature.

models as having agency. While the latter can be a fruitful lens for studying scientific practice, it goes against my version of explanatory particularism. On my view, agency pertaining to models (and thus to explanation) is localized to scientific communities and their members. The next chapter addresses that issue in detail.

[7] The model–target relation need not involve similarity.
This expands on point [1]. Given (M)'s reliance on Giere's ideas, it's worth highlighting key contrasts. Giere (1988, 2010) characterizes the model–target relation in terms of similarity in relevant respects and to certain degrees, determined by agents' purposes. But this goes against explanatory particularism, as discussed in point [1]. I reject any single substantive analysis of the model–target relation in scientific practice. The relation between a model, a target, users, and their purposes may be best understood as similarity in some cases and not others.[11]

To sum up: the sense of "model" at issue for explanatory particularism draws on a range of philosophical resources. The theses included in (M) are for the most part uncontroversial in philosophy of science today. Of course, accounts of models and modeling go further than the minimal schema of (M), aiming to answer questions that are not the focus of this book. The meaning of "model" here should be clear, familiar, and uncontroversial for many philosophers of science. That is all that's needed, alongside fit with EP. To that issue, I now turn.

2.2 Integrating M and EP (EP_M)

Particularist explanations are not just models in the above sense; (M) and (EP) aren't synonymous. Schema (M) is much more inclusive; not all models are explanatory, but some are. The next stage in developing my particularist framework is to specify that class of models. Such a specification provides some analytic purchase on the notion of particularist explanation; part of a conceptual framework for studying particularist explanation. Specifying EP-explanations in this way does not imply or presuppose a full and exhaustive classification of all possible varieties of models in science. That's all to the good, because it's unlikely that philosophers will ever produce such a taxonomy. There is not even consensus about the basic types of models across all the sciences. Instead, different classification schemes emerge, depending on whether the basis of classification is model structure or use (surveyed in Downes 2020, ch. 3). The diversity of models in scientific

[11] Other features of Giere's agent-based view are notably absent here; e.g., a general hierarchy of models. EP rejects the idea of general hierarchical relations among different explanatory models (though the theory–data continuum may figure in construction of an explanatory model in a particular field).

practice is such that even generic classifications (e.g., Weisberg's 2013 typology of models as concrete, mathematical, or computational) do not map onto important distinctions in practice (see O'Connor and Weatherall 2016). These trends and results support particularism about models; i.e., their diversity in practice precludes any but the thinnest general definition or functional characterization of models in science. However, although such a view is consistent with EP, this book is not a defense of model particularism. The more modest task is to characterize the kind of model identified with particularist explanation. Modifying the four-part relational schema (M) offers a characterization of this special case, without presuming any fixed taxonomy or typology of models in science.

Recall the basic thesis of explanatory particularism:

> (EP) A scientific explanation is a product of a particular scientific community that is significant, specialized, and advances understanding for that community.

Explanations so construed are specialized epistemic achievements of particular scientific communities, which advance understanding for members of those communities. As products constructed by scientists, designed for and succeeding in an important epistemic role, particularist explanations are fruitfully conceived as models. This conceptualization does not itself pick out explanations in scientific practice. As discussed in Chapter 1, identifying explanations in scientific practice is in the first instance a matter of listening to scientists from a particular community. A significant scientific achievement is highlighted and communicated as such by the relevant community. The next question is whether such an achievement advances understanding—and, if so, of what, in what ways, and in virtue of what features? These questions are directions for generating a particularist account of explanation in a specialized context.

These basic ideas and directions both fit with and constrain the values of variables in (M)'s four-place schema of agent, model, target, and purpose. Agent S is, in the first instance, a particular scientific community. Model M is the explanation as such—whatever form it takes. Purpose P is to advance understanding of some target X. (Note that this does not exclude S's having other purposes for the same model; advancing understanding is just the one that individuates EP-explanations.) So, S is comprised of members committed to M as an achievement of understanding. The explanatory model M relates to some target X about which understanding is achieved, in a way that satisfies conditions for explanation in that particular scientific community. Target X is entirely unrestricted; it can be anything whatsoever that scientific community S is interested in understanding. For now, I leave the notion of understanding as a placeholder for whatever epistemic benefit is associated with explanation in a particular community S. That is, understanding (at this stage of my account) is an actor's category, context-dependent and community-relative. In general, all that can be said of explanatory models is that they advance understanding in the way endemic to the particular scientific context in which they are constructed and used. In empirical study of any such context, answers to 'understanding of what?' and 'how so?' reveal features of M and the M-to-X relation that satisfy P according to S. The explanatory role of the class of models that are particularist explanations, in virtue of satisfying EP_M, is in principle unrestricted. The same goes for the idea(s) of understanding in P, and target X. Any particular scientific community will aim to advance understanding of *something* (probably many things). The community, and its members, will have other aims as well. But as long as there is some target, whatever it may be and however conceptualized by members of the community, then there is a purpose and target toward which an explanatory model can be constructed. If successful, the model is used to advance understanding of the target, whatever that amounts to in the particular context. The purposes for which a model is used, and the heterogeneous resources that go into its construction, vary widely across different scientific communities.

The basics of EP thus constrain values of S and of P in (M)'s four-place schema, leaving the model, target, and their relation all unrestricted. Therefore, an explanatory model is any M that fits the schema above, where S is a particular scientific community, and P is an achievement of understanding (however that epistemic purpose is cashed out in S). An explanatory model (M) relates to the target (X, explanandum) about which understanding is achieved, in a way that satisfies conditions for explanation in that particular scientific community. Beyond this sparse characterization, little can be said about explanatory models in general. The features in light of which a model is an achievement advancing understanding of some target vary with context; i.e., the particular scientific community in which an explanatory model is constructed and used. Summarizing all this, we can recast EP in terms of models:

> (EP_M) A scientific explanation is a model (M) constructed by a particular scientific community (S), recognized in S as significant and successful with respect to the purpose (P) of advancing understanding of some target (X).

Importantly, specialization for M and S is presumed, given the sense of "model" at issue. Model-construction is context-dependent and heterogeneous, just as in the models as mediators account. Moreover, the path of explanatory model-construction may be wandering and indirect in practice. Untangling that complex history so as to articulate the ways a model M accomplishes P for X in S is a major task for explanatory particularists. Any epistemic product that satisfies EP thereby advances understanding in a particular community, and is something members of the community recognize as an achievement. Some process leading to that achievement must have taken place in that community—although the details vary widely across cases.

Implications

What does the modeling framework add to EP? Identifying particularist explanations as models has a number of important implications for my view. First, explanatory models are *non-exclusive*. The same target can be modeled in multiple ways, and there is no necessary conflict between prima facie incompatible models of a single target. Particularist explanations work just this way. Different scientific communities can offer different explanations of the same target—whatever it is they aim to understand.[12] Different explanations of the same thing can coexist in science; they need not conflict with one another. A target of understanding is not exclusive to, or exhausted by, one explanation over another. This feature captures the open-endedness of scientific goals as well as the many ways understanding (and advance thereof) can be conceived across the sciences.

Second, and relatedly, models in the sense at issue are dynamic, constructed by particular communities, and continuously refined and adapted as features of the community context change over time. Like modeling, explanation is an ongoing project that succeeds by degrees, not an all-or-nothing affair. EP_M highlights the processual aspect of explanation, framing it as ongoing model-construction. This affords closer engagement with ways of understanding advances in scientific practice. Within a particular community, it's common for ideas about explanation, and major achievements, to change over time. EP_M allows for such change, although (at this stage) the particularist account doesn't include criteria for judging explanations apart from those 'internal' to a particular scientific community. This does not bar philosophers from critiquing scientific explanations, but it does put such critiques on par with (and requiring the same expertise as) critiques by members of that community.[13]

A third implication of EP_M is that explanations are inherently incomplete and partial. Practical limitations of specialized scientific contexts for solving real-world problems are rooted in more general epistemic limitations. Context-specific goals, methods, and norms yield scientific knowledge that is unavoidably partial or incomplete, relative to the real world. This insight is one of the major contributions to date of philosophy of scientific practice, articulated most thoroughly in philosophical studies of models and modeling and feminist epistemology. It is now

[12] And similarly for the same community at different times. Also, a single scientific community can encompass different explanations of the same target at the same time—but this kind of case has some unusual features (see Part II).

[13] One judgment excluded by EP_M is that a community is *mistaken* about an explanatory achievement. The epistemic product is a construct—a model—and recognition of it as an achievement is a community phenomenon. A community can disagree about, revise, or retract ideas about the nature and significance of an explanatory achievement. All such judgments come from the community of specialized researchers who construct and use the explanatory model in question. As Kuhn said long ago, there are no better grounds for judgment about these matters than the (relevant) community of scientists. Philosophers can also be members of such a community, but, like anyone else, they need to acquire expertise to do so. Thanks to an anonymous reviewer for raising this issue.

widely recognized that representation by models is partial and selective, and that model-users' purposes guide such selectivity. Explanatory particularism makes explicit the social dimensions of such selective, goal-directed modeling: different scientific contexts impose very different methods and norms on practices of model-construction and evaluation. Conceptualizing explanations as models, and so inevitably partial and selective in ways that accord with norms of a particular context, undermines the idea that any scientific explanation can be, on its own, complete and absolute.

Fourth, many kinds of model can in principle contribute to constructing explanatory models. Scientific practice is replete with models, in the sense of (M). According to EP, explanations are specialized products of particular scientific communities; each a complex construct reflecting goals, values, and presuppositions of its community of origin. In scientific community S, there may be many other models that contribute to the construction and/or use of an explanatory model M. That's not a problem for explanatory particularism. Models can contribute to scientific practice in many different ways, and can contribute to a scientific explanation in ways other than being such an explanation. The particularist conception of explanatory models allows for many (indeed, unlimited) other roles for models in science. Therefore, models are even more diverse than explanations; a single scientific community may construct and use a great many different models that fit the schema <S, M, X, P>. However, EP's significance condition remains in force. An explanation, then, according to EP_M, is something special: a significant achievement that advances understanding in the community in which it is constructed and used. Particularist explanations are models that satisfy that extra condition. There are not so many of those, in any particular scientific community. Explanations stand out as major achievements of understanding. Their features, as models, reflect the epistemic norms and values of the community in which they are constructed and used. They are, so to speak, the highest expression of those values. (And by that same token, specialized.)

This is an important result in its own right, though from EP rather than the models framework. Individuating explanations in scientific practice is not straightforward. There is a tendency among philosophers to take the category of explanation for granted, as though it is obvious what explanations are across all the sciences. But identifying explanations in scientific practice is not so straightforward. In many fields, scientists don't use the term very often (and not at all consistently). Of course, just because scientists rarely apply the term "explanation" to their work, does not mean that explanations are absent. It does mean that philosophers interested in scientific practice need some way to identify scientific explanations; they cannot be read off scientists' literal statements, in a way accessible to outsiders. There is a prima facie vexed relation between philosophy and scientific practice as regards explanation. It's not obvious what exactly counts as an explanation in science. Conceptualizing explanations as models offers a way to face and respond to this issue without making anti-particularist assumptions. This is because models in science are similarly amorphous, in ways philosophers studying scientific practice have learned to deal with. There is no generally-accepted philosophical account of what a model is, nor any typology or classification of kinds of model. And so there is no account of how models are individuated. Trying to count the models involved in a particular scientific episode is an exercise in futility. Models are continually revised, adjusted, and altered, as scientific inquiry proceeds. Although it is hard (perhaps impossible) to individuate models rigorously, philosophers with experience in a particular scientific field can recognize significant models in its practice, and draw many insights from studying their construction and use. Scientific explanations can be identified in just this way. According to EP_M, explanations are scientific models recognized as achievements that advance understanding of some target of interest in their communities of origin. The reasons why any particular model is deemed such an achievement depend on features of the epistemic situation; the particular context in which the explanation is constructed and used.

The multiplicity of models points to a fifth EP_M implication: just as different models can relate to one another, so can different explanations. This raises the question of what those relations are. In what ways do distinct explanations relate to one another, in scientific practice? Insofar as philosophical accounts of explanation have considered relations among different explanations, those relations have been assumed to be reductive or otherwise hierarchical (that is, one scientific explanation subsumes another).[14] Explanatory relations are often presumed to track tacit social epistemic relations among different sciences—relics of Oppenheim and Putnam's stratified unity of science (see Chapter 1). Another way to put the point is that philosophical accounts of explanation tend to extend the directionality of explanation (explanans to explanandum) to relations among different explanations (one explanation subsumed by, or otherwise made dependent on, another). This tendency is very entrenched, rooted in Hempel's D-N account and many of its successors. But conceiving of explanations as models breaks this connection. The prospects for relating *models* are much more varied and less laden with hierarchical assumptions.[15] If explanations are a kind of model, inevitably partial in the way of models, then it is natural to enquire how models can complement or augment one another, as well as conflict. Models can relate to one another in myriad, multifarious ways. I'll return to this point in Section 2.4.

A final result is dialectical. Chapter 1 introduced EP as a break from tradition in philosophical studies of explanation. Breaking with that tradition not only offers an alternative approach to studying explanation, but also reveals how much work along those lines has already been done. Many studies of models in scientific

[14] Potochnik (2017) offers a rebuttal and alternative to this view.

[15] Although EP_M identifies the directionality of explanation with the model–target relation, I reject any parallel asymmetry in relations among different branches (or fields) of science.

practice do exactly as EP prescribes, illuminating the nature and values of specialized explanatory models in a wide variety of scientific contexts.[16] So although explanatory particularism is a non-traditional approach to explanation, it aligns nicely with recent studies of models and modeling in scientific practice. EP_M makes this connection explicit. This opens the door to admitting many more case studies of models in particular scientific fields as contributions to philosophical study of explanation.

2.3 Example: embryo models

To illustrate the modeling framework for EP, I'll add to this wide-ranging collection of cases. My example is a family of concrete models that are also living things. "Embryo models" (aka embryo-like models, embryo-like structures, synthetic human entities with embryo-like features) are a recent achievement of stem cell research.[17] They are produced by growing (mouse or human) stem cells in specific artificial culture conditions. Artificial cell culture is a century-old method for growing cells of a multicellular organism outside that organism's body (Landecker 2007).[18] Human embryonic stem cells (hESC) were first cultured "ex vivo" by James Thomson and an international team of colleagues (1998). The method begins with extracting interior cells from a very early human embryo and "seeding" those cells on a two-dimensional plate bathed in a cocktail of specific chemical factors. Under such conditions, some embryo-derived cells take on a characteristic morphology and form rapidly growing colonies. An hESC line is made by continuously selecting cells from such colonies and culturing them on new plates. Embryonic stem cells not only "self-renew" indefinitely in culture but can also, in other environments, develop to produce any cell type of the adult body.

Embryo models are produced by changing the culture conditions of (human or mouse) ESC: their spatial arrangement, geometric organization, mechanical forces/constraints, chemical environment, and (sometimes) cellular context.[19] Certain combinations of these factors induce stem cells to form "embryo-like structures" resembling one or another stage of embryonic development in the species from which the source stem cell line originated (Figure 2.1). Resemblance of products supports an inference to process. In this case, what is inferred is that the

[16] For example, Marchionni (2008, 2013) on explanatory network models in economics and sociology.

[17] See, e.g., Warmflash et al. (2014), Aach et al. (2017), Rivron and Fu (2021a, b). This case builds on my earlier work focused on stem cells (Fagan 2016b, 2017a, 2020a, 2021).

[18] Bacterial cell culture goes back further, and is foundational for microbiology and many related fields. See O'Malley (2014).

[19] The last involves combining ESC with other early embryonic stem cells. Thomson et al.'s original hESC concept has since been refined into an armamentarium of distinct stem cell lines, each derived from a different embryonic developmental stage or location.

developmental process in culture is similar to normal embryonic development. Stem cell researchers state this in terms of "recapitulation" or "re-enactment." For example:

> ... embryo models are generated by combining pluripotent stem cells, sometimes along with extraembryonic stem cells, in precisely controlled numbers, geometries, and physicochemical environments mimicking developmental cues. This triggers cells to re-enact aspects of embryogenesis.
>
> (Rivron and Fu 2021a, 1011)

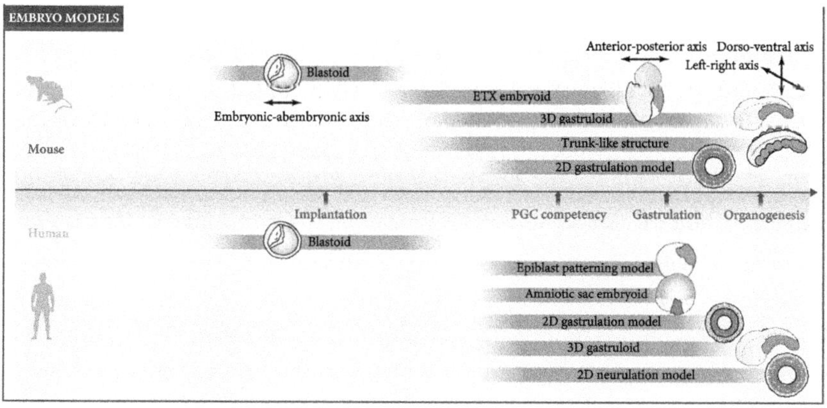

Figure 2.1 Sketch of new embryo models
Source: Rivron and Fu (2021b).

The resemblance to normal embryos is imperfect, of course. For example, if grown in small circular disks with BMP-4 (a molecule that stimulates cell differentiation), hESC differentiate in a "center-to-periphery" pattern, with stem cells at the center and more mature cells toward the periphery (Warmflash et al. 2014, 847). This process is likened to human embryonic development up to gastrulation, the stage when human embryonic cells commit to form a single multicellular organism ("individuation"), germ layers are distinguished, generic cell fates are fixed, and basic body plan polarity is established. The models' simple radial structure does not, of course, resemble a human embryo anatomically.[20] But it does exhibit several aspects of embryonic organization, such as robust timing and order of appearance of cell types, a reproducible spatial arrangement of germ layers, and polarity along

[20] Another kind of embryo model, the gastruloid, does more closely resemble the corresponding human embryonic stage.

the radial axis. The robust appearance of differentiated cell types (germ layers) in a particular spatial orientation is attributed to "the intrinsic tendency of stem cells to make patterns"—self-organization (Warmflash et al. 2014, 847).

Having sketched the example, let's analyze it with EP_M's schema:

> (EP_M) A scientific explanation is a model (M) constructed by a particular scientific community (S), recognized in S as significant and successful with respect to the purpose (P) of advancing understanding of some target (X).

Embryo models are, eponymously, models (M); I'll discuss their main features shortly. The scientific community that constructs them (S) is the community of stem cell research. That community is highly interdisciplinary, with contributors ranging from clinicians to bioethicists to computer scientists.[21] Even so, familiar sociological markers limn the community's boundaries. Stem cell research assumed its current configuration soon after the 1998 announcement of cultured hESC—notwithstanding that most stem cell research, before and since that watershed, does not involve embryos. (Most stem cell research and applications focus on cells found in blood or bone marrow.) An annual international meeting with an associated society was instituted in 2002, with specialized journals appearing a few years later (e.g., *Cell Stem Cell* in 2007, *Stem Cell Reports* in 2010). The topic of stem cells cuts across many other aspects of social organization in life science and medicine—there are many, many different contributing research programs. Being so radically diverse and decentralized, the community of stem cell research is unified by scientific interest in the eponymous entities, and little else.

Significant achievements in this sprawling community are quite varied. Embryo models are one such, as evidenced by recent high-profile presentations, webinars, and editorials in the field's specialized venues.[22] In this case, however, success is largely promissory. Stem cell researchers are interested in possible uses of embryo models, while research is still in early stages. That doesn't undercut the example's purpose here, though. Speculation about hoped-for achievements can be very revealing of a scientific community's views on explanation. Several recent editorials and essays discuss hoped-for uses of embryo models. These are mainly medical applications: reproductive medicine, treatment for infertility, treatment for genetic disease, regenerative medicine, drug screening.[23] However, alongside and supporting these practical goals is another: advancing understanding of embryonic development. Many remarks attesting to that purpose appear in recent discussions of embryo models. For example:

[21] "Gathering a community comprising stem cell biologists, embryologists, synthetic biologists, computational biologists, physicists, engineers, and ethicists will be key" (Rivron and Fu 2021a, 1013).

[22] A good selection of these appears in Rivron and Fu's special issue on embryo models (2021a, 2021b).

[23] See further discussion in Fagan (2020a, 2021).

> As a bottom-up approach to embryology, embryo models could help reveal *the first principles of development*... (Rivron and Fu 2021a; italics mine)

> Their utility in providing experimentally tractable, high-throughput, and *human-specific developmental insights* makes them amenable to many different approaches. (Moris et al. 2021; italics mine)

It seems evident that "first principles" or "insights" about embryonic development would advance understanding of that process (P). Most of their other hoped-for uses would be supported by this explanatory role. So embryo models fit EP_M's schema, and so qualify as (potential) particularist explanations.

The precise target (X) of embryo models is harder to pin down. What, exactly, is it hoped that embryo models will help us understand? Because their explanatory role is speculative, statements about embryo models' explanatory goal are less constrained than for realized explanatory achievements. The explanatory target of embryo models ranges widely in scope and 'grain': development in general, embryonic development, mammalian embryonic development, human embryonic development, early stages of human embryonic development, etc. Overall, though, explanatory focus is primarily directed toward early stages of embryonic development, especially in humans. If one considers each kind of embryo model separately, a more narrowly-defined target makes sense: "[d]ifferent culture formulations capture different stages of developing [cell] lineages" (Posfai et al. 2021, 1118). The directed relation to a target stage of embryonic development is explicit and central. Collectively, though, embryo models aim at understanding the more holistic process of early embryonic (human) development. Embryo models' significance is often articulated in terms of providing observational and experimental access to that naturally inaccessible process.[24]

Three insights about this case emerge from applying the EP_M framework. First, the explanatory target is diffuse and open-ended, not fixed and definite. Second, although embryo models are material—indeed, biological—objects, their material aspect alone doesn't amount to an achievement of understanding (i.e., an explanation). The explanatory model isn't the concrete experimental system alone, but the system interpreted in a way that amounts to an explanatory advance. (I'll discuss the main features of that interpreted system next.) Third, the explanatory target is understood by way of multiple models, rather than by partitioning models

[24] This way of presenting matters elides the obvious (yet often unstated) fact that human development naturally takes place inside a living human person, and is an intensely interactive process that cannot as such be observed and experimented on in the way stem cell systems are. Embryo model research does in a way address that fact, by modeling placental development (in a few research programs, at least) and aiming at improvements to reproductive medicine.

and targets in one-to-one ordering. The question of how these multiple models relate to one another thus arises. These features of the case are likely to extend to many other explanatory models in experimental life sciences—although more particularist studies are needed to support that extension. Here I'll only gesture to that particularist project. For now, I'll look more deeply at key features of embryo models (M), their relation(s) to targets (X), and the familial ties—and other relations—among them.

In current stem cell research, embryo models are interpreted as follows.[25] Stem cells are seeded in a three-dimensional arrangement, which sets up interactions among cell surface molecules. Together with physical forces and biochemical factors, those interactions commit cells to develop along particular pathways and to adopt specific positions relative to one another. To the extent that these patterns lead to recognizable features associated with normal embryonic stages, they comprise "self-organization" in embryo models. Self-organization, as currently understood, is a pattern of dynamic change initiated by interplay between cells' internal mechanisms and multiple interacting environmental factors ("support cells" secreting signal factors, the extra-cellular matrix (ECM), mechanical forces, oxygen levels, pH, and more; see, e.g., Yin et al. 2016, 28). This emerging account indicates key features of embryo models qua explanations of embryonic development. The concept of self-organization is central. The emerging explanation takes the form of a description of how stem cells self-organize in a specific context to produce embryo-like structures. The description includes many different components: stem cells' gene expression, morphology, spatial arrangement, binding interaction, and movements, but also various physical and chemical factors, as well as other macromolecular components and overall geometry of the culture system. It's not a simple interpretation, but complex and multiscaled. There are no equations or other elements of mathematical modeling; the description is qualitative. Of particular interest—though as yet uncharacterized—are aspects of coordination among components of the developing system, as well as embryonic regions that develop more or less autonomously. Researchers seek an account of self-organization that shows how body patterning, axis formation, and regionalization of the embryo all result from the interactive, stem cell–driven processes, such that developmental milestones in normal embryonic development are "recapitulated" in embryo models. Relating the mechanisms of self-organization to robust and reproducible features of normal development is another important theme.[26]

[25] This treatment is simplified and omits most technical jargon. It is thus less specialized than the scientific treatments it's based on. This kind of 'accessible translation' is one function of philosophy of science in practice (see Part III).

[26] For example, "How this self-organization proceeds and which mechanisms determine the reproducibility and robustness of development are still key questions" (Gupta et al. 2021, 1104).

At this point, the heuristic role of earlier philosophical accounts of explanation comes into play. The emerging embryo model explanation conforms in some ways to accounts of mechanistic explanation, interventionism, and other causal accounts. All these—and more—are valuable tools for philosophical analysis of this case. I won't work through these applications here (see Part III for more examples). It's important, though, that attachment to one or other substantive philosophical account not obscure the ongoing work of constructing embryo model explanations in scientific practice. As the embryo model interpretation comes into clearer focus, a debate about its explanatory role is also emerging. At stake are embryo models' key explanatory features and their specific relation to targets. The debate is this: should embryo models aim to accurately represent normal embryonic development, or to embody core mechanisms that underlie key features of that process? Although most of the current scientific literature on embryo models opts for the former, several commentators have raised concerns about its feasibility and desirability. I'll summarize this debate and show how it bears on features of M and the M-to-X relation in this example.

The 'accuracy' view characterizes embryo models' role as "recapitulating" normal human embryonic development. Taken literally, this means that the *same process* unfolds in both embryo models and normal development. Emphasis on recapitulation attests to a powerful norm in the stem cell research community, that embryo models should accurately represent their target. Satisfying that norm makes for an advance in understanding human embryonic development, because many aspects of embryonic development are species-specific. This limits the utility of traditional model organisms for understanding human development. We can learn a lot from observing and experimenting on (for example) mouse embryonic development, but there are a lot of differences, known and unknown. Embryo models get around this limitation, being constructed from embryonic stem cells, which are in turn constructed from a human embryo. (Mouse embryo models are constructed too—so there's more to discuss there too.) Molecular and cellular material from that organismal source is transmitted to embryo models by stem cells' developmental capacities (Fagan 2013, 2021). So there's extensive material continuity between stem cells, embryo models, and (some aspects of) their target: normal development in the source organism species. A continuous lineage of experiments connects every embryo model to a stem cell line, and thereby to a source organism. In conjunction with stem cells' characteristic generative developmental abilities, this material continuity establishes key similarities between embryo models and their targets. Important genetic, molecular, and cellular components of human organisms are literally built in. These continuities provide grounds for inferring further similarities; analogical inferences from model to target. Models that include many of the same molecular, genomic, and cellular components as are present in the source organism, can be reasonably expected to combine those components in a similar way during development. All this suggests that embryo models advance understanding of normal development via model–target similarity, and that further advances will be made by increasing embryo models' verisimilitude.

However, there are significant limitations to the project of increasing embryo model verisimilitude, which some scientific commentators have noted. Most obviously, stem cell researchers can't observe and experiment on normal human embryonic development: that process is inaccessible, and "ex vivo" human embryos are scarce. Although researchers can grow human embryos in vitro for a short time, there are ethical barriers to extending this to later embryonic stages.[27] Inaccessibility and scarcity of human embryos are the main motivations for embryo models in the first place. But if their role is to be as similar as possible to that (unknown) process, how can embryo models be evaluated? Stem cell researchers refer to this as the "bench-marking" challenge. (This challenge arises in many other modeling projects too, as philosophers well know.) One proposed strategy is to take a comparative approach, diversifying stem cell–based models to cover a broad range of species, and compare these to identify general rules or principles that plausibly extrapolate to (inaccessible) human embryos. But, if the project of verisimilitude leads stem cell researchers to infer general principles of embryonic development, why not make such principles the explanatory focus from the start? This leads to an alternative view of embryo models' explanatory role: as idealized experimental systems for discovering "principles of development." That view too is well-grounded in stem cell research practices. One of the great advantages of embryo models is their accessibility to many forms of data collection—we can learn much from them precisely because they are ex vivo, visible, and tractable objects of experiment. Their relative simplicity and accessibility allow for experimental manipulations that are not possible or ethically permissible in normal human development. Rather than perfectly mimicking in vivo processes, they display certain aspects of embryonic development in a simplified way. On this view, embryo models advance understanding not by accurately representing normal development but by revealing "principles" of embryonic development.

Philosophers of science might expect this approach to embryo models to include norms directed toward discovering something like traditional scientific laws. But that is not what stem cell researchers mean by "principles." In their community, the term means something close to "mechanism" in philosophy of science—notably, a system with a lot of specific moving parts that together yield a reproducible and desirable outcome. The role of embryo models, on this view, is to identify robust mechanisms initiated by specific combinations of stem cells and culture conditions, such that the result exhibits important features of embryonic

[27] Recent calls to reconsider the bioethical fourteen-day limit for culturing human embryos are largely a response to embryo models.

development that can be exploited for medically and practically useful outcomes. On this view, an advance in understanding takes the form of principles for robust and reproducible developmental outcomes; i.e., describing how a set of key properties robustly bring about one or more developmental outcomes of interest. Candidate properties include stem cell type and number, stem cell gene expression patterns, stem cells' spatial organization and morphological structure, cell–cell and cell–matrix contacts and resulting signals, and physical properties (e.g., size, shape, stiffness). The goal is to identify the key relevant properties that yield the developmental outcome of interest (e.g., three germ cell layers wrapped around a lumen).

This second approach strikes me as more in line with the practices of stem cell research than the first. Debate in the community is ongoing, however:

> Some would argue that this [the project of verisimilitude] defeats the point of such models: rather the goal is not to perfectly recapitulate development but to have a suite of tools that allow specific mechanistic questions to be addressed . . . The field will continue to debate whether the future focus should be on making embryo models "less wrong" and whether this necessarily equates to "more useful."
>
> (Posfai et al. 2021, 1133)

So, for now, the explanatory features of embryo models and associated relation to target(s) are still unsettled. The contours of this debate will be familiar to philosophers studying scientific models and modeling. There's a clear opportunity here for philosophers of science to contribute, using accumulated insights about models. Those insights will be most relevant if the implications for explanation and understanding are made explicit in a particularist framework.

One further insight can be gleaned from the embryo model example. Family relations among stem cell models are very intricate, and these relations have explanatory import in stem cell research. Human development is an extremely complex phenomenon, which no single, tractable model can adequately represent. Understanding of this target (and aspects thereof, such as early embryonic development) is advanced via multiple models. Different models complement one another, highlighting different aspects of the in vivo process. This is a familiar point: different models 'radiating' from a common target. But the various models also relate to one another. Embryo models are constructed from stem cell lines, which are themselves models of development. Stem cell lines in turn beget more stem cell lines, as well as other embryo- or organ-like constructs. Relations among these diverse models are central for understanding their shared complex target. Human embryonic development is represented by a sprawling family of concrete models: cultured stem cell lines, (relatively) pure populations of stem cells found in organisms, more elaborate constructs such as embryo models and organoids, etc.. These models relate to one another via material continuity as well as comparisons and contrasts discovered by scientists. A great deal of epistemic activity in stem cell research involves working out these comparisons and contrasts, determining relations among different stem cell-derived models, and then triangulating these comparisons and contrasts into a picture of the overall developmental process—i.e., understanding of the real-world phenomenon that is the models' collective shared target. In this way, embryo models are part of a larger interconnected fabric of diverse experimentally-produced models. Conceptualizing embryo models as part of this broader 'fabric of models' offers insights into their modeling role, limitations, and epistemic potential. There is much more to analyze about this epistemic organization and the forms of understanding that emerge from it.[28]

2.4 Model–model relations

The preceding example shows that model–model relations sometimes figure prominently in the explanatory practices of particular scientific communities.[29] This brings us to the second task: developing the EP-models framework to analyze relations between distinct models. As noted above, focus on model–model relations makes it easier to jettison traditional assumptions of hierarchy among the sciences. Philosophers tend to give explanatory priority to physical theories of the basic constituents of matter and fundamental forces, or mathematical abstraction/purity, or both. But these 'explanatory world-pictures' aren't shared by working scientists in their various communities. (Of course, some individual scientists likely do have this view, but traditional hierarchies don't figure in the education and practice of most working scientists today.) Yet these assumptions continue to influence philosophical studies of explanation, often working in the background. EP_M clears away all such hierarchical background assumptions. Multiple models of a single target are different, but there is no presumed order to them. The same goes for models of different targets. Here I develop a positive, non-hierarchical framework for conceptualizing model–model relations.

That is not to say that any particular scientific model (in the sense at issue here) must or should relate to any other. Multiple models of a single target phenomenon constructed in the same community often do relate to one another in some way—embryo models, and other stem cell models, are cases in point. Here the basic assumption of a dynamic array of scientific communities comes to the fore again. Scientific communities are not isolated from one another. As introduced in Chapter 1, my version of explanatory particularism takes the community structure of science as basic, with explanations localized to social contexts of construction

[28] See Chapter 5 and Fagan (2020a, 2021) for more discussion of these issues.

[29] Other case studies corroborate this result: climate models (e.g., Lenhard and Winsberg 2010, Morrison 2021).

and use. This social epistemic organization is dynamic; scientific communities continuously engage in boundary shifts, splits, and mergers. A research community may fragment into distinct specialized groups, while distinct communities may form connections. Relations between explanatory models often figure prominently in these social epistemic changes. Philosophical accounts of models and explanation, if they attend to the social epistemic organization of science at all, tend to presume a stable classification of scientific fields and correspondingly orderly division of explanatory activities. But in practice, scientific fields continually interact. So explanatory activities in scientific practice do not always partition neatly into a clear-cut division of labor coinciding with sociological community boundaries—still less with traditional disciplinary divisions. In some cases, connections between distinct scientific communities initiate an *interdisciplinary context for explanation* (see Part II.) But it's also common for different models to have no connection whatsoever in practice. For models constructed by different communities, 'no relation' is the default. In scientific practice, often there is no connection, just a diverse collection of explanatory models from different fields. However, there are plenty of cases in which models do (or can, or should) relate to one another in practice. Here, I develop a conceptual framework for analyzing these relations, as a further development of EP_M. This framework is extremely general, and can be applied to models constructed by the same or distinct scientific communities.

A taxonomy of model–model relations

My framework for conceptualizing model–model relations consists of two crosscutting distinctions and two continuous axes.[30] The two distinctions yield a simple four-part taxonomy of model–model relations, while the two axes afford a simple space in which to locate those relations. I'll set out each element of the framework in turn, beginning with the continuous axes.

[A1] Process: When distinct models relate to one another, the relation must be established in scientific practice. Models don't interact with one another in some paradise of abstracta, independent of practicing scientists.[31] Relations between them are intentional, like models themselves. Another way to put this point is that model–model relations are part of the dynamic process of science, the same sociological and historical perspective that serves as the backdrop of explanatory particularism. So model–model relations (if they obtain) are established at some point

[30] Earlier versions of this taxonomy appear in Fagan (2017b, 2020b).

[31] More precisely, if such interactions are possible, they are outside the scope of my particularist account of explanation.

in the model-construction process. That process begins with initial sketches and plans and continues up to a recognized achievement that satisfies all or most of a model's purposes. This is not to say that all model-construction processes follow the full trajectory (many, perhaps most, are abandoned as unsuccessful). Rather, the process of model-construction is an axis along which any model can be located. (Because model–model relations involve two distinct models, this amounts to two axes. I'll pass over this complication for now.)

[A2] Degree-of-connection: If two models relate to one another, their relation must involve *something* about each of those models. At one extreme are purely formal relations, such as simple additivity. The latter involves no substantive connection between distinct models. Statements pertaining to each are simply strung together in a conjunction, making the connection between models one of bare logical consistency.[32] A more substantive, but still thin, relation between distinct models is input–output; i.e., the output of one model serves as input for another. Unlike formal relations, input–output relations are important in at least some scientific communities. Yet they involve a very slight degree of connection. Other relations between models are more extensive or intricate (e.g., stem cell–derived models). One way to conceptualize higher degrees of connection is to consider how many elements of each model comprise the relation between them. Any model consists of one or more elements—whatever it is that makes up that model. Relations more intimate than formal or input–output connections must involve at least one element of each model. The relation could be much more extensive than this, involving many elements. For example, the various stem cell models are related to one another with high degrees of connection, due to the material continuities and stem cells' own generative capacities. At the extreme of this axis would be pairs of models whose elements are all related to one another: mutual dependence, or reciprocity, between all elements of each distinct model. (For example, van Fraassen's (2008) view of ongoing "joint construction" of theoretical and experimental perspectives.)

[D1] Direct vs. indirect: Traditionally, different models of the same target phenomenon are construed as competing alternatives—models in conflict. Conflict between models is a direct relation. If different models make genuinely incompatible claims about the same target, then the models directly contradict one another. Studies of scientific models and modeling show that many other relations among models are possible, however. For example, apparently competing models are often rendered compatible by each being subsumed by a more fundamental model (Morrison 2011, 2015) or underlying structure (Chakravartty 2010). Models that appear to make contradictory claims about the world are reconciled as different

[32] This axis endpoint is based on Chakravartty's account of models from different perspectives (2010).

ways of elaborating the same set of basic principles. Each is an alternative idealization based on a shared common structure (e.g., Potochnik 2017). This reconciliation is an indirect relation between models (or some elements of those models). Distinct models are reconciled—rendered consistent with one another—by each being coordinated to some other thing; an underlying structural description. The different models do not directly relate to one another but are all subsumed by a more abstract or general model. Indirect relations of this sort involve hierarchical assumptions, with the abstract shared feature being privileged. This privileging is often associated with explanatory power. Direct relations among models do not require hierarchical assumptions of this sort. More generally, relations between models are either direct or indirect: models (or elements thereof) either relate to one another directly or via some further mediator. Another familiar direct relation between models is overlap or coincidence: different models have certain features in common. These points of coincidence, or zones of overlap, can serve as "joints" that connect distinct models, often into families (e.g., Giere 1988, Lloyd 1994).

[D2] Similarity vs. difference: As noted, models can relate to one another in virtue of similarities. Shared or common elements are maximally similar; areas of 'overlap' between distinct models. But models can also relate to one another in virtue of differences. That is a less familiar idea, and may seem counterintuitive. Direct conflict is a relation of difference between models—but usually treated as unstable, demanding that one or the other be chosen, or that their respective uses be kept separate. However, there is another way to conceptualize the relation among directly conflicting models from different perspectives: namely, as complementarity. The idea is illustrated by occlusion in visual perspective. Occlusion is a kind of representational limitation on model-construction in visual perspective: selecting one feature to represent rules out or otherwise compromises successful representation of other features of the same target. A feature that is blocked or distorted by representation of another in a model is occluded in that model. One model's occlusion can be another's successful representation—this is one reason multiple models of one target are sought in the first place. But models that complement one another in this way cannot, in general, simply be combined into a single coherent model of their shared target. They directly conflict by representing the target in mutually exclusive ways. This is an example of a direct difference relation between models: they relate to one another by contrast. More abstractly, such models are complements or "negative images" of one another. To return to the visual metaphor: that which one model casts in shadow, the other presents in full light. I will term this relation N-complementarity (for "negative"). This immediately suggests another way in which models from different perspectives can relate to one another: namely, by complementing one another more directly. In this case there is no occlusion; different models just supply what is missing from one another's representations of a shared target. This kind of relation is epitomized by the seamless fitting-together of jigsaw puzzle pieces, via complementarity in shape. Such "lock and key" complementarity (LK-complementarity) is a form of mutual dependence among distinct models. LK-complementarity is a direct relation: different models represent different features of the target, compensating for one another's partiality. N-complementarity is indirect; the properties that N-complementary models represent the target as having cannot be combined into a single coherent model. Such models can be integrated only via mediating relations such as van Fraassen's "duality" (2008) or subsumption. Other indirect relations of difference are part-whole hierarchy relating different levels of description, or cause-effect relations representing elements of one model as causes of elements of another.

The two distinctions are orthogonal. Similarity-vs.-difference concerns features of models in virtue of which they connect with one another. Models can relate to one another via having features common to both. Or they can relate to one another through difference: a contrast leveraged into a positive link. Most philosophical discussions about integrating models focus on similarities across distinct models: shared structure, properties depicted, and so on. But models can be interestingly related in virtue of their dissimilarities as well. Orthogonally, distinct models can relate to one another directly or else via some additional mediator. These two cross-cutting distinctions, provide a general comparative framework for analyzing relations between distinct models. The two axes provide a general graphical framework for characterizing relations between distinct models in terms of the process of model-construction and the extent to which the models are related (Figure 2.2). Alongside this graphical framework, the two cross-cutting distinctions yield a four-place taxonomy of relations between models (Table 2.1). In Part II, I'll explore the idea that combining different, local explanatory models yields better (in one sense) understanding of a phenomenon. Comprehensive explanations of this kind are valued in practice, but we lack the vocabulary to

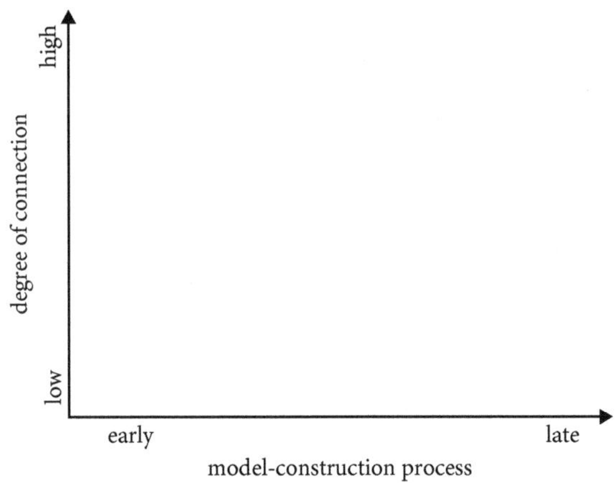

Figure 2.2 Graphical framework for classifying model–model relations. The horizontal axis indicates the stage or interval in the process of model-construction at which models are related in practice. The vertical axis indicates the degree of connection between two (or more) models.

Table 2.1 Four-way classification of model–model relations

	direct	**indirect**
similar	overlap (shared element)	subsumption unification (traditional)
different	conflict LK-complementarity	N-complementarity causal nexus part-whole structure

articulate why this is, and its scientific significance. This and the next chapter develop concepts and standards for doing so, to make explicit this aspect of scientific practice and understanding. Part II applies this framework to an important strand of scientific practice: interdisciplinary research.

2.5 Conclusion: toward social understanding

This chapter has argued that particularist explanations are a kind of model, in one important sense of that term. That sense is related, but not identical, to models as characterized by Giere, Morgan, Morrison, and other philosophers focusing on models in scientific practice. Adapting insights from these philosophers' accounts, I've argued that particularist explanations should be conceptualized as models, in a loose sense of the term. An explanatory model (M) relates to some target (X) about which understanding is achieved, in a way that satisfies conditions for explanation in a particular scientific community (S). More precisely:

> (EP_M) A scientific explanation is a model (M) constructed by a particular scientific community (S), recognized in S as significant and successful with respect to the purpose (P) of advancing understanding of some target (X).

Features of M and X in virtue of which an explanation is recognized in S as significant and successful with respect to P are left unspecified. These are revealed piecemeal, through study of explanatory practices in particular scientific communities. Philosophical theories of explanation can be valuable tools for such study, but as heuristics not elevated to universal monistic requirements. However, more can be said of S and P. Philosophical theories of social action offer a deeper analysis of scientific communities and their purposes. That is the task of the next chapter, which articulates a framework in which EP's social epistemic view of explanation is genuinely normative and critical. To preview: explanatory values are social values aimed at overcoming fragmentary understanding, the narrowness and myopia of scientific viewpoints. An important kind of understanding aims at transcending community boundaries; fitting together pieces of a larger puzzle into a more comprehensive view. There are many targets of scientific interests that we'd like to understand in this comprehensive way—via a linked system of different explanatory models. Construction of these models by particular communities is a species of social action, a special case of collaboration. Chapter 3 uses insights from studies of collective activity and intention to extend the particularist modeling framework.

This chapter does important groundwork for that account. I've illustrated EP_M's ideas with an example from stem cell research: embryo models. This case highlights the importance and intricacy of model–model relations within a diverse scientific community. I've proposed a very general framework for characterizing such relations, comprised of two cross-cutting distinctions and two continuous axes. The two axes locate relations between distinct models with respect to stage in the process of model-construction and extent to which the models are related (degree-of-connection). Alongside this graphical positioning, two cross-cutting distinctions provide a general comparative framework for analyzing relations between distinct models. Relations between elements of models may be direct or indirect, involving similarity or difference. Part II puts this framework of model–model relations to work. Emphasis on model–model relations, and the social epistemic organization of science more generally, distinguishes my account of explanatory models from other recent work on this topic.

3
Social Action

This chapter develops the second part of explanatory particularism's philosophical framework. The starting-point for this part of the framework is the idea that *constructing a scientific explanation is a collaborative activity.* I first clarify the meaning of that claim, using a detailed example (Section 3.1) and theoretical resources from philosophy of social action (Section 3.2). Philosophy of social action is the sub-field of philosophy concerned with analyzing normative foundations of human social behavior. Building on those theoretical resources, I propose an original account of collaborative activity, articulating constraints on participant attitudes (Section 3.3). This account is then integrated with the main results from Chapters 1 and 2. This yields the full account of my version of explanatory particularism, insofar as it can be stated in general terms (Section 3.5).[1] Section 3.4 examines the normative ground of my account of collaborative activity, emphasizing aspects of understanding that come to the fore in Part II.

Philosophy of social action has few points of contact with philosophy of science. These two sub-fields of philosophy are quite distinct.[2] My reasons for venturing into this new philosophical terrain are three-fold. First, EP's background assumptions about scientific communities and explanatory practice are well-suited to analysis in terms of action. My account of scientific communities is dynamic, processual, and historical, conceptualizing scientists' explanatory practice as active. This orientation toward actions is further supported by insights from philosophical studies of models and modeling, highlighting the importance and intricacy of model-construction processes. That explanatory practices are *social* activities follows from EP's basic assumptions and concepts. Given EP and the modeling framework developed so far, social action theory is a natural place to look in further developing the framework. Second, it's appropriate to look outside philosophy of science to develop the social aspect of EP's conceptual framework. The general framework for studying particularist explanation needs to be neutral, so to speak, as to substantive features of scientific explanation. Here

[1] Recall that much of the account's substance is to emerge from detailed studies of particular scientific communities, collation and summary of which is beyond the scope of this book (see Chapter 1).

[2] I am not, of course, the first to connect philosophy of science and of social action. Kristina Rolin (2008), K. Brad Wray (2003, 2007), Hanne Andersen (2010), and Deborah Tollefsen (2002, 2015), for example, appeal to social action theories to analyze social aspects of scientific knowledge and conceptual change. A larger literature concerns the existence and nature of "collective scientific knowledge" (see Section 3.2). However, none of these accounts, to my knowledge, focuses on explanation.

Explanatory Particularism in Scientific Practice. Melinda Bonnie Fagan, Oxford University Press.
 DOI: 10.1093/9780191998874.003.0004

I seek to characterize social conditions for particularist explanations, without pre-judging features of those explanations or the ways they advance understanding. Philosophy of science does not have these theoretical resources, having traditionally very little to say about the community structure of science and its implications for knowledge and understanding, and indeed about social aspects of scientific knowledge more generally. So it makes sense to look to another area of philosophy for social conditions for particularist explanation. Philosophical theories of social action, which target the full range of human social behavior, are suitably inclusive and generic.[3] Third, and most straightforwardly, I have experience in philosophy of social action, having studied and published a few works in that area (Fagan 2011, 2012a, 2014). Specialization and understanding being key ideas of this work, it's reflexively important to be aware of one's own areas of expertise and their limitations. The conceptual framework of my version of explanatory particularism is based on two areas of philosophy in which I have professional expertise. For all three reasons, a journey into philosophy of social action is appropriate here.

There's a thematic connection as well. In science, a lot of boundary-work is concerned with explanation (e.g., "what *we* do is explain, what *they* do is stamp-collecting"). Boundary-work, to reprise, draws a line between one's own group and outsiders in a way that epistemically privileges insiders.[4] Boundary-work around explanation is a mark of the latter's importance as an aim of science; it's territory worth defending. This book aims to counteract boundary-work focused on scientific explanation, proposing a more inclusive, variegated, and pro-social view of explanatory practices in science. That is, the book as a whole is designed to combat in-group tendencies in both science and philosophy. Ideas from philosophy of social action, adapted to my purposes, are central to that project. To preview: this book advocates a certain kind of connection between scientific fields as vital to science, and conceives the latter as a pro-social human activity. Although that argument unfolds in later chapters, this chapter's foray into social action theory reflexively embodies the kind of connection I defend in Part II.

3.1 An example

Now, a brief introduction to social action theory. Although philosophical subfields are not as highly specialized and differentiated from one another as scientific

[3] Social epistemology of science (e.g., Longino 1990, 2002, Solomon 2001, Zollman 2007, O'Connor and Weatherall 2019, Dragos 2021) and recent studies of scientific collaboration (e.g., de Ridder 2014, Dang 2019) also offer resources. However, those epistemological frameworks are not as well-suited to develop the social aspect of explanatory particularism's core ideas. See Fagan (under review) for connections between EP and Longino's social epistemology.

[4] See Gieryn (1999), also Chapter 1.

communities, boundaries and barriers pertaining to explanation do appear in philosophy. It would be reflexively obtuse for this book to ignore them. So let me be very clear about how this project, a contribution to philosophy of science, engages philosophy of social action. The engagement is modest and largely one-way. I'm not offering a general explanatory theory of social action, and (as with models) I remain neutral with respect to most ongoing debates in that sub-field. My strategy is, rather, to survey the main philosophical theories of social action, identify key concepts and a rough theoretical consensus, and then adapt that rough consensus for my own purposes.

Philosophy of social action, as noted, is broadly concerned with general human experience of social activity and behavior. To begin navigating this theoretical terrain, an illustrative example will be helpful. Here's one from personal experience as a novice scientist.[5] Just after graduating from college, I spent a summer studying behavioral ecology at the Duke University Marine Lab on Pivers Island, North Carolina. The island was flat terrain, ringed with oil refineries and other industrial infrastructure. In morning and evening the beaches shone gold with scrims of shallow warm water, metal towers looming hazily behind. My final project was on the fiddler crabs who swept across the beaches at those gently-lit times. The waving pugilators—their descendants probably strutting the same beach now—were the focus of my attention for many hours that summer. I would follow one crab for a few minutes, noting the number of times it raised its larger claw at neighbors. Then I'd shift and follow a different crab. And repeat. Intermittently, I'd rest my attention and take in the whole delicately side-stepping wash of crabs along the curve of the beach.

The reason I'm indulging in this gold-tinted memory is to note a contrast—indeed, the central, fundamental contrast that motivates general philosophical theories of social action. The anecdote above is about me alone, alternately recording crab data and taking in the beach, the light, the scuttling rows of crabs, and background industrial filigree. That's one experience of watching. But it's not the experience I usually had, that summer. The behavioral ecology final project was, like most scientific work, a team effort. Most of the time when I watched the crabs, my project partner Alex was watching them too. (That made sense, because the pugilators only strutted in numbers at morning and evening; two of us could get twice the data in those peak times.) Both of us watched the crabs, counting their pugilistic waves, twice daily for a couple of weeks. The scene was just as described above. My note-taking was too. Alex used the same method; both of us

[5] The style of writing in this example uses the kind of first-person reflection on the character of their experience that is common in social action theory. The example itself, in this way, effects a modest inter-community connection within philosophy.

were novices learning the basics of behavioral ecology fieldwork. (That, together with getting acquainted with data analysis and write-up methods, was the purpose of the final project.) The contrast with the opening anecdote is just that Alex and I watched the crabs *together*. And this wasn't the same experience as watching them by myself. While attending to one or another individual crab, or the overall scene, I was throughout aware that Alex was watching them too—that he was also following one individual crab after another, counting their claw-waves, and intermittently taking in the sidelong water-waves, the light on the beach, the overall scene. I don't know that his experience mirrored mine, and he certainly couldn't have had the same web of background associations and memories coloring the experience. But without any such exact matching, I was all the time aware that we were doing the same thing—*we* were performing one of the most basic scientific activities, observing things in the world. (Our joint observing took in more than the crabs. I still remember the overall scene more than any of the data-point pugilators.)

Throughout those data-collection sessions, I was all the time aware of my research partner and of the pugilator crabs. But it wasn't the same kind of awareness; Alex wasn't to me part of the scene like the crabs or the beach or the industrial shadows. Or if another person had come walking by—I don't remember that happening, it would have been rude to scatter our to-be-collected data-points by treading on them—but anyway, a passer-by would have been part of the overall scene like the rest of the world. Alex wasn't, though. We were the observers; the world was the rest of the scene. Furthermore, I was aware of that scene in a different way than I would have been if observing it alone, albeit in a way that's hard to pin down. The closest I can get is that during those data-collection intervals, I saw the world as *what we were both seeing*—as the intersubjective world, to use a philosopher's term of art. Summing up: in several ways, some vague and difficult to articulate, the experience of *our* watching the Pivers Island fiddler crabs differed from the experience of *me* watching those same crabs by myself. Clarifying those differences, and locating them within a general theory of human social activity, is a core task for philosophy of social action. A satisfactory general theory should be able to clearly articulate the differences gestured at in my example, reveal their deeper grounds, and answer a range of related questions: How should we understand the experience of group activity? What's the significance of that experience, for human social life? What norms govern human social activities, and why?[6]

[6] This and other references to a "general theory of social action" may suggest that I am not a particularist about social action. That's not entirely correct. My views are congenial to particularism (and pluralism) in that theoretical domain, but as with model-particularism I don't argue for it here. This book is focused on particularism about explanation, which I've argued is warranted by the history and state of philosophical debate on that topic (Chapter 1). Tracing the implications of this version of particularism for theories of social action is a different task.

3.2 Philosophical theories of social action

Now to sketch the theoretical terrain more systematically. The central phenomena that theories of social action aim to explain are activities and attitudes attributed to groups. Everyday examples abound:

- The Dave Brubeck Quartet recorded "Take Five" for their 1959 Grammy-winning album.
- The opposition leader of Venezuela, Juan Guaidó, was recognized in early 2019 by the United States, Canada, Brazil, and Argentina.
- Scientific societies worry that Europe's Plan S will make science journal publishing unsustainable.

As these examples indicate, groups of various sorts are commonly referred to as engaging in activities (e.g., the Quartet recording) and as having intentional attitudes (e.g., nations recognize a political leader, scientific societies worry). Philosophical theories of social action encompass both kinds of case. (Indeed, they tend to blur the boundary between them, for reasons that will shortly be obvious.) To make sense of these everyday attributions, philosophers of social action seek an illuminating analysis of the basic unit of social activity; "the social atom" constituting the fabric of human social life and institutions.[7] Philosophical efforts along these lines draw on a wide and diverse array of resources, with important influences stemming from sociological theory, psychology/cognitive science, and philosophy of mind, as well as general action theory.

Background and basics

Two background assumptions unify these diverse theoretical efforts:

(1) A theory of social action must explain how it differs from individual action (i.e., action of an individual agent, typically assumed to be a human person).
(2) Social action is distinguished from individual action by a mental attitude, not physical (external) actions or events.

Putting these ideas together yields the theoretical starting point for much philosophy of social action. The difference between a random group of people and a group engaged in social action is assumed to be localized not to behaviors the

[7] Phrase borrowed from Margaret Gilbert. There is interesting recent work on the non-human origins of social activity (e.g., Tomasello 2009) and on naturalistic approaches to same, but philosophy of social action remains largely focused on human activities.

people perform—what they actually do in the world—but in how they think about what they are doing (e.g., Jankovic and Ludwig 2018, 1).[8] The central theoretical task is to analyze this intentional difference. The analysis should apply to both group activities and attitudes; behaviors and beliefs alike.

The distinctive attitude of social action is characterized as an *intention*; i.e., a conative state had by an agent (or agents) that has propositional content, is directed toward the state of affairs represented in that content, engages the agent's will, and is implicated in the agent's practical reasoning. Accordingly, a great deal of work in philosophy of social action is aimed at analysis of "collective intention" (aka "joint intention," "we-intention," "shared intention")—i.e., that which unifies multiple individual activities and attitudes into a single social action.[9] Collective intention is what makes the difference between genuine social activity and "a mere summation or heap of individual acts" (Roth 2017). So a theory of social action must explain what collective intention is, and how it relates to individual intentions. The sought-after explanations have conceptual, ontological, and psychological aspects, which are differently weighed and blended by different theorists. Although not scientific explanations, the explanations provided by a theory of social action have the same specialized and local character. As noted, theorists usually take for granted that social action is to be understood on the model of, and in relation to, individual action.[10] Debate centers on the issue of *reduction*: Are social activities and attitudes reducible to activities and attitudes of individuals, or not? Is collective intention reducible to individual intentions, or not?[11] Reductive theories identify a collective intention with a collection of suitably related individual intentions. Different reductive theories posit different ways that individual intentions are woven together to constitute a collective intention. Nonreductive theories, in contrast, posit some irreducible feature had by collective intentions. In this debate, four theories have been particularly influential.[12] Bratman (1999, 2014) argues that a complex of interdependent individual intentions to perform an action under conditions of

[8] But see Epstein (2016) for an important dissenting view, from the perspective of social metaphysics.

[9] For simplicity, I will use "collective intention" throughout. The term was introduced by Searle (1990). The *Stanford Encyclopedia of Philosophy* entry of the same name provides a thoughtful discussion of historical antecedents and recent literature (Schweikard and Schmid 2013), and the new *Routledge Handbook of Collective Intentionality* offers a broad interdisciplinary overview of the main theories and arguments on the topic (Jankovic and Ludwig 2018). Roth (2017) also provides a detailed overview of philosophical theories of shared agency.

[10] See Chant (2018) for criticism of this approach.

[11] Reduction is at the core of what Schweikard and Schmid (2013) term "The Central Problem" for collective intention, as well as what Roth (2017) terms "the traditional ontological problem" for shared agency. See also Alonso (2018). Studies of collective scientific knowledge are also organized around this debate. Some have argued that scientific groups have knowledge in some irreducible sense (e.g., de Ridder 2014, Palermos 2022). Dragos (2021) proposes an insightful, inclusive taxonomy for theories of group belief and knowledge, organized around oppositions of individual/group, internalism'externalism, and epistemic autonomy/extension.

[12] Jankovic and Ludwig (2018) refer to these four accounts as "field-shaping" (see this work for further references and interdisciplinary influences).

common knowledge and mutual responsiveness is sufficient for collective intention. Gilbert (1989, 2000) defines collective intention in terms of irreducible joint commitment to believe or bring about p "as a body"—a "plural subject." Searle (1990) analyzes collective intention as an attitude had by individual agents in a "we-mode" that is irreducible to the "I-mode" of individual action. And Tuomela (2005) analyzes collective intention as a multi-agent, concatenated attitude had by individual agents in an irreducible we-mode under conditions of mutual belief.

In earlier work, I've argued against the reducible/irreducible dichotomy, emphasizing instead connections linking individual and group beliefs (Fagan 2011, 2012a, 2014). As an alternative, I've advocated an "interactive" approach, analyzing a group's belief in terms of its members' beliefs in interaction with one another.[13] On that approach, group (or community) S interactively believes that p just in case S's members' beliefs about p interact, via some characteristic set of social epistemic relations, to yield S's belief that p. This is a process view of group belief, in that a group S's belief that p is analyzed in terms of a process mediating between individual members' beliefs concerning p and the group belief. Theoretical focus thus shifts from the question of whether groups can have beliefs (or knowledge) to the processes by which beliefs (or knowledge) commonly attributed to groups are produced. The mediating process consists of interactions between individual group members' attitudes. So, instead of "Who has (scientific) knowledge?" the central theoretical question becomes "What interactions produce (scientific) knowledge?" Group belief-producing interactions, taken together, comprise a group's social epistemic organization. Exactly which social epistemic relations are involved varies according to the case; e.g., relations of epistemic authority, trust, and pedagogy. Rather than dichotomizing mediating processes as summative or collective, the interactive account aims to articulate the social epistemic interactions that mediate between individual and group belief. These mediating interactions vary widely across groups—e.g., particular scientific communities. Reductive theories are a limiting case, in which no epistemic relations among members are involved in producing group belief. Such a group is effectively a single iterated individual; an aggregate of interchangeable individuals. But in most cases, a group belief does not result from simple aggregation of individual beliefs. Group belief often involves aspects of social life that knit the members together into a group - interactions involving trust, authority, and the like. The interactive account allows for reductive cases, but also for those in which

[13] Interactive belief: Group S interactively believes that p if and only if S's members' beliefs concerning p and epistemic interactions among those members together result in S's belief that p. For other accounts of group belief involving similar ideas, see Corlett (1996), Ernst and Chant (2007), Kitcher (1991), Niiniluoto (2003), Palermos (2022), Tuomela (2005), and Wray (2007). The last is presented as an account of irreducibly collective belief, but Wray's concepts of epistemic interdependence and functional organization, not to mention his focus on scientific practice, better fit the interactive view. De Ridder's epistemic dependence account (2014) resembles Wray's in these respects.

a group's social epistemic organization is involved in generating group belief. The simple dichotomy of reducible vs. irreducible group beliefs is replaced by a plurality of analyses of group belief.

In this way, interactive belief offers a path beyond—or through—the reduction debate. A group's epistemic organization is constituted by epistemic practices involving synchronic or diachronic relations among members' beliefs: trust, authority, and decision procedures for integrating individuals' beliefs to yield a group-level result (e.g., voting). The interactive account allows for the full range of such interactions, highlighting connections between individual and group belief. As will be seen shortly, this interactive, processual approach comports well with basic ideas of EP and EP_M. Whatever the merits of my interactive account for philosophy of social action, it's well-suited for the task of articulating a conceptual framework for explanatory particularism. This is because it fits with key features of social action theory on the one hand, and with tenets of explanatory particularism on the other. It forges a clear conceptual connection between accounts from different areas of philosophy.

Theoretical consensus

The interactive account of group belief is easily extended to social action more generally. For general theories of social action, this amounts to setting aside the reduction debate. Once this is done, commonalities among the major theories come to the fore.[14] Although there are many different analyses of collective intention on offer, they generally involve the same elements in similar relations (e.g., Bratman 1999, Tuomela 2005, List and Pettit 2006).[15] The main points of contention among philosophical theories of social action (aside from the reductionism issue) concern the nature of mental attitudes and norms of practical reason. These are substantive concerns specific to philosophy of social action. For my purposes, they can be set aside. What remains is substantial consensus on social action. There are four consensus elements: multiple participants, interdependent participant intentions, a shared goal, and common awareness. *Multiple participants* are obviously required for social action. And their attitudes—individual participant intentions—must be somehow conjoined so as to constitute a collective intention. That is, a collective intention represents *interdependence* among participants' intentions regarding the social activity. Most theorists explicate this interdependence in terms of "other-regarding" attitudes; e.g., each intends to participate partly because each

[14] See also Pettit and Schweikard (2006) Ludwig (2007), List and Pettit (2011), Theiner (2018). The terminology here follows Theiner's.

[15] Precursors from somewhat different philosophical traditions include Walther and Collingwood (discussed in Schweikard and Schmid 2013). There are also parallels with Andersen's (2010) notion of heterogeneous conceptual consensus.

believes others so intend. This other-regarding aspect is central to what *Stanford Encyclopedia of Philosophy* terms the "majority view" of collective intention:

> For a complex event to be not just a coincidentally matching pattern of actions by individual agents but an intentional joint action, it needs *a complex structure of attitudes in which the agents involved refer to one another.*
>
> (Schweikard and Schmid 2013; italics mine)

So some reference to others' attitudes within an individual participant's intention is another consensus element of social action theories. (The next section proposes an analysis of this interdependence, for the special case of collaborative social activities.)

A third consensus element of social action theories is the concept of a *shared goal.*[16] This idea needs a bit more introduction, being a term of art in philosophy of social action. A goal is a focus for agency; the object, destination, or state to which means are taken (e.g., Roth 2017: "commitment to an end [is] implicated in any instance of intention or intentional action"). A *shared* goal is a motivational focus of this kind for multiple agents and/or a group. For example, consider a group of people exiting a movie theater. Suppose each individual person in the group has an intention with the content 'I intend to get up and move to the exit.' And then each person acts on their intention. Their individual intentions are 'shared' in the sense that each has the same content. But their exiting isn't a social activity. Each person is just doing what they intended, separately. It just happens that multiple individuals act on their intention at the same time (or near enough). What's missing is a shared goal: some activity that each person in the group intends they all do together. For many theories, this shared goal figures in the content of a collective intention; e.g., 'I intend that we all get up and move to the exit.' Group members coordinate so that each has this intention, and then they exit as a group: a social activity.

In a social activity like this, each participant is aware of the others' counterpart intentions and beliefs regarding the shared goal. Some form of *common awareness* is a fourth consensus element of social action theories. The motivating idea is that one cannot participate in social action without some knowledge or belief about the others involved. But there's no consensus on exactly what this requirement amounts to. Bratman and Gilbert explicate it in terms of "iterative common

[16] This is at least partly because many theories of social action use the means-end relation to explicate the connection between an individual's own intended role and the overall social action. That is, the individual-group relation is analyzed as coinciding with a means-end relation; individual participants in a social activity each take their own actions to be instrumentally related to a shared goal, which is the social action itself. This allows theorists to bring norms of rationality to bear on their analyses of collective intention. It also creates puzzles, which have spurred a lot of theoretical work. I set the latter aside here.

knowledge," a rather demanding condition that weaves epistemic attitudes of different agents into an ascending structure of transparency (i.e., A knows that p, B knows that p, A knows that B knows that p, B knows that A knows that p, A knows that B knows that A knows that p, B knows that A knows that B knows that p, and so on, ad infinitum). Weaker conditions along the same lines are epistemic openness; mutual awareness, or a "public context" such that no participants are necessarily epistemically aloof from others (e.g., Kutz 2000, Ludwig 2016). Roy and Schwenkenbecher (2021) provide a thoughtful recent overview of different accounts of "common knowledge" for social action. They propose a pluralistic, context-dependent approach, such that different awareness conditions are conducive to (rather than constitutive of) different kinds of shared intention. Amidst this theoretical diversity, however, there is consensus that some form of *common awareness* is involved in social action.

To sum up the argument so far: philosophical theories of social action center on the attitude of collective intention, as the 'ingredient' constituting the difference between individual and social action. A central theoretical task for this area of philosophy is therefore analysis of collective intention. A number of theories have been offered, four being notably influential. Debate about these theories is mainly organized around the issue of reduction. However, taking the interactive approach shifts focus from that issue toward explicating relations that weave together individual intentions into a single collective intention. This shift in turn highlights four common elements across theories of social action: multiple participants, a shared goal, interdependent individual intentions, and common awareness. The rough theoretical consensus, then, is that social action consists of multiple agents who form interdependent intentions under conditions of common awareness to participate in a social activity (the shared goal). Different social action theories offer different analyses of the four consensus elements and their relations to one another. But the basic theoretical picture is fairly robust.

EP and social action

The theoretical consensus about social action fits with and builds on tenets of explanatory particularism. To reprise: the task is to identify conditions for construction and use of particularist explanations. Such framing conditions are at once epistemic and social. Their epistemic aspect is articulated in terms of models and modeling (Chapter 2). Their social aspect is articulated in terms of social action theory. The key move for the social side of the framework is to conceptualize the process of constructing a particularist explanation (explanatory model, in EP_M's sense) as a social activity. That view comports well with the basic ideas of explanatory particularism. According to the latter, a scientific explanation is a specialized achievement of a particular scientific community, which advances understanding for members of that community. It follows that a particularist explanation is a form of knowledge attributed to a group: the scientific community that produced it. Therefore, a particularist explanation achieves an important goal for a scientific community. It does so by combining diverse ingredients into a model that community members recognize as advancing understanding. That complex process is framed by EP_M as model-construction. But it's not a stretch to conceive it as social activity. Particularist explanations are specialized, constructed out of diverse resources to achieve a kind of harmony of values and constraints in that community. The constructive process is complex, involving an array of practices performed by multiple agents in the community. The combination of diverse community resources into an explanatory model mirrors the combination of participant actions and attitudes into a social activity.[17] It's straightforward to conceptualize the process of constructing a particularist explanation as collaborative social activity in a scientific community.

This social action framing offers guidance for particularist studies of explanation in scientific practice. Practices contributing to an explanatory achievement vary widely across communities, so there's little to be said of them in general. Philosophers studying explanation in scientific practice should, on a particularist approach, begin by learning the epistemic ways of a particular scientific community—its boundaries, requirements for membership, major accomplishments, work-in-progress, values and goals, methods, and technologies, and more.[18] This empirical task is demanding—without some guiding framework, intractable. While substantive philosophical theories of explanation can be valuable heuristics for such study, there is no shortcut for doing the empirical work of characterizing what makes for a significant achievement of understanding in a particular community context. Social action theory offers a framework for that task; a thin but non-vacuous general analysis for studies of explanation in particular scientific communities. Whatever the activities in a particular scientific community that contribute to an explanatory achievement, explanatory modelers and research teams are participants in an overall collaborative process. Diverse individual activities and attitudes combine to produce a significant explanatory achievement for the whole community. The interactive view highlights social epistemic interactions involved in that process. Alongside details of model-construction, philosophers should also track patterns of epistemic trust, authority, openness, etc.—features of a community's social epistemic organization that mediate between the activities of individual scientists and the community's explanatory achievement. Alongside epistemic practices of model-construction, relations of trust, authority, and power

[17] This is not to say that explanatory model-construction is nothing over and above social activity in a scientific community. Many other resources are involved, material and otherwise. EP_M's framework is not displaced, but augmented by the social action framework.

[18] Also broader institutional differences; more the area of sociology of science. Particularist study of scientific explanation is thus allied with, and closely akin to, sociology and history of science.

mediate the process by which a new model is recognized as an explanatory achievement in a scientific community. So a scientific community's social epistemic organization is implicated in scientific explanation. Philosophers of science should first map these interactions—sketch the social epistemic organization; the 'anatomy and physiology' of a scientific community.

A key point highlighted by the interactive approach is that producing a particularist explanation is not the sole task of individual modelers/research teams. Of course, the latter are vital—and looking into the details of these practices reveals a great deal about a scientific community's explanatory values (see Chapter 2). But constructing an explanatory model, in EP_M's sense, is not limited to direct model-building contributions. There are several reasons for this. First, a scientific community's major achievements are, by definition, imbued with and exhibit that community's values. Significance of an epistemic product is judged by the community, not just the members who directly construct an explanatory model. Second, among a community's significant products, explanations are distinguished by their epistemic role: advancing understanding for members of that community. What counts as such an advance, and the features of an explanation that afford it, are also relative to a community—not determined by a few individual modelers or a research team. Third, ideas, constraints, and values that are widely shared in and characteristic of a particular scientific community are centrally involved in constructing and using an explanatory model. These background ideas, values, and commitments, which guide and stabilize the process leading to a particularist explanation, are prevalent within the community. Individual modelers make use of them, as community members. For all these reasons, the social activity of producing a particularist explanation involves members beyond the researchers/teams that directly construct a particular model.

Furthermore, all four consensus elements of social action theory figure in that constructive process. Evidently, constructing an explanatory model involves multiple participants. Individual members of a scientific community work together to produce various results, epistemic and practical. Any explanatory model's construction will involve some community members as central, key participants, while others play more of a supporting role. Even a lone modeler must use a range of community resources to construct an explanatory model. Recognition of the result as a significant achievement requires input from other community members. So constructing an explanatory model, in EP_M's sense, always involves multiple participants. The activity also has a shared goal. In social action theory, the shared goal featured in a collective intention is performance of the social activity: enacting what is intended. So the shared goal is the social activity of constructing an explanatory model. However, this case is more complex and outcome-driven than the examples of social action that ground the main theories. Those exemplars tend to be simple, e.g., walking together or dancing a tango. Explanatory model-construction is more evidently aimed at an outcome than these classic theoretical exemplars.

As EP_M specifies, explanatory models have the purpose (P) of advancing understanding of some target (X) for members of a particular scientific community (S). That purpose, or end, is a major influence on the process of constructing a particularist explanation. So not only is there a shared goal in this case—constructing an explanation—but the use to which that construct is to be put—advancing understanding of some target explanandum—serves as a grounding purpose or end of the collaborative activity.[19]

Interdependence among participants in explanatory model-construction is more difficult to characterize in EP's terms. Across different scientific communities, the specifics of such interactions vary widely. The next section offers a general analytic framework for conceptualizing the interdependence of scientific community members in the collaborative activity of constructing a particularist explanation. Finally, common awareness among members is an established feature of scientific communities. Within the social epistemic organization of a scientific community, members are neither aloof nor hidden from one another. This doesn't mean that all members of specialized scientific field know one another as individual persons (i.e., *de re* common awareness). Of course, any member of a scientific community will personally know some (even many) researchers working in their specialization. But it's unlikely that any member of a scientific community will know every other member in this way. Rather, members of a scientific community are aware of others' attitudes *de dicto*; i.e., they know there are other members of the same specialization, whoever they are. That is, each member of a scientific community is aware that their community includes other members with a substantial shared epistemic, methodological, and practical background. Moreover, contributions to an explanatory model are public within a scientific community.[20] To meet the requirement of significance, an achievement of understanding has to be exhibited and accessed widely within a scientific community. Publicity in this sense is required for scientific knowledge-production more generally, implemented in familiar norms such as the need for peer review, publication, and venues for criticism of new results.[21]

So, all four consensus elements of social action theory figure in the basic ideas of explanatory particularism (interdependence being a promissory note). These four theoretical elements expand EP_M's conceptual framework. A scientific community S includes multiple participating members—working scientists comprising that community. Their shared goal is to construct and use an explanatory model M. M's purpose (P) is to advance understanding of some target of interest, X. To

[19] See Section 3.4.

[20] This does not imply that every aspect of these contributions is public; there is a lot of propriety secrecy among research groups, institutions, laboratories, and projects. But *explanatory* models, in EP_M's sense, are necessarily public.

[21] This characterization follows Longino's social epistemology of science (1990, 2002). Connections between the latter and explanatory particularism are examined in Fagan (under review).

accomplish P, M and the M-to-X relation must exhibit features conforming to (expressing) S's specialized epistemic values, alongside other constraints bearing on explanation in that community (e.g., available technologies). Constructing M is a social activity guided and constrained by S's ideas and values bearing on P. The interactive account offers deeper analysis of explanatory model-construction within a scientific community S. Participating scientists work together to construct an explanatory model. Their individual attitudes toward that shared goal are interdependent—woven together in some way that can yield an advance in understanding for (members of) S. Common awareness of each another's explanatory efforts is the norm for members of scientific communities. So the case at issue includes all four elements featured in theories of social activity.

3.3 Structures of collaboration

This section builds on general theories of social action, sketching the structure of relations among the four consensus elements. The key task is to specify interdependence and common awareness of multiple participants' attitudes toward a shared goal in *collaborative* social activity. So what's offered here isn't a general theory of all social action. Not all social activities are collaborative. Those that are, are distinguished by greater interdependence, tighter integration, and/or closer cohesion among multiple participants aiming at a shared goal.[22] I propose an original analysis of the interdependence of collective intention in collaborative social activity. That analysis grounds further results about particularist explanation and understanding. I don't want to overstate the originality here, however. The main theories of social action, with their consensus elements, inform what follows. All specify a structure of relations among individual intentions and associated attitudes, implicated in if not constitutive of collective intention.[23] For example, Gilbert's plural subjects theory posits interdependent participatory commitments ("a web of mutual obligations and rights") constrained by a group's irreducible joint commitment. Some theorists, notably Bratman, characterize this structure in more detail (1999, 2014). The analysis that follows builds primarily on those theories.[24] My account is thus an elaboration of the general consensus view

[22] See León et al. (2019), Roy and Schwenkenbecher (2021) on pluralism about collective intention. There are different forms of shared intention, more or less "tightly connected, interdependent, or interlocking" (Roy and Schwenkenbecher 2021, 4537).

[23] Ludwig's (2016) account is an exception, as he dispenses with the requirements for openness and interdependence. Yet his account does include analogs of the three layers discussed here, though the resemblance is not as exact as for, e.g., Bratman and Tuomela.

[24] See also Satne (2021). The main inspiration for this part of the framework is Bratman (2014), specifically the interlocking intentions of his "modest sociality." However, I omit "by way of" relations and layering via plans/sub-plans, so the structure articulated here is not identical to Bratman's.

of collective intention, the main contrast being tighter, denser interdependence relations than would be appropriate for social action in general.

Three-layer interactive structure

Interdependence is analyzed as an integrated complex of attitudes of participants in some collaborative social activity. The shared goal is performing the collaborative activity. Participants' attitudes toward that goal are intercalated via three layers of relations. 'Layering' refers to the fact that successive relations take the previous as one of their relata. The interactive structure is intercalated, with reciprocal intentions of each participant combining to comprise the overall structure. This tripartite set of intercalated attitudes with respect to a shared goal comprises the *interactive structure* of collective intention for a collaborative activity. This structure coordinates multiple participants in their pursuit of a shared goal, thereby enabling performance of the social activity corresponding to that shared goal. I'll work through each layer in order.

First layer

For simplicity, consider a collaborative activity involving two participants, A and B. Their shared goal is the social action that participants intend to perform; call this collaborative social activity C.[25] The first layer of relations is between a participant and the shared goal: A intends to perform C, and similarly for B.[26] But these first-layer relations are not directly "other-involving." The shared goal is like a maypole with multiple lines streaming from the top, each grasped by an individual dancer. Each grasps a line tethered to the same center, but there is no other connection among them (see Figure 3.1a). Additional layers are needed to weave a collective intention. The first-layer relation has a dual aspect, however. So there is some weaving at this layer, of singular and plural perspectives for each agent; A relates to C as both 'I' and 'we,' as does B. That is, each agent intends C in a way that involves both singular and plural self-regard.[27] Singularly, each intends to do her part in C, as a participant. Plurally, each also intends that 'we' (the group A-and-B) perform C. This dual aspect affords interplay in each agent's self-identification. Each

[25] It is important to distinguish the shared goal from the intended outcome or end-state of the social activity to be performed. The collective intention is to perform some social action. The outcome of a collective intention that is successfully executed in action is outside the scope of this analysis. It does come into play for particularist explanation, however.

[26] Social action theorists typically express the content of this intention as "I intend that we J" (e.g., Bratman 1999).

[27] Social action theorists demand further analysis of this dual identification, and rightly so (see Schweikard and Schmid 2013 for an incisive breakdown of theories). However, for my purposes this analysis can be bypassed. It is enough to assume that some satisfactory analysis can be provided, without specifying its exact nature.

identifies, with respect to C, both as an individual participant and as the group as a whole.[28] That is, both A and B are aware of their own participation in C, at once as individuals and as a group.

Figure 3.1a Interactive structure of collective intention for collaborative social activity: first layer. Each participant (A and B) relates to shared goal C.

Second layer

The second layer relates each agent (A and B) to the other's first-layer relation. This introduces the "other-regarding" aspect of collective intention. Both A and B relate to the other's relation to the shared goal. The relation is one of regard, or recognition: A regards B as intending to C, and B regards A similarly. So the attitudes of individual agents regarding C are reciprocal. To return to the maypole metaphor, it is as though individual dancers each grasping a tether also grasp hands with one another (Figure 3.1b). The two-layer structure entails common awareness of each agent's intention to C. This awareness imposes a kind of symmetry on each agent's intention to C. Each agent regards the other as a participant in the group's activity, just as she identifies herself. Even if the specific roles they intend as participants diverge greatly, A and B recognize one another as standing in the same relation to C. As participants, they stand in relation to C as equals. The group-level aspect of the first-layer relation further effaces boundaries between A and B. Each agent regards the other as identifying with the group, just as she does. In this sense, their agency is combined. Yet their experiences aren't merged or fused. Agents A and B remain distinct individuals, aware of one another as connected via C yet not the same. Schmid (2014) refers to this as "the difference condition" for collective intentions associated with shared emotion. On his view, A and B share an attitude (a we-intention), yet retain their own phenomenal experience. Their distinct experiences match or complement one another, giving rise to an overall unity.[29]

[28] A number of philosophers have attempted to capture the phenomenal or theoretical features of this dual experience, acting as an individual and as a plural subject (e.g., Schmid 2014, 2016; Høffding and Satne 2021).

[29] An individual's experience (of an activity or attitude) is "socially extended in the sense of standing in a relationship of constitutive interdependence with the phenomenal character" of another's experience, "without this entailing a breakdown of the individual boundaries." Differences between individuals "far from jeopardizing the integration, sustain[] it" (Léon et al. 2019, 4856).

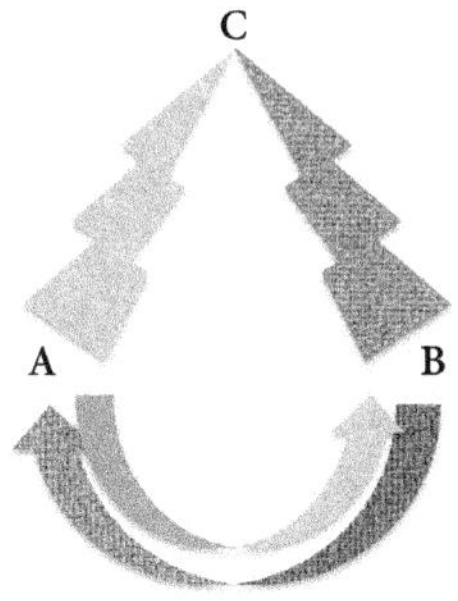

Figure 3.1b Interactive structure of collective intention for collaborative social activity: first and second layers. Each participant (A and B) relates to shared goal C *and* regards the other as so related (A regards B as intending to C, and B regards A as intending to C).

Third layer

The third layer of relations closes the loop, effectively weaving together individual attitudes of A and B into their collective intention to C. In this third layer, agent A relates to B's second-layer relation, and B relates to A's second-layer relation. This relation too is one of regard, or recognition: each agent regards the other as other-regarding, with respect to C. That is, A regards B as regarding A as intending to C, and B regards A similarly (Figure 3.1c). This is clumsy to state, but the basic idea is simple: one sees oneself, in relation to C, through the other's eyes.[30] The reciprocal third-layer relations deepen agents' common awareness of their intention to C. Not only does each regard the other as a participant and as identifying with the group, just as she identifies herself—each also regards the other as regarding her so. Not only do they recognize one another as participants, standing in the same relation to C, but each recognizes the other as recognizing this. Each agent is aware of the other's other-regarding attitude with respect to C.[31] Each recognizes the other not only as another participant in C, but a *fellow* participant; a partner. This third layer of relations completes the 'back and forth' intercalation of agents' attitudes, weaving individual intentions into a collective intention.[32] Each layered triplet of relations forms a loop from one agent to the other and back. A intends to C, B recognizes that A intends to C, and A recognizes that B recognizes A's intention. And similarly for B: B intends to C, A recognizes that B intends to C, and B

[30] The visual metaphor should not be pushed too far, however. The maypole analogy from the first two layers breaks down here also.

[31] Higher 'layers' of common awareness are possible, of course, but these seem otiose for collective intention. Most theories of social action today do not require common knowledge in Lewis' sense (full "epistemic ascent"). Three layers suffice to interweave individual intentions, structurally speaking.

[32] E.g., "a bridge intention to mesh my J-related plans and intentions with yours" (Roth 2017).

recognizes that A recognizes B's intention. The two loops are mirror-images of one another, each tracing the same connections with the agents' roles reversed. Agents' recognition of one another as partners in C prevents collapse of boundaries between A and B into a 'we.' That the other is distinct from oneself is highlighted in the third-layer relation—yet A and B are more deeply connected through this 'looping' effect, which bears some resemblance to influential conceptions of empathy.[33]

Figure 3.1c Interactive structure of collective intention for collaborative social activity: first, second, and third layers. Each participant (A and B) relates to shared goal C, regards the other as so related, *and* regards the other as other-regarding, with respect to C. This interactive structure has the form of two interlaced feedback loops.

To sum up: the three layers of relations effectively interweave the attitudes of participants in social action, as each regards the shared goal and their individual role therein; the corresponding position of others as attending to the shared goal and their individual role therein; and the corresponding attitude of others as analogously other-regarding. The three-layer interactive structure is symmetric, reciprocal, and interlaced. Reciprocal awareness of oneself and others, as distinct individuals and as comprising a single 'we,' transforms each agent's experience in C, relative to a singular counterpart. A visual analog is the motif of "interlace" found in illuminated manuscripts of the British Isles and Ireland before AD 1000.[34]

[33] See León et al. (2019) for a similar account of joint intention, stemming from a very different theoretical perspective.

[34] See for example the carpet page from the Lindisfarne Gospels, held by the British Library. The manuscript's interlace motif captures key ideas of the interactive structure of collective intention, which interweaves participants' intentions into the defining attitude of collaborative activity.

Norms for collaborating agents

Social action theorists argue that collective intentions entail norms for individual participants, which constrain their individual attitudes and mental states.[35] Following this line of thought, the three-layer structure of collective intention has conceptual implications with normative significance.[36] The interlocking structure weaves individual attitudes into a collective intention to collaborate in some social activity. What does it mean to participate in a collaborative activity characterized by that intentional structure? What does it mean to partner with others in a community, in some collaborative task? The three-layer framework offers (partial) answers to these questions. Each layer of relations in the interactive structure constrains participants' attitudes in specific ways. The interactive structure spirals outward from the shared goal in three layers, which weave the attitudes of different individuals into a single collective intention. This structure constrains participants' attitudes, in ways that have normative (indeed, ethical) overtones. I'll discuss these implications in order.

The first layer connects each agent to the shared goal. A shared goal implies multiple participants, with whom the goal is shared. So an agent committed to a shared goal cannot, on pain of inconsistency, intend to perform the activity as a solo project (except insofar as that agent identifies with the group as a whole; as "we"). From an individual-level perspective, each participant must acknowledge that others are involved in the intended activity. A participating agent can't conceive of the activity as somehow exclusively theirs. Each must be open to other participants in the activity; participation is *non-exclusive*. Relatedly, an individual agent's intended contribution to a shared goal isn't sufficient to accomplish that goal. The part an individual agent intends to play is just that—a part, which combines with others in the overall performance of the activity. Each participating agent should recognize that their individual contribution is *incomplete* on its own. Recognizing this incompleteness requires humility. The dual aspect of a participant's relation to C, which affords interplay in each agent's self regard, is also significant. With respect to C, both A and B self-identify as an individual participant and with the group. Interplay between group and individual identification results in uptake by individuals of values and interests associated with the group.[37] The latter, as discussed above, are major determinants of the features of an explanatory model constructed in a particular scientific community.

[35] These include internal consistency, consistency with other beliefs, means–end coherence, and stability (see, e.g., Bratman 2014).

[36] The next section discusses grounds for these norms.

[37] The reverse may also occur; the values and interests of an individual diffusing throughout a group.

The second layer of relations connects each agent with the other's relation to the shared goal. This reciprocal regard for one another's intentions implies *equality* among participating agents. There are several aspects to this equality. First, all participants bear the same relation to the same shared goal; they are on equal footing with respect to performing the social activity, although their roles in it might differ widely. Moreover, insofar as each identifies herself with the group (the plural subject "we") boundaries between individuals are elided; their respective identifications with the group coincide. This equal standing in regard to the activity should be reflected in agents' attitudes toward one another; they are all part of the same thing, broadly speaking. In addition, participants' attitudes towards themselves and one another should be symmetric. That is, each participant treats the other's intention to participate in C as she treats her own, as parts in the same overall activity. Their enacted parts may differ, but their status as parts is the same. Finally, common awareness implies a form of epistemic equality. Participants are aware of one another's intentions regarding C. No participant is hidden from others, pulling the strings as 'the man behind the curtain.' Within the group associated with C, these intentions are 'out in the open'; i.e., *public*. Participants are, in this rather narrow sense, epistemically equal.

The third layer relates each participant to the other's other-regarding attitude, closing the intentional loop. Each agent recognizes the other as a full partner in the intended collaborative activity. That is, each should recognize the other as making a contribution analogous to their own.[38] The norm associated with this attitude is *respect*. Participants in a collaborative activity should respect one another as participants with equal standing in the project. These implications of the interactive structure, taken together, suggest norms for constructing particularist explanations. Stated as guidelines for collaborating agents, these are as follows:

- The activity is not one's own exclusively; others are involved.
- One's own contribution to the activity is incomplete; exercise humility.
- Participation is public; no one acts behind the scenes.
- Other participants have equal standing in the activity; respect them as full partners.

None of these guidelines should be surprising; all are fairly commonsense maxims for a collaborative project. However, applied to the case of explanatory model construction, they have further, less obvious, implications.

[38] Wylie (2015) articulates this norm for collaboration in archaeology: "At a minimum, what is at stake in demands for accountability is that archaeologists *respect* the interests and sensibilities of descendant communities, even if they don't credit them with epistemically compelling norms of justification" (194–195; italics mine).

Norms for explanatory modeling

I'll now apply this interactive structure, with associated norms, to the case of constructing an explanatory model within a particular scientific community. The theses discussed so far, EP and EP_M, characterize an explanatory product rather than a process: explanatory model M advances understanding of some target X for members of a particular scientific community. M is a significant achievement, according to community members. It follows that target X, the advance in understanding thereof, and the M-to-X relation that secures the advance, are all important to that scientific community; i.e., they exhibit the community's values pertaining to explanation. M expresses, harmonizes, and epitomizes those values. Those values, along with details of M's construction and resultant features of M, vary widely across particular scientific communities. However, participants' attitudes toward one another in that process are constrained, albeit minimally, by the collaborative norms presented above. How do these constraints operate in scientific community S, constructing a model M with the purpose P of advancing understanding of target X?

The first layer of relations connects each agent to the shared goal: constructing explanatory model M. An individual member of the research team engaged in doing just that must conceptualize it as part of a group effort, not as their own exclusive solo activity. Others in the community are interested parties, even if not as directly involved in M's construction. Constructing a model to advance understanding of X is a task open to others in the community; it is not exclusive to any individual modeler in that field.[39] Even if it happens that no other community members are working to understand target X at the time, a lone modeler must allow that they could do so—that their solitude is a mere accident (and unlikely, if understanding X is considered important in that community). Relatedly, no individual modeler can single-handedly construct M in a way that satisfies EP_M. For one thing, most of the resources for constructing M are 'in the community'; i.e., any individual modeler's activities are scaffolded and supported by those resources. Second, recognition of M as significant and successful is a community judgment, not up to any individual modeler (or team of modelers). Judgments of significance and success are necessary for M to accomplish its purpose. Third, those judgments are relative to the purpose of advancing understanding of target X. Such understanding is public and shared, not limited to any individual modeler. So the purpose relative to

[39] Compare the sense of "we" grounding collective intention: "the experience of persons being specifically related to each other, called *the sense of us*, renders a core feature of a we-perspective, which is instantiated in collective attitudes" (Crone 2021, 11829; italics mine). Using empirical findings from psychology and cognitive neuroscience; "basic intersubjectivity, forms of nonverbal interaction [micro-interaction] and a feeling of binding are conditions" on a "minimal sense of us" that functions as a precursor (conceptual and causal) to full-fledged plural subjecthood (Crone 2021, 11829). See also Martens (2018), Schmitz (2018).

which M is constructed and evaluated belongs to the community—it's not an end that can be had or accomplished by any individual member of that community. An individual modeler constructing M is working toward that end, which guides the constructive process throughout. The most a lone modeler can do is propose a model to the wider community: an explanation that is judged successful or not by the wider community. An individual effort is incomplete; an explanatory model, in EP's sense, isn't constructed by any single researcher or group.

The second layer of relations connects each agent with others' relation to the shared goal. In most scientific communities, no individual member will be directly acquainted with all others. So they will relate to other members' intentions *de dicto* (i.e., the intentions of other community members, whoever they are) rather than to the intentions of individual persons known to them. How does an individual modeler in community S conceptualize other community members' attitudes to constructing an explanatory model? All have the same stake in the outcome: an advance in understanding of X. The value placed on understanding X is common to all members of S—even those who aren't actively working on that project are committed to the values, standards, and methods that characterize S. In this sense, all members are on equal footing with respect to constructing M. Relatedly, each individual agent should see their own relation to those community values, standards, and methods as symmetric to others' relation to same: the community features bearing on explanation are intersubjective, neutral with respect to community members. Judgments of explanatory significance and success are symmetric (neutral, "objective" in one sense of that term) among community members. Finally, the community's values, standards, and methods bearing on M's construction are public and 'out in the open.' Community members thus share common awareness of these contextual features; no one's intentions are hidden from others, with respect to constructing M. In this sense, community members are epistemically equal in this activity.

The third layer constrains each modeler to recognize others in the community as making contributions analogous to their own. Participants in a collaborative activity should respect one another as participants with equal standing in the project. Relatedly, each participant's intended part should fit with the others'. This excludes cases in which agents intend to do something together, but are settled on ways of doing it that are not compatible. This norm is put to the test in communities that include different, competing ideas about explanation. I'll address this issue in Part II, deferring further explication for now. Overall, explanatory model-construction as a collaborative activity entails several constraints on individual researchers:

- Explanatory practice is non-exclusive; others are involved.
- One's own contribution to explanation is incomplete; exercise humility.

- Contributions to an explanatory model are public and accessible to other members of the scientific community.
- Community members have equal standing to recognize and use an explanatory achievement.

These four requirements apply to participating members of S, and to model M. The details of M and its construction process are, of course, context-dependent. So they're filled in case-by-case, via study of particular scientific communities. But insofar as constructing a particularist explanation is a social activity, the general theoretical framework for analyzing social activity applies.

These collaborative norms develop several implications of the modeling framework set out in the previous chapter. Models are *non-exclusive* with respect to their targets. The same target can be modeled in multiple ways, without any conflict between different (even prima facie incompatible) models. If particularist explanations are models, then different explanations of the same target can unproblematically coexist in science; they need not conflict with one another. The social action framework extends this non-exclusivity to modelers' attitudes. It is not up to any single researcher whether or not an explanation is successfully constructed in any particular scientific community. Constructing an explanation is a collaborative activity involving multiple members of a community. A scientist who viewed it as a solo activity would be proceeding inappropriately; failing as a participant in scientific inquiry. In addition, EP_M implies that explanations are inherently incomplete and partial. Model-to-target relations are partial and selective, with model-users' purposes guiding such selectivity. The social action framework builds on this insight. Not only should individual scientists recognize the inevitable partiality and incompleteness of all explanatory models; they should also see their own work as contributing to a larger project. The purpose of an explanatory model is such that no single agent can achieve it.

Aspects of understanding

These results flesh out the view of understanding associated with EP and EP_M. The purpose of an explanatory model is to advance understanding of some target. The social action framework reveals additional aspects of understanding in the process oriented to that end; i.e., aspects of understanding involved in explanatory model-construction. These aspects supplement rather than supplant the understanding advanced by an explanatory model for a particular scientific community. That is, these other aspects provide a more nuanced picture of understanding in the social activity of explanation, building on EP_M's analysis. To see this, it's helpful to recast EP_M's ideas in terms of the collaborative process:

(EP_{MC}) Constructing an explanatory model (M) is a collaborative activity (C) performed by members of scientific community (S), for the purpose (P) of advancing understanding of some target (X).

This thesis integrates the two strands of EP's conceptual framework. Importantly, EP_M's significance and success conditions are assumed. The activity of model-construction (C) is further analyzed in terms of the three-layer interactive structure and associated norms, as set out previously. This intentional structure entails three further aspects, or modes, of understanding for participating members of S:

- understanding of *oneself* as a participant in C;
- understanding of *other members of S* as collaborative partners in C; and
- understanding of *model M* as distinct from target X, with respect to purpose P.

I'll discuss each in turn, then consider how they are integrated as means to achieving the end of explanatory collaborative activity; i.e., advancing understanding of X. An example will be helpful, so I'll return to my opening reminiscence about the Duke Marine Lab research project.

That research project didn't include an explanatory model. That didn't matter in the earlier presentation (Section 3.1), which aimed to intuitively capture the difference between individual and social activities, as experienced by an agent. That same perspective—an individual agent's experience—is here applied to characterize the three contributory modes of understanding. But the example needs to be modified. Suppose an explanatory model did result from that long-ago fledgling research work. In this wildly fictional example, Alex and I discern in our data a pattern that we attempt to capture in a predictive formula—say, something like Volterra's mathematical model which explained patterns in fish populations under different predation conditions (among other targets). Suppose that our potential explanatory model accounts for pugilistic waving in Pivers Island crabs, and many other populations. Eventually, a refined and improved version of our model is recognized by the behavioral ecology community as importantly advancing our understanding of, say, aggressive displays in a wide range of populations. What are the key features of the collaborative process that links our novice-level data collection and analysis, to the community's explanatory model?[40] And how is understanding implicated in that process?

First, our two-person research team needs to expand and connect with the social epistemic organization of behavioral ecology. Alex and I had at the time just

[40] Of course, I would hope to receive credit for the explanatory model, alongside Alex. For purposes of assigning professional scientific credit, the generalized-crab-formula is "our model." For the purpose of advancing understanding, however, the model is not "ours" in any meaningful sense. That is, as an explanation, it is had by and benefits the entire behavioral ecology community.

received our bachelor's degrees in biology. We were well-positioned and qualified to successfully complete a final project for a summer course at Duke Marine Lab, but not to make a major explanatory contribution to behavioral ecology. Indeed, neither of us would have been able to judge the significance of our result for that community. The imagined crab-formula emerged out of data-analysis for our very circumscribed research project, carried out over a few weeks on a single population of jauntily-waving fiddler crabs. The intended audience was the rest of our summer program class, and the professor—there was no expectation of a wider audience or continued research after July 1992. For our model to be even considered as a behavioral ecology explanation, it must go beyond that intended institutional setting. The most straightforward route would be to involve our course instructor, a professor from Princeton, and work with him on crafting our project into a paper and/or presentation for a behavioral ecology journal or conference. Our two-person research team thus expands to include more senior, expert participants in the field: our professor and his own research network.

This (counterfactual) effort toward an explanatory purpose involves a change in my own self-understanding—not of myself as a person, full-stop, but rather my understanding of what I am doing. Completing a final research project with another student is a very different activity from proposing an explanatory model to a wider scientific community. The latter requires that I position myself within that community, engaging with its social epistemic networks. As a novice, of course, I'd have very little idea of those social epistemic patterns. My point of entry is clear enough, however: our professor, his research group, his colleagues at Princeton, extending to a constellation of similar research groups spread across the world. Even as a novice, the shift would be obvious: from sequestered (protected, scientific low-stakes) coursework to a working research community comprised mainly of those with greater expertise. I'd be entering as part of a smaller team, riding on more experienced members' trust and authority—but also bearing my-and-Alex's own contribution, to start making our own place in the community. In so doing, I understand myself as a member of that community (albeit an entry-level one). As a participant in the explanatory project, I understand myself as connected to existing, ongoing community networks in behavioral ecology. It's an important shift in attitude and identification—an aspect of understanding, of oneself in relation to the behavioral ecology community. More generally, interplay between group and individual identification affects a participating agent's "self-understanding." This is another facet of the difference between social and individual activity. Part of one's equipment as a practicing scientist is a map, so to speak, of the social epistemic landscape and one's own position within it. Who are the best researchers, whose work is the most innovative, who's a likely collaborator, whose over-hasty publications can be discounted? Getting a sense of the terrain requires becoming acquainted with patterns of epistemic authority and trust in one's community; becoming involved with the field's social epistemic organization. This is part of what members learn as they're educated and socialized into a specialized scientific community, through mentoring, graduate societies, and attending regular conferences.

The next step is refining our group's model, to articulate its full significance and relations to targets. This is done by our expanded research team preparing the paper and/or presentation to show our proposed model to an audience of working behavioral ecologists. We'd refine our model, look for other populations in which it applies, and so on. Then our small (but augmented) research team would make the case to the wider community that our model should be recognized as explanatory; i.e., that it's based on satisfactory methods (including our jointly-acquired pugilator data) and exhibits features that confer understanding, according to the standards of the field. Neither myself nor Alex would be in a position to lead these efforts, not having much background. But we could follow along and become more expertly specialized through this participation.[41] Our research team would be demonstrating (or trying to) that our model satisfies criteria for an explanation in behavioral ecology. This requires making the target explicit and showing that advancing understanding thereof is important for behavioral ecology. In addition, we need to show others in the community what our model can do—exhibit its key features, and how they yield understanding of the target, in a way that satisfies the community's explanatory standards. That is, we need to understand M as a model, distinct from its target X, and show that M and X are related so as to satisfy criteria for explanation/understanding in behavioral ecology. By our efforts, others in the community come to understand our model too. In this way, understanding of M qua model is a means to the end of constructing an explanation, thereby advancing understanding of target X for the community.

Alongside understanding of oneself and a model, a participant's understanding of others in the community is transformed by the explanatory process. Our paper/presentation is an invitation for others in the field to judge that our model is suitably explanatory, not a decree to that effect. It isn't up to me, or Alex, or our professor, or the Princeton behavioral ecologists, or all of us together, to determine that our M is an explanation—though of course we'd believe that it is. Nor is that up to any single other member or team in the behavioral ecology community. Rather, the explanatory status of M is settled by the community as a whole. So, alongside my newly-acquired self-understanding as a participant in the community, is awareness—even if vague—of others in the community, of equal standing with respect to our model's explanatory status. This follows from the constraint that a successful explanatory model is not a solo production; it must be evaluated and recognized as such in the community more broadly. Other community members are of equal standing, in determining this. No modeler is working in isolation; their resources and values are shaped by their community context. So some

[41] An extremely important part of scientific mentoring (not usually from coursework, though).

understanding of others is implicated, in the process of advancing understanding of some target X. Crucially, those others are understood as having equal standing with oneself with respect to explanation; participants' relations to the model and its purpose are symmetric. Members of a scientific community understand one another as engaged in the same explanatory activity. They understand one another as partners in the overall explanatory project.[42] (Boundary-work denigrating the explanatory capacities of other communities is the obverse of this community-building feature.)

These three aspects of understanding are means to the end—purpose P—of advancing understanding of some target X for members of community S. All three are implicated in the collaborative activity of constructing an explanatory model M. The purpose, in turn, guides and shapes the collaborative activity of model-construction. These results deepen EP_M's conceptual framework, offering more nuanced characterization of scientific understanding. From these different aspects of understanding, a thin general notion of that concept emerges. Very roughly, understanding involves fitting that which is understood into a larger whole. A participating agent locates herself and others within a scientific community; proponents of a model figure out how its features bear on community values associated with explanation. The larger whole is the scientific community; the basic analytic unit for particularist explanation.[43] That understanding involves fitting things together into a larger whole, overall pattern, or system has precedent in the philosophical literature—I'll return to this matter in Chapter 6. For now, my modest point is that the various aspects of understanding implicated in the general social action framework for scientific explanation suggest a shared, general concept of understanding that is relevant for science. Of course, the specialized notion of understanding advanced by an explanatory model in a particular scientific community is not exhausted by that general notion. Just as with explanation, particularism admits only a thin general notion of understanding; substantive features emerge from careful study of particular scientific communities. That thin general notion, however, has important implications for explanatory particularism (see next section).

I'll here summarize the particularist account of understanding so far. Advancing understanding of some target X is the purpose P of an explanatory model M. Implicated in the process of constructing an explanatory model that succeeds in its purpose are three other aspects or modes of understanding: of oneself, of

[42] Satne's (2021) account of direct participatory understanding is an important source for these ideas; e.g., "individuals understand not only their own individual [intentions] but those of others involved in the same activity, this is part and parcel of understanding and being directed to their own activity" (S520); how social interaction "provides or constitutes ... some—albeit basic—forms of mental understanding" (S525). Satne is concerned with a very different class of cases, however (early human cognitive development and understanding of other minds).

[43] Woody's functional account of explanation (2015) anticipates this point.

others in the community, and of the constructed model. These different aspects of understanding, like the corresponding elements of the social activity, don't collapse into one another but are integrated into an explanatory process. Becoming a specialized expert in a particular scientific field involves triangulating among all three—one's self-understanding as a community member, understanding of others as collaborative partners in explanatory practices, understanding of model M, and understanding of target X thereby. All these elements are distinct but interrelated in explanatory practice in a particular scientific community. Philosophers studying explanation in such a community should attend to them as well. The image of interlace is again relevant: intricate relations between distinct elements. Ties between self, others, models, and target emerge as part of explanatory practice; the collaborative activity of constructing an explanatory model in EP_M's sense.

3.4 The end of understanding

The previous section articulated generic norms for agents participating in collaborative activity, and for the special case of explanatory model-construction. Left unstated (so far) is the ground of normativity. What, if anything, gives these norms their force? Philosophy of social action offers several mutually compatible answers. In addition, ideas about scientific objectivity suggest normative grounds specific to the case at issue. This collection of normative grounds is another source of "added philosophical value" for the particularist framework as developed in this work.

One ground for social action norms is instrumental. Collective intentions, according to many social action theorists, are necessary for practical planning and effective social activity.[44] So they are instrumentally valuable, enabling accomplishment of various practical ends. In order to play this instrumental role, collective intentions must be stable; i.e., hold together over time and across multiple participating agents.[45] In this way, the instrumental value of collective intention grounds a stability requirement for same. The intercalated, looping structure of collective intention proposed in the previous section satisfies that requirement. The norms implied by that structure are thus grounded as doubly instrumental: requirements for a collective intention to be a stable scaffold for planning and practical efficacy. More simply, these norms constrain participants' attitudes so as to stabilize a collective intention. Epistemic openness, alongside equality and parity among agents, afford a stable basis for those agents' planning around C. The dual aspect of the first-layer relation between agents and a shared goal also stabilizes a collective intention in at least two ways. First, recognition of both self

[44] See, e.g., chapters 1 and 2 of Routledge's *Handbook of Collective Intentionality*; Bratman (1999, 2014)

[45] See, e.g., Gilbert (1989), Tollefsen and Gallagher (2017), Castro and Pacherie (2021).

and other as identifying with the group counters an ego-centric perspective and further stabilizes the concept of "we" as the agent intending to C. Second, interplay between group and individual identification leads to uptake by individuals of values and interests associated with the group.[46] Such uptake offers a basis for ongoing collaborative activity, as shared values can become normative standards for individuals acting in a group. More broadly, shared background context, reasons, and values among individual agents contribute to the stability and effectiveness of collective intention. The more background beliefs and values bearing on an activity that are shared, or held in common across participants, the more stable the shared intention, and so the more useful for planning/practical reasoning. Overall, then, the interactive structure and associated norms contribute to stability of a collective intention, which is instrumentally valuable for accomplishing practical ends. In this way, the social action side of the particularist framework is grounded in instrumental normativity.

A related instrumental requirement is means-end coherence. The latter is necessary for an action to be practically effective.[47] Means wholly mismatched to an intended end are vanishingly unlikely to lead to practical success. It follows that means and ends involved in an intention (collective or singular) must be coherent with one another. Requirements for coherence, as well as the exact meaning of the latter term, are notoriously vague (see, e.g., BonJour 1985). Consistency, or absence of contradiction, is thought to be necessary but not sufficient. Practical consistency is plausibly involved as well; one shouldn't adopt means that undercut one another, nor construe ends in such a way that one's chosen means cannot accomplish them. If one commits to an end, one should also commit to appropriate means. In social action, means-end coherence has several facets. For one, participants' means are constrained to be coherent with one another; i.e., each participant's intended part should fit with the others' in the performance of C ("meshing sub-plans," in Bratman's terms).[48] This excludes cases in which agents intend to do something together, but are settled on ways of doing it that are not compatible, and cannot be jointly pursued so as to accomplish our shared goal.[49] The latter requires that participants' individual intentions be coordinated, to make up a coherent overall plan of action. At the very least, our respective means should

[46] The reverse may also occur; the values and interests of an individual diffusing throughout a group. Many conceptual and methodological innovations in science originate in this way.

[47] Means–end coherence is also cited in theories of practical rationality (e.g., Bratman 2014, Castro and Pacherie 2021, Roy and Schwenkenbecher 2021). Although nothing in this chapter excludes such theories, the concept of rationality is ill-suited to connect philosophy of science and of social action. So I will not pursue that strand of the latter.

[48] "[E]ach group member considers it possible that, first the joint activity is successful, and that it is so by way of the intentions and meshing sub-plans of the group members" (Roy and Schwenkenbecher 2021).

[49] For example, if each of us intends that we climb Half Dome, but I plan to hike up using the chains and you plan to climb the technical Regular Route with rock-climbing equipment, then our collective intention is incoherent.

not conflict with or undercut one another. Positive coordination and mutual support among participant means enhance coherence in a social activity. In the case of explanatory model-construction, achieving coherence or even absence of conflict among means is a substantive requirement. An explanation is a significant achievement for a community in part because it effectively harmonizes the diverse values and constraints associated with understanding/explanation in that community. A particularist explanation exemplifies such coherence, reflecting and exhibiting a community's values associated with explanation and understanding.

Means-end coherence has a further aspect, because coherence cuts both ways. Of course, a chosen end imposes normative requirements on means taken to it.[50] But we can also ask: what must the shared goal of our scientific practices be like, given the means we take to it? Means-end coherence has normative force on either side of the instrumental relation. Applied to particularist explanation, this constraint yields further insights about scientific understanding. To review: EP_M states that an explanatory model M's purpose (P) is to advance understanding of some target X for members of a particular scientific community S. The social action framework yields a deeper analysis of the process of model-construction, conceptualized as a collaborative activity. That conception, as shown above, fits with and builds on EP's basic characterization of scientific explanation. Explanatory practice in a scientific community is collaborative, with participating members' attitudes constrained by the interactive structure of collaborative activity. These collaborative interactions are mediated by a community's social epistemic organization; e.g., relations of epistemic authority and trust among members. Engaging in this activity transforms an agent's understanding of herself, other participants, and the explanatory model. These interlaced modes of understanding contribute to the activity's purpose: an advance in understanding of some target X. In all these ways, then, means by which a scientific community constructs an explanatory model are collaborative. On pain of incoherence, the end (or purpose) must be such as to possibly result from those means. That doesn't guarantee success, only that the end be possible to achieve by means taken to it. Just as means to an end can't directly thwart its achievement, symmetrically, an agent's end must be such as to be achievable by the means they take.

For particularist explanations, as noted, those means are essentially collaborative. Explanatory practices in science are non-exclusive, incomplete, public, and involve multiple community members acting as full partners in constructing models that advance understanding of some target X. In social action, one conceives the means one takes as one's part in a larger action involving coordination with others, rather than as an isolated or singular action. These features of explanatory

[50] This instrumental normative aspect is also built into EP_M and philosophical accounts of modeling more generally, in that a model's purpose determines many features of the model and its evaluation criteria.

practice are reflected not only in models thereby constructed (products of collaborative activity), but also the epistemic purpose for which they are constructed. That is, features of the means are reflected in the end, due to coherence constraints. Understanding conveyed by explanatory models has characteristics traditionally associated with objectivity: non-exclusive, incomplete, public, and available to all (or at least most) community members. Of course, these means-end constraints don't determine other, more substantive features of understanding and the model-target relation for a particular scientific community S. Howsoever a scientific community conceptualizes and values understanding—whatever the term means, in that particular context—means-end coherence implies that understanding of target X is such as to be *shared* among community members, not limited to any individual modeler or member of S. In this way, requirements of means-end coherence afford a thin, non-exhaustive characterization of understanding advanced by particularist explanation.

This result suggests another ground for particularist norms: pro-social values associated with human flourishing. This idea needs some introduction. To start, consider the thesis that shared understanding is a constitutive aim of science. What I mean by a constitutive aim is captured by Henry Richardson's concept of a final end. Very briefly, Richardson defines an end as "something for the sake of which an action is to be done" and a *final* end as one "for which the action would still be done even if no other advantage flowed from it" (1997, 6). That is, a final end is valuable in itself. It isn't, however, the terminus of practical reasoning nor the unique source of justification for such reasoning. Deliberation about final ends is possible, both because reflection can reveal new final ends and because efforts toward practical coherence can bring out new relations among an agent's diverse ends. The latter notion is an ideal, on Richardson's view, constituted by relations of mutual support among an agent's diverse ends, values, and other commitments.[51] Mutual support among elements of an agent's practical reasoning is accomplished through specification of how an end is to be achieved and what its value consists of. Mutual support relations are bidirectional, revisable, holistically assessed, and stable upon reflection (176-177). Although Richardson doesn't apply these notions of mutual support and practical coherence to epistemic communities or explanatory models, there are parallels with my particularist account. The key point I want to emphasize here, alongside these parallels, is Richardson's main thesis: human agents can reasonably deliberate about final ends. That is, normativity in the practical sphere isn't restricted to instrumental reasoning about 'given' ends. Agents often reflect on their ends, discovering and refining their understanding of new final ends in the course of life. Indeed, part of a good life is achieving practical coherence among

[51] "Finding or constructing intelligible positive connections or links or mutual support among [one's norms, means, ends] and of removing relations of opposition or conflict" (Richardson 1997, 144).

one's various ends—so Richardson argues, anyway. Philosophy of science too can contribute to such reflection.

My account of explanatory particularism, as developed here, is a contribution along these lines. This addresses concerns about philosophical "added-value" (Chapter 1). One objection to particularism about explanation—or about any methodologically significant aspect of science—is that it collapses into empirical history and sociology of science. That is, to endorse EP is to give up on philosophy of science as a distinct line of inquiry; a specialized epistemic community in its own right. So the objection runs. The social action framework and its normative implications rebut this objection. I am proposing, based on that framework, that *shared understanding* (of targets of interest to members of scientific communities) is a final end—a constitutive aim—of scientific activity. If explanatory practices are collaborative activities, then they are pro-social—that is, social in a positive way, not re-inscribing social relations of political power and dominance. According to EP, an explanation reflects and exhibits human values in the context of a particular scientific community. So one implication of explanatory particularism, as developed here, is that *explanatory power is social power.* This is not an endorsement social "brute force" in science, however. The means-end coherence requirement that understanding of target X be shared among community members moves my account in a pro-social direction.[52] Explanatory virtues as means to shared understanding are social virtues: norms of tolerance, inclusion, and respect. It's a familiar point, these days, that science's social practices and organization are epistemically significant. But the pro-sociality of explanatory practices in science is less recognized. The value of pro-social behavior is a final source of normativity for the particularist account as developed here. Pro-social behavior and attitudes are intrinsic to human flourishing. My account identifies a subset of pro-social values as explanatory. Explanatory virtues form "part of our idea of human cognitive flourishing"—and social flourishing (Putnam 1981, 134).

Relatedly, explanatory particularism satisfies some philosophical desiderata for an account of explanation, albeit in a non-traditional way. Rice and Rohwer (2021), for example, "would like an account that is able to: (1) distinguish explanations from non-explanations in a principled way, (2) help us understand why explanations are valuable, and (3) provide normative guidance to scientists who aim to provide explanations" (1026).[53] Explanatory particularism as developed in these chapters manages all three. The principled distinction between explanation and non-explanation is made in the first instance by scientists themselves. Philosophers can then examine scientists' explanatory models in terms of the community's values and other contextual features. This opens the way to

[52] This requirement is the starting-point for a new, pro-social conception of scientific objectivity (see Chapter 6).

[53] See also Woody (2015) "Five challenges."

criticism.[54] The value of explanations, on this account, stems in part from pro-social, collaborative values central to human social life and instrumental for practical success. More specialized values in particular scientific communities are the substantive focus of any particularist study. The pro-sociality of science given EP is worth emphasizing: the scientific significance and social value of explanation go hand-in-hand. In contrast, many stock philosophical examples of explanation obscure or efface both. Third, normative guidance to scientists is provided by the social action framework: the four norms summarized above, and the final end of shared understanding.

So far, shared understanding (of some target X) as a final end is limited to members of a particular scientific community. But explanatory particularism allows for the extension of shared understanding beyond those boundaries. Part II takes up this challenge; what follows is a brief preview. EP in the first instance conceptualizes explanations as thoroughly enmeshed in their contexts, particular scientific communities. Yet it's not uncommon for explanatory models from different scientific communities to make contact with one another. Such interdisciplinary contexts afford opportunities to connect explanations originating in different scientific communities. In such a context, members of different communities can be partners in constructing an explanatory model. Analogs of the norms for collaborative activity within a community apply to interdisciplinary contexts (see Chapter 5). The view of science as a dynamic assemblage of distinct, variously connected epistemic communities suggests a further argument for extending the notion of understanding beyond community borders. It's a truism that scientists don't know exactly where their investigations will lead. (If they did, the whole enterprise of scientific inquiry would be otiose.) One aspect of this unpredictability is the pattern of connections between different scientific communities. This pattern is highly contingent. Who could predict what interdisciplinary interfaces will arise, as science proceeds and the world presents us with problems and opportunities? We can't predict which interdisciplinary connections will form, just as we can't predict the results of a particular research program in advance. Indeed, particular connections that form between different specialized communities, depend in part on the results of those communities' inquiries. Therefore, the pattern of interdisciplinary connections cannot be specified in advance. We can confidently predict, however, that new connections will form, in some pattern or other. As scientific specialization accelerates, interdisciplinary contacts also increase.

Due to this unpredictability, no member of a particular scientific community can specify in advance the other communities theirs might in future connect with. Those in-principle unpredictable connections might involve explanation and understanding, alongside other core elements of scientific practice and method.

[54] See Parts II and III.

At least, the possibility of connecting explanatory models with those of other epistemic communities cannot be ruled out. Here considerations of means-end coherence come into play. Means-end coherence, as discussed above, demands that participants in a social activity believe or accept that the shared goal of that activity is possible to achieve by means taken to it. More simply, the shared goal must be achievable by means taken by participants. Explanatory practices in a scientific community must, on pain of means-end incoherence, be directed towards an end achievable by those means. Those means are collaborative within a community and, in the longer term, include connections with other communities that cannot be predicted in advance. Understanding sought must be such as to possibly result from these means: shared understanding of some target X. So thoroughly community-specific understanding is not an achievable end, from a within-community perspective. Scientific explanations that count as such only for members of a specialized community are means-end incoherent, if one conceptualizes scientific activity along particularist lines. Although explanatory models are deeply and significantly specialized, it's an important aim of science to transcend such specialization. These means-end constraints yield a thin specification of an ideal of scientific understanding; a conception of a final end for the scientific community as a whole. The final end of explanatory model construction is understanding such as to possibly satisfy standards not specifiable in advance. Explanatory practices, conceived as social action satisfying normative constraints, ultimately aim at understanding that is so in virtue of satisfying standards not limited to any specifiable scientific community.

How can this ideal be reconciled with tenets of explanatory particularism? The key thesis is that explanatory models can *connect with one another*. Scientific understanding doesn't aim at isolated islands of knowledge, but entire continents. A mark of understanding is not only acceptance within a particular epistemic community, but also the potential to connect with explanatory models in other epistemic communities. Part II explores this idea, proposing normative standards for an interconnected 'fabric' of scientific understanding. Shared understanding of this sort is itself a scientific achievement, requiring that countervailing impulses and tendencies be overcome. Collaborative explanation, and the associated concept of shared understanding, is an ideal—but one we can work toward.

3.5 Conclusion and summary

I'll conclude by summarizing key results and theses of Part I. The basic idea of explanatory particularism is that a scientific explanation is a product of a particular scientific community that is significant, specialized, and advances understanding for that community (EP, Chapter 1). Explanations so construed are specialized epistemic achievements of particular scientific communities. Science as a whole is a dynamic network of diverse epistemic communities. These communities are socio-historical entities, changing over time but sociologically bounded at any given time. Scientific inquiry is a dynamic landscape of interconnected and ramifying lines of inquiry, clustered into specialized scientific communities. This social epistemic organization is fundamental for studying explanation. Particularist explanations are not given or static, but continually revised as the community context changes. Furthermore, scientific communities are dynamic and variously interrelated—some recently split, others connected or merging, but as a whole disarticulated and distinct. To develop EP's basic ideas, I've used two main philosophical resources. The first resource is from philosophy of science: conceiving epistemic products as models. The second is from social philosophy: analyses of collective action and intention. These two resources correspond to epistemic and social aspects of particularist explanations, respectively. The epistemic aspect is explicated in terms of models (Chapter 2). The social side is explicated in terms of collaborative activity (this chapter). The social action and modeling parts of EP's framework dovetail. In EP_M's terms, an explanatory model M has the purpose (P) of advancing understanding of some target (X) for members of a particular scientific community (S). A scientific community S has multiple participating members—working scientists comprising that community. Their shared goal is to construct and use an explanatory model M. M's purpose (P) is to advance understanding of some target of interest, X. To accomplish P, M and the M–X relation must exhibit features conforming to (expressing) S's specialized epistemic values, alongside other constraints bearing on explanation in that community. So constructing M is a social activity guided and constrained by S's ideas and values bearing on P. In this way, EP_M's model framework and ideas from social action theory are brought together in an analysis of background requirements for particularist explanation. Other versions of explanatory particularism are possible. Different philosophical resources, or different interpretations of these same resources, would yield different versions of explanatory particularism.

Looking ahead, relations between explanatory models can be classified according to the framework introduced in Chapter 2, consisting of two continuous axes (stage of process, degree of connection) and two cross-cutting distinctions (direct/indirect, similar/different). Model-model relations are established at some point in the model-construction process, and these relations involve varying degrees of connection ranging from very thin (formal) to extensive and intricate; mutual dependence, or reciprocity, between all elements of each model. Relations between elements of models may be direct or indirect, involving similarity or difference. Direct relations are just between the two models (elements thereof), while indirect relations include some further "bridging" element. Models can also relate to one another in virtue of similarities (e.g., common elements), or differences. The latter are various: direct conflict, N-complementarity, LK-complementarity, part-whole hierarchy, cause-effect relations, and more. The two axes provide a general framework for characterizing relations between distinct models in terms of the process of model-construction and the extent to which the models are related. The two cross-cutting distinctions provide a general comparative framework for analyzing relations between distinct models. Part II applies this framework to the problem of explanations that transcend the boundaries of specialized scientific communities: *interdisciplinary explanations.*

However, models do not relate to one another in and of themselves. They are constructed, used, and (sometimes) related to one another by epistemic agents working in scientific communities. To these activities, norms of collective intention apply. The social action framework, applied to the case of scientific explanation, yields a set of norms for constructing such an explanation. These are, briefly: non-exclusivity, incompleteness (humility), publicity, and respect for others as full and equal partners in explanatory activity. These norms constrain the attitudes of participants constructing an explanatory model. In this way, social action theory offers a normative framework for analyzing explanation. The centerpiece of this framework is a three-layered interactive structure of collective intention. That structure in turn implies norms for collaborative activity. These constraints are grounded in the requirement of stability for collective intention—and, in the case at issue, on shared understanding as a final end of science. These collaborative, pro-social ideas ground the norms of my version of explanatory particularism. Part II aims to show the potential of a "pro-social," collaborative account of explanation that engages deeply with explanatory practices across the sciences. The basic idea is that social action norms also apply in contexts where there is a shared goal of connecting two (or more) distinct explanatory models from different communities. Those contexts are a class of interdisciplinary research. Just as within a scientific community, agents A and B from different scientific communities should see their respective models as contributing to, but not wholly accomplishing, the overall goal. Each should recognize, in accordance with the norm of humility, that his own specialization's explanation is not absolute, complete, or final on its own. There is more explanatory work to be done than one field's model can provide. This is so even if members of one community add some feature to their model enabling connection with others. 'Outreach' of this sort must be met on the other side as well. A suitably humble attitude is encouraged (perhaps even required) by recognition of one's disciplinary explanation as a *model.* Models are not exhaustive of the targets to which they are directed, and do not exclude other models of the same phenomenon. Recognition of one's own explanation as a model affords openness to connection with different explanatory models.

A final disclaimer, to end Part I. The basic ideas of explanatory particularism are not wholly my invention. Rather, the methodological and substantive commitments encapsulated by EP are already at work in many areas of philosophy of science, including (through recent pluralist efforts) studies of scientific explanation. However, this and preceding chapters are the first sustained defense of a particularist approach to studying explanation, presented as an explicit alternative to traditional and prevailing approaches. Chapter 1 synthesized and articulated ideas already in circulation. Chapter 2 developed those ideas, building on insights about models and modeling—another important topic in philosophy of science. This chapter looked outside philosophy of science, using theoretical resources from philosophy of social action to articulate a framework for analyzing particularist explanations as products of collaborative activity within a scientific community. My conceptual framework for EP is at once social and epistemic, the latter side analyzed in terms of models and modeling, the former in terms of the structure of collective intention and associated norms.[55] The particularist account so developed offers deeper and more nuanced characterizations of understanding, explanatory model-construction, and norms for explanation as the product of collaborative activity. The idea that shared understanding of significant targets is a constitutive aim of science, or final end, suggests a way to extend the normative account proposed here, engaging explanatory practice in interdisciplinary contexts. This completes my account of explanatory particularism, insofar as the account can be presented in general terms.

[55] This dual aspect of scientific practice has been noted by other philosophers of science, but has not been studied with respect to explanation specifically (e.g., Andersen 2016).

PART II
EXTENDING THE PARTICULARIST FRAMEWORK

4
Interdisciplinarity and Explanation

In 2016, the US National Science Foundation announced new funding dedicated to "Convergence Research" (NSF 19-551). "Convergence" here refers to the mingling of "knowledge, theories, methods, data, research communities and languages" via ongoing interaction between researchers from different fields. This funding aims to encourage collaborative research "at the nexus of traditional disciplines." Anticipated benefits of such "deep integration" include new "research vistas," "convergence culture," and "long term scientific vision," as well as practical solutions addressing societal needs. Although key terms are left vague and open-ended, the underlying motivation is plain. Research spanning multiple scientific disciplines is thought to be practically and epistemically valuable. NSF 19-551 is just one example. Funding calls for interdisciplinary institutes, educational programs, and research projects have been widespread for some decades. It is widely thought (notably by policy-makers, but also many scientists) that contributions from multiple disciplines are necessary to address complex problems, spur theoretical innovation, and make scientific progress.

Despite the popularity of its slogans among administrators and funding bodies, the meaning and value of interdisciplinarity in scientific practice is somewhat murky. The particularist framework from Part I, together with insights from empirical and philosophical studies of interdisciplinarity, lights a path through the murk—not the only possible way, but one fruitful avenue. From the welter of issues associated with interdisciplinarity, explanatory particularism charts a throughline in terms of scientific modeling, collaboration, and social aspects of understanding. This contribution to interdisciplinary studies also answers the challenge raised at the end of Part I. To reprise: explanatory particularism assumes that science is epistemically fragmented, siloed into many specialized subgroups. Different specializations impose different norms and standards on explanatory models' construction and use. So explanations are localized to particular specialized communities. But if explanations are so thoroughly context-dependent, tied to their fields of origin, then how can explanations and understanding transcend those communities?[1] Relatedly, how could an explanation be constructed from two (or more) distinct fields? How is *interdisciplinary explanation* possible?

[1] A traditional, ontically-inflected response is that scientific explanations are independent of researchers' values and interests. But that response is unavailable to an explanatory particularist.

Explanatory Particularism in Scientific Practice. Melinda Bonnie Fagan, Oxford University Press.
 DOI: 10.1093/9780191998874.003.0005

The latter question concerns one kind of interdisciplinary research: pooling explanatory resources from different specialized fields toward the shared goal of an interdisciplinary explanation (IDE). I show that the framework from Part I, applied to this strand of interdisciplinarity, yields an account of IDE that accords with explanatory particularism. This shows how IDE is possible, in a particularist framework. The more general question is also answered, albeit more briefly. How can scientific explanations and understanding transcend their communities of origin? Often, *they don't*. Those outside a particular scientific community do not gain understanding from that community's explanatory models—at least, not in the same way as community members do. Transcending community boundaries doesn't come for free and oughtn't be assumed as a default, in the matter of explanation. "Extension" is a more appropriate term; less lofty, more oriented to the work it takes to move an explanatory model beyond the boundaries of its original context. That work is interdisciplinary research. This chapter and the next show how and under what conditions explanatory models can be extended via interdisciplinarity.

This chapter proceeds as follows. Section 4.1 surveys the background for this application of EP/EP_M: empirical and philosophical studies of interdisciplinarity. Section 4.2 shows that particularist ideas from Part I comport well with key results from that background literature, as well as offering several new insights. Section 4.3 goes deeper in applying Part I's framework, focusing on EP_M and the taxonomy of model–model relations from Chapter 2.[2] The latter classifies interdisciplinary research projects relevant to explanation/understanding, setting up both the challenge of IDE and my particularist response. Section 4.4 concludes. To preview the main results: explanatory particularism reveals an important role for explanation/understanding in interdisciplinary research, provides an analytic framework for assessing that role, and turns the IDE challenge into an advantage for particularism as an approach. On the last: features of scientific practice that seem incompatible with explanatory particularism in fact support my view. But to bring out these challenges and the particularist response, some background is needed.

4.1 Interdisciplinarity in scientific practice

What is interdisciplinarity, exactly? The question is surprisingly hard to answer. Conceptual confusion around interdisciplinarity has several sources.[3] For one, the term and its cognates apply variously to people, programs, projects, departments,

[2] The social action framework comes into play in the next chapter.

[3] There is much more to say about interdisciplinarity than the narrow swathe that explanatory particularism touches upon. A full treatment is beyond the scope of this book. What follows is a very incomplete sketch, within which I position my particularist account. For a more inclusive overview, see Frodeman et al. (2017).

and more. Because such a wide range of things are called "interdisciplinary," that term's meaning becomes extensionally stretched. Second, and relatedly, the root term "discipline" is hard to pin down precisely. "Interdisciplinary" both inherits this vagueness and compounds it—the latter because disciplines (whatever they are) can be imagined combining in many ways. So the core idea is vague and ambiguous from the outset. A third source of confusion arises from scholarly usage: "interdisciplinary" has both broad and narrow senses. In its broad sense, the term is a generic umbrella covering a very wide range of activities, situations, entities, and processes. Its narrow sense, however, marks contrasts with trans-disciplinarity, multi-disciplinarity, meta-disciplinarity, and the like. One needs to know which sense is intended, in any scholarly discussion. Last but not least, different scholarly fields highlight different meanings of "interdisciplinarity"—exemplifying the specialization that looms so large for particularists. For all four reasons, "interdisciplinarity" and its cognates resist clear and simple definition. Researchers in sciences, humanities, and administration use them in many different ways, often without explicit definition. Discourse around interdisciplinarity is confusing, polysemous, and (it seems) sorely in need of clarification.

Attempts at the latter have been made, of course, by social scientists and a few philosophers. The most inclusive definition, to my knowledge, is Uskali Mäki's: interdisciplinarity refers to "whatever relevant relationship between two or more disciplines or their parts" (2016, 331). This covers a huge range of scientific and philosophical activities, including even abstract comparison of different disciplines for any reason. (And what counts as a discipline?) Another broad definition is from Katri Ilona Huutoniemi and colleagues (2010): "interdisciplinary" refers to "a genus of integrative research activities that combine more than one discipline, field, or body of knowledge" (309).[4] Although this construal restricts interdisciplinarity to "integrative research activities"—immediately raising the question of what "integration" amounts to—the sociological unit of analysis is left open. The United Nations also uses a broad definition covering a wide range of social phenomena, from "the simple communication of ideas to the mutual integration of organizing concepts, methodology, procedures, epistemology, terminology, data, and the organization of research and education in a fairly large field."[5] Other definitions build on one or another specific interpretation of "discipline": epistemic cultures, social entities, self-perpetuating institutions, economic units, communities of discourse, and more.[6] Canvassing the array of proposals, J. Britt Holbrook (2013) distinguishes three basic meanings

[4] Porter and Rafols (2009) and Weingart (2010, 11) offer similar definitions.

[5] The definition is from the Office of Economic Cooperation and Development, part of the United Nations Education, Social and Cultural Organization. Quoted in O'Donnell and Derry (2005, 54).

[6] See, e.g., Klein (1990, 2005), Boix Mansilla (2006), Jacobs and Frickel (2009), Frodeman et al. (2017). Related studies of scientific collaboration include Thagard (1997), Wray (2002), and Wylie (2015).

of "interdisciplinary": integration, dual-expertise, and synthesis. The first is exemplified by a kind of scientific practice; i.e., meeting with others in a conference room and together producing an epistemic product. Dual-expertise is exhibited by a researcher becoming proficient in all relevant fields. Synthesis is associated with innovation, understanding the problem in a completely novel way. The three are not exhaustive, and other scholars carve up the conceptual terrain differently.[7] All definitions on offer are, of course, influenced by the particular scholarly specialization in which they arise. Overall, there is no standard definition of "interdisciplinary" or its cognates. The most one can do is be transparent about the concept one uses and its associated scholarly context. This situation is congenial to a particularist approach.

Empirical studies

Conceptual vagueness aside, some features of interdisciplinarity in science are empirically well-established. Science has become increasingly interdisciplinary since the 1970s, by several metrics. The trend is seen in scientific research, education, administration, and funding. Bibliometric analyses show that co-authorship of scientific publications spanning multiple fields increased more than 50 percent between 1975 and 2005, while cited sources in article references showed similar increase in diversity (Porter and Rafols 2009).[8] A more extensive bibliometric analysis of more than thirty-five million journal articles in the Web of Science found a similar result between 1950 and 2010 (van Noorden 2015).[9] Van Noorden also reported increased usage of the term "interdisciplinary" and its cognates in the same interval. Other measures corroborate the overall bibliometric trend. The number of PhD dissertations in interdisciplinary subject areas steadily increased in the US from 2003 to 2012 (American Academy of Arts and Sciences 2016). The number of interdisciplinary programs in US four-year colleges grew steadily between 1975 and 2000 (Brint et al. 2009). Although these measures do not capture the features of disciplines and interdisciplinary contexts highlighted in proposed definitions, the empirical results paint a compelling picture of interdisciplinarity on the rise.

[7] For example, O'Rourke et al. (2016) distinguish three major themes or foci in the philosophical literature on interdisciplinarity: research questions; activities and practices; and relations of subjugation and power.

[8] This study measured "the number of disciplines cited (variety), the distribution of citations among disciplines (balance), and, crucially, how similar or dissimilar these categories are (disparity)" (721).

[9] Van Noorden subdivided this large dataset into fourteen "major conventional disciplines" (e.g., chemistry, psychology) and 143 specialities (e.g., genetics, virology). Interestingly, while citations outside a paper's discipline rose, citations to nearby specialities declined slightly. Van Noorden further breaks down the incidence of extra-field citations by discipline/specialty and national context and reports a lagging but significant impact of co-authored interdisciplinary publications.

Some of this increase can be attributed to institutional encouragement. Funding for interdisciplinary research is emphasized at national and international levels. For example, the German National Science Program since the 1990s has explicitly encouraged "fruitful new mutual relationships between hitherto separately working disciplines . . . [or] cooperation beyond the boundaries of disciplines, institutes, departments and faculties" (DFG 1996, 202). The US supports a number of initiatives aimed at developing technologies in "nano-, bio-, and information" areas, as well as cognitive science. These areas were conceptualized as "convergences" of multiple science and engineering fields, even before the 2016 funding call.[10] Another recent call from the US NSF offers funding to establish "Research Coordination Networks" and/or educational programs in four "deeply interdisciplinary" areas: Harnessing the Data Revolution, Navigating the New Arctic, the Next Quantum Revolution, and the Human–Technology Frontier.[11] The EU's Horizon 2020 program is not discipline-based but "challenge-based," meaning that its considerable funds are targeted to broad issues such as food security, health and well-being, clean energy, and environment—challenges that require contributions from multiple specialized disciplines, if they are to be addressed.[12] These and other funding initiatives encourage interdisciplinary research, although, as Mäki (2016) notes, it is important not to confuse top-down policies with actual scientific practices.[13]

As the above examples show, interdisciplinary research (IDR) is seen as desirable, even necessary, from the perspective of many funding and science policy agencies. For worldwide practical challenges such as climate change, sustainable energy, and global health, the need is obvious (Ledford 2015, Koskinen and Mäki 2016). These are complex problems that cannot be solved using epistemic resources from just one scientific specialization. But knowledge transcending the narrow confines of disciplinary expertise is valued in many "small science" cases as well. This seems an inevitable consequence of increased scientific specialization: the more distinct specialized scientific communities exist, the greater the need to combine their epistemic products to solve any problem that arises from outside a community's own specialized approach. Unsurprisingly, then, "interdisciplinarity" has become a buzzword; a theme accorded epistemic and institutional priority, especially for research on complex real-world problems.

Implicit in the funding and policy emphasis on interdisciplinarity is the idea that IDR *needs support*; i.e., it does not happen without institutional intervention,

[10] E.g., http://www.nano.gov/about-nni/what/funding

[11] NSF 17-065 "Dear colleague letter," https://www.nsf.gov/pubs/2017/nsf17065/nsf17065.jsp

[12] See <http://ec.europa.eu/programmes/horizon2020/en/h2020-section/societal-challenges>

[13] See also the US National Science Foundation (2008), Impact of transformative interdisciplinary research and graduate education on academic institutions. Workshop report; National Academy of Sciences (2006) Facilitating interdisciplinary research. Report; European Union Research Advisory Board (2004) Interdisciplinarity in research; University of Bielefeld's Centre for Interdisciplinary Research (ZiF), MIT 'Convergence' (2016).

at least not at the scale or in the specific ways required to solve Grand Challenges. It's therefore unsurprising that many empirical studies of IDR are directed toward strategies for making this kind of research (however loosely defined) more efficient and successful. Sociological studies of IDR focus on institutional and social factors that structure and constrain individual researchers' activities. For example, Wesley Shrum and colleagues (2007) conducted a "meso-scale" study of fifty-three multi-organizational collaborations in the physical sciences, identifying major characteristics of collaborations in their sample, and relationships among these characteristics.[14] Their data-driven organizational sociology approach yields a complex social framework for analyzing interdisciplinary research at a coarser grain than most philosophical approaches (including mine). Another sociologist of science, Elihu Gerson (2013, 2021), uses the concept of "metawork" to characterize the organization of interdisciplinary research in particular contexts.[15] Drawing on sociological case studies, Gerson emphasizes the obstacles to fully collaborative research across specializations, highlighting the role of contingency and the need to adjust many aspects of scientific practice. These and other sociological accounts of interdisciplinarity use a wider lens than most philosophers, emphasizing forms of organization that connect academia, industry, governments, and the public. Philosophical studies of interdisciplinarity tend to be more narrowly focused, in this way complementing bigger-picture sociological accounts.

Another major takeaway from empirical studies is that IDR is challenging. Successful projects must overcome numerous obstacles, which many fail to clear. This has been obvious since early interdisciplinary studies in the 1970s, when Hugh Petrie dolefully observed that "the importance of interdisciplinary work has seldom been matched by its fruitfulness" (1976, 9). A few decades later, Yvonne Rogers and colleagues reflected that "[t]he added value of such [interdisciplinary] efforts can all too often fail to materialize, and instead a mishmash of ideas, methods, and theory may be the ensuing result" (2005, 269). Although Porter and Rafols (2009) found a thirty-year "sustained and important increase" in interdisciplinarity by bibliometric criteria, most of these connections were between neighboring fields. When measurements were calibrated to reflect "disparity" between fields, the bibiliometric data "show a more modest growth" in interdisciplinary research (mostly around 5%) (Porter and Rafols 2009, 721). These social scientists concluded that "science is indeed becoming more interdisciplinary, but in small steps . . . only modestly increasing the connections to distant cognitive areas" (719). A 2015 *Nature* review observed that "research that transcends

[14] E.g., hierarchy, formality, and specialization are inversely related to interdependence among fields/sectors.

[15] Gerson's view of science as comprised of many different specialized communities continuously proceeding along lines of work that split, intersect, or otherwise interact, is shared with (and an influence on) my explanatory particularist account.

conventional academic boundaries is harder to fund, do, review and publish" (305).[16] The road to successful IDR, especially when disparate fields are involved, is littered with obstacles.

Philosophical studies

Philosophical studies corroborate this empirical result. For example, Miles MacLeod and Michiru Nagatsu (2016) argue that interdisciplinary research faces "cognitive barriers" stemming from "lack of shared concepts and language, educational background, and shared standards" (414–415). This lack of commonality, they claim, makes the already-challenging tasks of "grasping another's unfamiliar techniques and practices and coming to agreement on the epistemic standards to apply to new cases" extremely difficult.[17] Other philosophers have discussed interdisciplinarity as a counterbalance to scientific specialization. Thomas Kuhn put the point very strongly: "the evidence, once faced, is overwhelming: the development of human culture, including that of the sciences, has been characterized by a vast and still accelerating proliferation of specialties" (2000, 250).[18] Others have noted, relatedly, the fragmentation of knowledge into increasingly narrow spheres of technical expertise (e.g., Goldman 2001, Millgram 2015). This socially structured backdrop offers clear motivation for interdisciplinary research. It's often assumed, furthermore, that for collaboration across disciplines to succeed, many aspects of scientific inquiry need to be closely coordinated and integrated. This in turn requires new conceptual frameworks, methods, and evaluative standards—which have the added appeal of innovation and novelty. This general picture of the value of interdisciplinarity, as we have seen, underlies many policy and funding initiatives. Philosophers of science studying interdisciplinarity have been more cautious regarding its features and benefits. Several of these projects furnish important background for the particularist account that follows.[19]

Nancy Nersessian and colleagues conducted multi-year ethnographic studies of biomedical engineering research laboratories and integrative systems biology laboratories, focusing on collaborative strategies for producing physical and computer simulation models in these interdisciplinary settings (Nersessian 2006, MacLeod and Nersessian 2016, MacLeod and Nersessian 2019, Nersessian 2019,

[16] Nature Editorial Board (2015). See also Ledford (2015), Frodeman et al. (2017).

[17] See also Grüne-Yanoff and Mäki (2014), Grüne-Yanoff (2016), Mäki (2016). Others approach IDR in terms of social epistemology (e.g., Andersen 2010, Andersen and Wagenknecht 2013, Rolin 2015).

[18] See also van Noorden (2015), Koskinen and Mäki (2016).

[19] This short survey is far from a comprehensive review. Other philosophical studies that take a particularist approach to interdisciplinary research and/or integration include: Brigandt's problem-centered approach to explanatory integration in biology (2010, 2013), Bechtel's case studies of interdisciplinary relations (1986, 2006), Thagard's studies of "peer-different collaboration" (1997), and Mitchell's integrative pluralism (2002).

Nersessian 2022). This project was primarily concerned with cognitive and modeling practices in those laboratories. Nersessian and colleagues sought to characterize practices in particular laboratories or research groups that contribute to successful IDR. While these long-term case studies don't focus on explanatory models, the latter do play an important role (actual or aspirational). IDEs are a proper subset of the models examined in Nersessian's project, the results and ideas of which inform my account. Using a similar approach, Hanne Andersen (2016) analyzed interdisciplinarity in terms of individual scientists' research activities ranged along continua of epistemic dependence/independence (on one another) and cognitively convergent/divergent (i.e., drawing on one or multiple sets of cognitive resources). The latter requires "interlocking expertise," i.e., finding or constructing connections between cognitive resources drawn from different domains. The idea of "interlocking expertise," originating in Nersessian's work, has important parallels with my account.

One of Nersessian's collaborators, Miles MacLeod, subsequently applied the ethnographic approach to modeling in resource economics, an interdisciplinary field combining ecology and economics (MacLeod and Nagatsu 2016). One key result is "a relatively clear division of labor and distribution of tasks" largely preserving the epistemic standards and model-building methods of each field (428–429). Extrapolating from this case, MacLeod and Nagatsu propose three features of successful collaboration: models of one group are outside the other's area of expertise; models provide a clear structure of interactions to organize division of labor among groups; and disciplinary aims of each group are promoted. These insights (corroborated by other empirical studies) also inform the account developed here. MacLeod also contributed to the University of Helsinki's Institute for the study of interdisciplinarity (TINT), headed by Uskali Mäki from 2006 to 2018. Under the aegis of this Institute, several dozen researchers used a particularist approach informed by social science, to study aspects of interdisciplinarity, with emphasis on economic subfields, explanatory expansion and unification, scientific imperialism, and interdisciplinary model exchange (Grüne-Yanoff and Mäki 2014). EP's main ideas are shared by TINT researchers and their collaborators. Mäki and colleagues attend to the dynamic social epistemic organization among the sciences, as well as rankings of disciplines by prestige and the ways disciplinary boundaries are maintained and crossed. All these ideas and commitments are shared with my account. Ongoing TINT projects include Mieke Boon and Sophie Van Baalen's (2019) proposed "engineering paradigm" for teaching IDR. Their approach is particularist, centered on the idea that items of knowledge are epistemic tools—a view not dissimilar to the account of models in Part I. To counteract fragmentation and hyper-specialization, Boon and van Baalen propose "metacognitive scaffolds" as tools for examining knowledge-construction within and across disciplinary perspectives. Their account shares key tenets with EP, although focused on knowledge rather than explanation.

Another research program, the Toolbox Dialogue Initiative (TDI), has a similar approach. The TDI runs workshops on philosophical themes with the goal of improving communication among members of an IDR project on background assumptions relevant to that project (Eigenbrode et al. 2007, O'Rourke and Crowley 2013, Robinson et al. 2019). Collaborators engage in dialogue, structured by Toolbox questionnaires, about basic assumptions on methods, reality, values, and more. Both common ground and disagreements among participants are revealed, so researchers better understand one another's different scientific perspectives. Consolidating insights from hundreds of workshops, Toolbox researchers recently proposed a general framework for studying interdisciplinary relations (O'Rourke et al. 2016). The basic idea is to conceive relations between two (or more) disciplines as processes connecting inputs to outputs. Such processes may be integrative, preservative, or disintegrative, with further characterization in terms of scale, degree of conflict, and comprehensiveness. This general framework allows for comparison of many different kinds of interdisciplinary relation. My concern is with one variety: integrative relations between explanatory models. The taxonomy of model–model relations from Chapter 2 can thus be positioned as a refinement of one type of "input-process-output" relation bearing on explanation/understanding.

Overall, empirical and philosophical studies paint a picture of IDR as an increasingly prevalent feature of scientific practice, notably for projects aimed at solving complex real-world problems. In such research contexts, demands for explanation/understanding often transcend the boundaries of a single scientific community. Other epistemic benefits of interdisciplinarity tend to be assumed rather than argued for: conceptual, theoretical, and practical innovation, scientific progress, and the like.[20] Empirical and philosophical studies collectively indicate a formidable range of features of and challenges for IDR. The philosophical studies surveyed above (and others; e.g., Brigandt 2013), are explicitly particularist. Together, they form a background of accrued insights that fit well with EP and EP_M. I'll next apply these ideas from Part 1 to the case of IDR.

4.2 Interdisciplinarity: insights from EP

Explanatory particularism not only fits well with existing interdisciplinary studies, but also contributes new insights. The fit between explanatory particularism and interdisciplinarity is exhibited in several ways. For one, they are two sides of the same social epistemic coin. According to EP, specialized scientific communities construct and use explanations in accordance with features of

[20] See, e.g., Wray (2002); Jacobs and Frickel (2009) for a critical response.

those particular contexts. But scientific understanding does not always remain restricted to particular scientific communities. Often, as IDR studies show, there's a broader social goal: using science to address real-world problems. In such cases, researchers from different specialized scientific fields need to work together to advance understanding more broadly. As scientific specialization intensifies, so too does the need for IDR collaboration.[21] Interdisciplinarity and EP's specialization dovetail, in this respect. Unsurprisingly then, social scientists often define interdisciplinarity in ways consonant with EP. Broad definitions encompass not only interactions between traditional disciplines (e.g., biology, chemistry, physics) but also fields and subspecialties, such as "low temperature physics, molecular biology, developmental psychology, toxicology, operations research, and fluid mechanics" (Porter and Rafols 2009, 720).[22] This maps closely onto EP's particular scientific communities: collections of experts sharing a specialization, with distinctive patterns of communication, methods, aims, and standards of inquiry.

The basic image of science presupposed by EP also highlights the importance of interdisciplinarity between particular specialized communities. Because science's social epistemic organization is dynamic, distinct communities do not always remain so. At any given time, scientific practice as a whole consists of a multitude of diverse specializations, many proceeding more or less independently from others, some in the process of merging or splitting. Most philosophical treatments of the organization of sciences presume stable organization of fields, often hierarchically arranged in ways that track intuitions about the nature of explanation. But in practice, scientific fields are continually meeting with one another—and conversely, splitting to generate distinct specializations. Explanatory tasks do not always partition neatly for clear-cut division of labor coinciding with community boundaries. The latter, after all, are always in flux, and specialized scientific communities continually meet with and impact one another. With increasing numbers of researchers and more elaborate technological demands, specializations continue to proliferate. This in turn motivates more interdisciplinary research, with the aim of solving real-world problems or understanding phenomena of interest to multiple communities. The dovetailing of EP and IDR studies is not ad hoc, but grounded in EP's central ideas about scientific practice.

Interdisciplinarity involves more than explanation and understanding, so EP is concerned with only a subset of IDR contexts. Yet that subset is central. Interdisciplinary projects, as noted, tend to be motivated to solve real-world problems. Any epistemic products involved in solving such problems will be

[21] But see Beaver and Rosen (1978) for a dissenting view, critiqued in Wray (2002).

[22] Huutoniemi et al. (2010) also define the distinct contributors to interdisciplinary research as fields or specializations, as does Bechtel (1986) in the matter of integrating scientific disciplines.

significant—not only to the specialized communities involved, but also to wider publics. EP's significance requirement is thus satisfied in many IDR contexts. Moreover, understanding/explanation are likely to be instrumental for such projects' success, even if goals are practical rather than epistemic.[23] EP's conceptual framework casts IDR as a class of 'meetings' between different specialized scientific communities, motivated by significant problems, and spurring advances in understanding mediated by collaboration among members of different specialized epistemic communities. There is much common ground here with interdisciplinary studies, empirical and philosophical. EP's core tenets and assumptions highlight the need for and importance of IDR involving explanation. The particularist approach begins with the complex, dynamic social organization of scientific activity—the stage on which interdisciplinarity is enacted. So although EP in the first instance is concerned with scientific practices of construction and use of explanatory models to advance understanding within particular communities, the account extends straightforwardly to IDR contexts as well. Indeed, Part I's framework is well-suited to analyze these contexts. And this analysis yields insights for studies of interdisciplinarity, contributing to the literature surveyed above.

One such contribution is clarifying grains or scales of interdisciplinarity. A core assumption shared by EP and studies of interdisciplinarity is that science consists of many groups in complex social epistemic organization. Scientific practice involves many different groupings of scientists, which may overlap or crosscut one another. Individual researchers are organized into close working groups, both within and across laboratory boundaries. Large laboratories often include a number of such research teams. Laboratory groups themselves cluster into fields or sub-fields, working on related problems using similar methods and standards. Specializations emerge at this grain. Traditional disciplines are broader, encompassing multiple specialized fields. Other divisions, such as natural vs. social sciences, are coarser-grained still. The two extremes, as discussed in Chapter 1, are individual researchers on the one hand, and the scientific community considered as a whole, on the other. Questions about interdisciplinarity arise at every grain between these two extremes. EP focuses on the grain of specialized scientific communities with characteristic styles of explanation. But that focus doesn't occlude other grains or scales of interdisciplinary activity. In highlighting one important grain of social epistemic organization for IDR research, EP also recognizes the wider range of scales/grains relevant to interdisciplinarity. Attending to this range, and to the variable scales at which IDR can proceed, is a contribution to IDR studies.

[23] See Section 4.3.

Embryo models, revisited

An example will illustrate. Most philosophical and qualitative social science studies of interdisciplinary research are on the finer-grained end of the sociological continuum: small research teams or laboratory groups. The embryo models discussed in Chapter 2 originated in research at this grain.[24] Here I'll delve into the origins of one member of that interlaced family: "embryo-like structures." Embryo-like structures were initially constructed as part of a small-scale, long-term interdisciplinary project between a stem cell biologist (Ali Brivanlou) and a theoretical physicist (Eric Siggia) at Rockefeller University.[25] Their specialized epistemic communities are quite distinct. Brivanlou, with a background in biochemistry and molecular biology, aimed to dissect the "signaling network" of cell–cell communication underlying differentiation and morphogenesis in the early human embryo.[26] Explanatory models constructed by Brivanlou's research team take the form of molecular signaling networks. Siggia's research program is more explicitly interdisciplinary, applying ideas from condensed matter physics to aspects of biological development; e.g., techniques for quantifying morphogenesis. Brivanlou and Siggia's collaboration is not focused on explanation per se. Instead, their methods and ideas complement one another, and Siggia's physics graduate students occasionally work in Brivanlou's stem cell lab. One such transplanted physics student, Aryeh Warmflash, objected to the usual practice of allowing human embryonic stem cells (hESC) to grow randomly and unconstrained in cell culture. It was too irregular, he thought. Seeking greater order in stem cell growth, Warmflash cultured hESC in specific "micro-patterned" geometries: small circles, triangles, squares, etc. The results were unexpected.

In these geometrically restricted conditions, hESC don't simply reproduce, but develop in a robust "center-to-periphery" pattern: a central clump of stem cells surrounded by layers of more differentiated cells at the periphery. Brivalou's lab group interpreted these as "embryo-like structures," because the arrangement of cells resembles a classic layered cellular form in early human development (Pera et al. 2015, 917–918). Siggia's lab provided a high-resolution quantitative assay to measure results of Warmflash's experiments. So physics made a double contribution to this biological project. First, standards imported from physics imposed geometric order on hESC growth. And second, physics instrumentation provided quantitative high-resolution data for results.[27] The epistemic product, however,

[24] Arguably, all EP-explanations originate at this sociological grain. As this example shows, however, philosophical studies of explanation shouldn't be restricted to this grain of analysis, but should focus on specialized research communities, a coarser sociological grain.

[25] This setting is appropriate; the Rockefeller Institute was established as a site for physicists to begin investigating biological questions, central to establishing molecular biology as a new specialized field.

[26] See <https://www.rockefeller.edu/our-scientists/heads-of-laboratories/868-ali-h-brivanlou/>.

[27] Brivanlou (2018).

was a contribution to stem cell biology: a kind of embryo model, reproduced and adapted by that community as part of a work-in-progress explanation of early human development. In this case, physics contributed vitally to a project that advanced understanding in stem cell research. The process was interdisciplinary, but the epistemic product was constructed and used in just one specialized community.

This brief example shares many features with more extensive philosophical studies of IDR. The collaboration is small-scale, based on the choice of two individual researchers to combine their different research approaches.[28] Each is an established scientist in his own right, recognized within his specialization. The two collaborators work at the same institution—one with a long history of facilitating interactions between physics and biology—and have achieved a successful working partnership over a period of years. These features are common in successful interdisciplinary collaborations: small-scale, long-term, initiated by practitioners, and sustained by personal ties. Another noteworthy feature is the role of integrated pedagogy, with physics students like Warmflash encouraged to learn practices of biological experimentation.[29] Associations like that of Brivanlou and Siggia are highly contingent, local, and dependent on the vagaries of interpersonal connections. One goal of interdisciplinary studies is to make IDR success more reproducible and reliable; i.e., less contingent and dependent on interpersonal ties. A clearer account of what makes IDR successful could be used to plan and facilitate among larger-scale groups in response to specific problems, rather than "organically" arising over the course of individual researchers' careers. This is exactly what Nersessian and colleagues aim to do, for systems biology/engineering. That goal is shared by many social science researchers as well. But the latter use a wider lens, focusing on larger and more impersonal scales and different sectors of society (e.g., Shrum et al. 2007).

Explanatory particularism charts a middle way between these two grains of social epistemic organization: "micro" studies of one or a few individual research teams/laboratories, and "macro" studies involving multiple institutions and sectors of society. Most empirical and philosophical studies of interdisciplinarity take one or the other as their focus. However, EP-explanations are products of a wider community than a single laboratory or research team. To be sure, the Rockefeller team's original embryo-like structures were a significant achievement that garnered a lot of attention in the stem cell research community. But no one, including the original authors, claimed the initial achievement was explanatory, nor that it advanced understanding of human embryonic development (Warmflash et al. 2014). Explanation was not a primary goal of this IDR project at all. Rather, the goal was to construct a new experimental model for studying

[28] Durkheim's "organic solidarity"; see Wray (2002).

[29] See also Nersessian (2006, 2019), MacLeod and Nersessian (2016), and NSF 21-607 (NSF-Simons Collaboration on a National Institute for Theory and Mathematics in Biology).

human embryonic development. The goal of understanding that target is relevant, of course. But Warmflash's experimental constructs are contributions to that goal, rather than its accomplishment as such. In the stem cell research community, the major achievement is the more inclusive collection of embryo models, of which Warmflash's (and others') embryo-like structures are one variety. The stem cell research community is excited about these models, in part due to their potential ethical significance, but primarily for their explanatory value. These entangled, value-laden issues are worked out at the grain of the wider stem cell community, not within individual laboratories or teams. From an EP perspective, Brivalou and Siggia's interdisciplinary collaboration is one strand of the more inclusive process of constructing a fabric of interrelated models of early human development. The explanatory achievement in this case is still work-in-progress. Furthermore, advances in understanding to date don't rest on any one single embryo model, but the whole interrelated fabric of such models (see Chapter 2). That fabric of models, and associated advance in understanding, is still largely confined to the stem cell research community. This might be because the explanation is still under construction, or it might be a more permanent epistemic feature.

The explanatory target in the embryo model case is human embryonic development. That target is also of interest to other specialized scientific communities besides stem cell research. It is of broader social and political significance as well.[30] Connecting different explanatory models of one target is a straightforward extension of the particularist framework presented in Part I. In the case of embryo models, the advance in understanding so far remains localized to its original context, the stem cell research community. But the question of how this explanatory advance can (or should) extend beyond that community is very relevant. That extension is precisely the challenge for EP raised at the end of Part I. It coincides with, and clarifies, one of the main challenges for IDR. Far from excluding the possibility of interdisciplinarity, EP's core ideas make sense of its increased prevalence and at least one key challenge it faces.

The explanatory challenge

So far, I've argued that EP's ideas and background tenets fit well with existing scholarship on interdisciplinarity. But explanatory particularism does not just corroborate earlier work. Applying Part I's framework to IDR yields new insights about this

[30] I am writing this on December 4, 2021, just as the US is seeing erosion of women's reproductive rights, allegedly on the basis of the value of early human embryos. In my view, claims of full personhood for early human embryos are incompatible with models of mammalian/human development constructed and used in many biological and biomedical fields. Much of the current political discussion relies on carefully maintained ignorance of biological understanding. Deconstructing some of those fences for ignorance could be worthwhile.

mode of research. One was discussed already, stemming from EP's highlighting of specialized communities as foci for IDR. Such specialized communities are of intermediate social epistemic grain, between individual research teams and broader social institutions. In centering specialized scientific communities, EP shows an important dimension of variation for IDR projects in general: grain of social epistemic organization. Attending to this dimension, alongside others, makes for a richer and more nuanced general account of IDR.

Two more insights from EP form a pair: explanation is both a significant aim and a major obstacle to IDR projects. These two ideas set up *the explanatory challenge* for IDR, answered in the rest of Part II. Here I show how these ideas, which follow from EP's basic tenets, contribute to interdisciplinary studies. I then state the explanatory challenge in full. That explanation is a significant aim of IDR projects is unsurprising, but isn't emphasized much in interdisciplinary studies to date. Much of the empirical and philosophical interest in IDR stems from its integrative aspect. The challenge is to combine key aspects of scientific method into a single productive process: data collection practices, modes of inference or analysis, norms of model-construction, standards for scientific knowledge, and more. EP adds explanation, thinly but substantively defined, to the mix. For many IDR projects, an explanation spanning multiple specializations is either a primary goal, or required to achieve such goals. Explanations arising from interdisciplinary contexts are crucial for the success of many research initiatives, including nanotechnology, climate science, cognitive neuroscience, environmental studies, and systems biology. Even if goals are practical, "[i]n order to create a functional solution to a complex problem, the problem has to be understood comprehensively, and its attempted solutions should reflect the interests of all relevant stakeholders and employ all the available knowledge from different sources" (Koskinen and Mäki 2016, 429).[31] That is, an IDE encompassing the problem-situation is necessary for interdisciplinary problem-solving. So explanation is important for many IDR projects. EP affords a more precise and detailed examination of epistemic and social aspects of this important class of interdisciplinary contexts. In this way, EP complements the broader methodological focus taken by other philosophical studies of the issue.

Of course, not all interdisciplinary projects are concerned with explanation. But there is reason to think that many are. First, the concept of interdisciplinarity is often associated with integrated or unified theories or models. Julia Thompson Klein, a prominent scholar of interdisciplinarity, traces the origins of the concept to "ideas of a unified science, general knowledge, synthesis and the integration of knowledge" (1990, 19). Explanation is a traditional task of general, unifying theories. These associations persist even without the latter; biological sciences with

[31] See also Gorman's (2010) adaptation of Galison's notion of a "trading zone" to an ideal for multidisciplinary teamwork: organized around a shared mental model based on a common understanding of the task at hand.

few or no general theories nonetheless aim at integrative explanatory models: "diverse methods, bodies of data and models are brought together in order to gain a mechanistic and predictive understanding of biological systems" (O'Malley 2013, 552). Second, the epistemic benefits commonly attributed to IDR—theoretical innovation, reliable and accurate prediction—are also common functions of explanatory models (Mäki 2016, 331). It is a truism that IDR breeds innovation. While this assumption is open to question, it is clear that new explanatory frameworks are themselves an important kind of theoretical innovation. Many important scientific fields originate out of such connections, including molecular biology and cell biology (Judson 1979, Bechtel 2006). And although the idea that explanation as such is closely tied to prediction has fallen out of favor, reliable and accurate predictions are a hoped-for result of many explanatory models in practice. Finally, explanation is implicated in the main practical benefit attributed to interdisciplinarity: solving complex real-world problems. Putting these points together: the benefits of IDR are commonly characterized as being realized by or requiring explanatory models.

Focusing on explanation also yields an additional insight about IDR's challenges. A scientific community's commitment to a particular style of explanation (that is, ideas and values associated with explanation in that community) can itself be a major obstacle to interdisciplinarity. Although the problem has not been studied systematically, discrepant explanatory norms and goals seem implicated in many projects' failure to thrive. Of course, there is a strong bias against publicizing such cases. But a few philosophical studies hint at the scope of the problem. Rabinow and Bennett (2012), in a rare chronicle of scholarly defeat, describe the difficulties they experienced as "embedded humanists" attempting to collaborate with biologists, engineers, and social scientists at the Synthetic Biology Engineering Research Center in Seattle, Washington. This hoped-for interdisciplinary collaboration was stymied by unequal power relations, institutional wrangling, and—not least—conflicting explanatory goals. Another failed connection afflicts phylogenetics: despite the broad availability of phylogenetic tools and databases, prokaryote phylogeny remains largely separate from the rest of the tree of life. O'Malley (2013) diagnoses the exclusion of prokaryotes as a consequence of phylogeneticists' commitment to a model of a single branching tree of life representing every speciation event during the history of life on earth. The prokaryotic tendency to "horizontally" swap genomic sequences across species boundaries muddles the preferred branching structure. That is, practitioners in the field of prokaryotic phylogenetics aim at a different explanatory model—a tangled network rather than a branching tree. A third example is from my own earlier work, arguing that failure of stem cell and systems biologists to communicate is largely due to these fields' commitments to divergent explanatory models (Fagan 2016c; see Chapter 8). These few examples are the tip of an iceberg.

The core ideas of explanatory particularism show why this is so. Although the goal of explanation (in EP's sense, advance in understanding) is broadly shared across the sciences, it is conceptualized and accomplished in diverse, locally specific ways. Different scientific contexts impose different standards on construction and evaluation of explanations. Those norms provide specific constraints on what a successful explanation should be like—the features it should exhibit, as such, and the ways it represents or otherwise relates to some target of interest. Explanations are, therefore, among the most distinctive epistemic products of a particular scientific community.[32] They are achievements of coherence among diverse factors and resources within a particular specialized field.[33] This coherence is achieved over time and with hard-won labor, through the mutual adjustment of a specialization's objects of inquiry, goals, methods, and norms to one another. Concomitant with this mutual adjustment *within* a specialization is diversity *across* specializations. It follows that specializations are differentiated *from one another* in key aspects of scientific method bearing on explanation/understanding.[34] So, on a particularist approach, the various epistemic challenges for interdisciplinary research appear as aspects of a general problem. How to achieve harmonious fit among elements of a research program — methods, norms, and epistemic goals — with different provenance? The paradigm of such harmonious fit is the smooth joining of jigsaw puzzle pieces. But their complementarity is prefabricated. It's unlikely that such would be found among pieces shaped by independent, distinct sources. The fit must be accomplished in practice; it takes *work*. That work is IDR aimed at explanation.

Other features of EP-explanations are arrayed against those projects. EP-explanations are by definition significant; i.e., highly valued by members of communities in which they are constructed and used. A particular scientific community invests in its style of explanation, is committed to key features of its explanatory models, and attaches considerable value to those epistemic products. As Woody (2015) argues, scientific explanations function to define and maintain a community's identity and values. Accordingly, boundary-work and hierarchies of scientific knowledge are frequently invoked around explanation. That is, explanation often marks an "us vs. them" distinction that reinforces a particular scientific community's identity and epistemic aspirations. (Philosophy of science has traditionally been complicit in reinforcing physics-centric hierarchies and boundaries, rather than examining and critiquing them.) If researchers contributing to an IDR project are committed to different views of explanation, this very often leads to

32 This claim is the flipside, so to speak, of Woody's (2015) account of explanation as having the social function of binding together scientific communities.

33 Close fit (harmony) between different elements of scientific practice in successful lines of inquiry has been noted at least since Kuhn (1962); Hacking (1992) refers to this as "self-vindication" for laboratory sciences.

34 This is not universally true, of course. Specializations that have recently split, such as genetics and genomics, often overlap in explanatory style. Other specializations emerge at the interstices of others (e.g., molecular biology, cell biology) and thus have extensive overlap with their parent communities (Darden and Maull 1977).

tension that forestalls effective collaboration. Precisely because explanation is an important goal for many scientific specializations, yet understood in distinctive ways across them, IDE presents a stark challenge.

The challenge can now be stated in general terms. Research projects spanning multiple fields or disciplines are increasingly prominent in the scientific landscape. Projects that involve explanation spanning multiple fields or disciplines pose a heightened challenge, because explanations are among the most distinctive and specialized products of a scientific field. Ideas about what an explanation is, what norms it should satisfy, and the kinds of evidence required to construct and support it, vary considerably across scientific communities. Within a single specialization, such ideas are often left tacit. In a community with shared explanatory norms this is unproblematic. But when the explanatory project extends to another scientific community with a different view of explanation, problems arise. EP's core ideas highlight and clarify this challenge for IDR.

4.3 Interdisciplinarity and model–model relations

Explanatory particularism contributes to studies of interdisciplinarity in several ways: clarifying the various grains of social organization at which interdisciplinary activities can occur; highlighting the importance of explanation to a broad swathe of those activities; and articulating the challenge that disparate styles of explanation pose for interdisciplinary projects. All these follow from EP's basic ideas (Chapter 1). The models framework (Chapter 2) makes a further contribution, setting the stage for a particularist response to the explanatory challenge. The simplest situation for which this challenge arises is that of two independent scientific communities that aim to explain the same target (e.g., mechanical properties of solids under strain, human sexual behavior, pathways of embryonic development). By "independent," I mean that the communities involved have neither recently split from a single field, nor are they in the process of merging. An example of a recent split is genetics and genomics. Though distinguished from the former in core aims, rhetoric, and concepts, genomics shares a great deal with genetics; institutionally, their separation is incomplete. In the latter category are specializations that emerge at the interstices of others (e.g., molecular biology, cell biology) and thus overlap with those "parents" (Darden and Maull 1977). Splitting and merging do fit EP_M's framework, but it's simpler to begin with a more clear-cut (and practically challenging) example. So, suppose two different, independent scientific communities have been proceeding separately, constructing explanations in accordance with distinct goals, norms, and methods. For one explanation in each community, the targets converge. There are four possible outcomes for this simple scenario. Most commonly, the two explanations are isolated from one another; separate communities do not interact in regard to their explanatory models. (One or both communities may refuse to recognize others' as explanatory, relegating "outsider" models to mere description or the like.) Evidently, such non-interactions are not cases of IDR. So this possibility can be set aside.[35]

A second possible outcome is that explanatory models of the same target do interact, and are found to coincide in some or all key respects. A coincidence of this kind would afford smooth merging of distinct explanatory models into a coherent overall explanation. However, due to divergence across specializations, this kind of outcome is extremely unlikely for independently-constructed explanations. (It's much more plausible in cases of closely-related or recently-separated fields.) Communities that construct the same explanatory model to understand the same target phenomenon would only do so if those communities had very similar explanatory values, methods, technical constraints, etc. Unless the communities are near neighbors, such coincidence would be very surprising. And if fields are such near neighbors that their respective explanations are largely identical, the term "interdisciplinary" seems inappropriate. But that scenario, however unlikely, presents no challenge to constructing an IDE. So it too can be set aside—there's no challenge to answer. A third possibility is that two distinct explanatory models of the same target interact so as to conflict with one another. There's a large philosophy of science literature dealing with this scenario in terms of theory-choice or incompatible models. However, for projects aiming at IDE we can set this aside too. Direct conflict between distinct explanatory models of the same target is incompatible with the goals of (most) IDR projects. That isn't to say such cases never happen. But they are likely to be rare across interdisciplinary boundaries. The first scenario, reciprocal disregard of the other community's explanatory model, takes less effort. And it's hard to see how an IDR project involving direct conflict between distinct explanatory models could continue at all. This option amounts to embracing interdisciplinary discord, giving up on IDE.

The fourth possibility is that two distinct explanatory models of the same target interact, but neither coincide nor conflict. Instead the two are connected, or integrated, in some way wanting further analysis.[36] This fourth outcome is the only one germane to IDE: distinct explanatory models of the same target are integrated in some way, as part of an IDR project. Accomplishing this—achieving the shared goal of constructing an IDE—requires collaboration between members of distinct specialized communities. The next chapter analyzes this special case of collaborative activity using EP's social action framework. Here I canvas and analyze relations between models constructed in different communities—the range of possible products of such collaborative effort. To do so, I use EP_M's framework of model–model relations (Section 2.4). I'll first reprise that framework, then apply it to the case of IDR.

[35] Longino (2013) makes a pluralistic virtue of this situation, in the case of human behavior.

[36] See, e.g., Mitchell (2002, section 1.5) on strategies for integrating models—although most of this work concentrates on integrative efforts within particular scientific fields and disciplines. See also Grune-Yakoff and Mäki's (2014) taxonomy of interdisciplinary model exchanges, as a lens for comparing different scientific communities' treatment of the same representational device.

Model–model relations: reprised and applied

Chapter 2 proposed a general comparative taxonomy of the ways models might relate to one another. The sense of "model" at issue, recall, is more inclusive than mainstream philosophical accounts. Any construct by a particular scientific community S for some purpose P bearing on some target X counts as a model. (For explanatory models, P is specified as advancing understanding of some X for some community S, with EP's additional significance requirement.) In scientific practice, it's fairly routine for models to come into contact with other models, explanatory or otherwise. The model–model framework characterizes general ways distinct models can relate to one another, in accordance with EP. The framework consists of two continuous axes and two cross-cutting distinctions. The axes are (i) stage of the model-construction process (early vs. late) and (ii) degree-of-connection (low vs. high). Consider (i). Distinct models, if related to one another, must make that connection at some stage of model-construction. Orthogonally, (ii) models may relate to one another via many elements or few. These two continuous axes define a graphical space for representing integrative relations between distinct explanatory models (Figure 4.1). Alongside this graphical framework, connections between distinct scientific models are classified by two cross-cutting distinctions. Models can relate to one another in virtue of similarities or differences; directly or via some additional mediator. The two distinctions together yield a very general four-part taxonomy of model–model relations (see Table 4.1).

I'll now apply the model–model framework to interdisciplinary projects that involve multiple scientific communities, and aim at or require for success an explanatory model constructed via contributions of all those communities. Such IDR projects include IDE as a shared goal. According to EP, any particular explanatory model emerges as part of a complex process. Although details differ

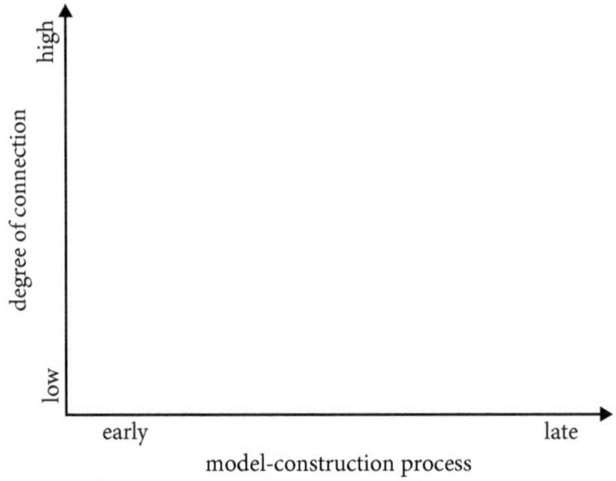

Figure 4.1 Graphical framework for classifying model–model relations, reprising Figure 2.2

Table 4.1 Four-way classification of model–model relations, reprising Table 2.1

	direct	**indirect**
similar	overlap (shared element)	subsumption unification (traditional)
different	conflict LK-complementarity	N-complementarity causal nexus part–whole structure

across particular scientific communities, each such process follows a path from early, sketchy proposals to a recognized achievement that advances understanding in the community. If two (or more) distinct models constructed by different communities are connected, this must happen at some point or interval along that path. We can distinguish three general categories of model–model relations in regard to this axis: early, late, and mixed stage relations between distinct explanatory models. How do these categories map onto interdisciplinary projects? I'll discuss each category in turn, showing connections with interdisciplinary studies.

Consider, first, explanatory models related early in their respective construction processes.[37] That is, all explanatory models are at an early stage, none having reached the status of an EP-explanation in their respective communities. This situation arises in IDR projects that bring together members of different scientific communities to innovate a new solution to some problem of interest. Projects of this kind are often motivated by complex real-world problems, which require multiple specializations to solve (e.g., a COVID-19 vaccine, bacteria that ingest toxic pollutants, better warning systems for earthquakes). Sometimes it's unclear which specialized communities are relevant to a problem. Funding agencies might leave this an open question, allowing for different combinations of specializations to try for solutions addressing one complex problem (e.g., the US NSF's call "Convergence Research"). A project of this kind effectively creates a new transdisciplinary community that persists for the duration of research, and perhaps beyond. Transdisciplinary research, as usually defined today, involves deep transformation of contributing specializations and is characterized by long start-up periods, continuous interaction across established specializations, conceptual

[37] For simplicity, I'll consider only the simplest case of two explanatory models/communities. More complex cases, as will be seen, consist of different pairwise connections between models, so the extension is straightforward.

innovation or novelty, and (potentially though infrequently) emergence of a new, enduring, specialized community.[38]

In such a project, all participants are out of their explanatory comfort zone, so to speak. The task is to construct an explanation nearly from scratch, rather than deploying a model already in hand among (some) community members. Even if the sought-after outcome is not an explanation per se, understanding of the problem itself is necessary for such transdisciplinary problem-solving to succeed. Understanding of a problem guides efforts toward its solution, outlining a template for future work. A template that yields such understanding is a model in EP's sense; e.g., of viral infection and vaccine impact thereon, of target pollutants and bacterial feeding patterns. Though not a full explanatory achievement, representation of a problem-situation is implicated in the former's construction, marking the starting-point of the process of explanatory model construction. In cases where explanation is not the goal, a model of the problem-situation is still needed. If its solution requires contributions from multiple scientific contexts, then the model of the problem-situation does also. One initial task of transdisciplinary IDR, then, is to construct an explanatory model of the problem (with an eye to its solution) that incorporates resources from all contributing specialized communities.

Subsequent transdisciplinary processes of explanation-construction involve explicit negotiation among participants about features of explanation (and other relevant aspects of scientific practice) that typically go without saying in an established scientific community. That is, constructing an IDE from multiple early-stage explanatory models requires new "bespoke" explanatory norms and standards for the new, transdisciplinary community. Although this can be accomplished in many different ways, involving various degrees of connection, a high degree of connection between explanatory resources from different communities of origin would be expected for projects of this sort. The framework of Part I, conditions for constructing explanatory models via collaborative activities in a particular scientific community, thus applies to IDR projects of this kind, just as for particular scientific communities. However, IDR's explanatory task is more demanding, because there is no shared history to draw on. Aspects of scientific practice contributing to the new problem-explanation are figured out (or negotiated) from scratch. To do this successfully requires a lot of time and effort, as well as frequent regular interactions among participants from the various original communities. It follows that transdisciplinary research is very labor- and resource-intensive.

The particularist framework of Part I has other interesting implications for this kind of interdisciplinary context. For this collaborative activity to succeed,

[38] See Leavy (2012). The term "transdisciplinary" originated in the 1970s, and was at that time associated with ambitions of a general systematic theory—the logical positivist dream of a unified science aimed at improving human lives. The standard definition in social sciences today is more down-to-earth, although traces of old ambitions remain.

participating members must consider one another on an equal footing with respect to their shared goal. That is, each sees themselves as contributing (but not solely accomplishing) that goal, and other members as full partners in that project. Members of a transdisciplinary community that arrogate to themselves alone the task of articulating the problem at hand—or who otherwise set themselves up as explanatory experts over other members—will scuttle the project. If participants are committed to background assumptions about hierarchies of scientific explanation, those ideas will undercut the transdisciplinary project from the outset. Although their roles in the collaborative activity will of course differ—as is the case in an established scientific community—all participants must recognize one another as being on an equal footing with respect to the explanatory task: understanding of the problem (and accompanying guidance toward its solution). Essentially, the account presented in Part I applies straightforwardly to transdisciplinary projects, offering a set of norms to guide such projects—or to abandon them as doomed from the outset.

The situation for explanatory models related at late stages is very different. In such cases, there are two or more explanatory models from distinct scientific fields, each recognized as an explanatory achievement (or incipiently so) in their communities of origin. Each specialized community has constructed an explanatory model of some target of interest. These models need not be complete, or full explanations—if indeed there are any such in ongoing scientific inquiry. They need only be models recognized as explanations, at a relatively late stage of construction, in their respective scientific contexts. Some inciting incident spurs some members of those distinct communities to combine their explanatory resources. This might be a real-world problem on which both are thought to bear, the siren call of "Convergence" funding, recognition by community-members that the models' respective targets coincide or overlap, etc. What differentiates this kind of interdisciplinary project from transdisciplinarity is not its impetus, but that the relevant explanatory models stand as separate achievements in their communities of origin. The task is then to relate these models, so as to yield a more comprehensive overall explanation of some target of interest—an IDE.

This brings us to the second axis of EP's model–model relation framework. This axis represents the degree to which distinct models are connected: how fully their elements are related. At one extreme are purely formal relations between each model taken as a whole (e.g., conjunction, input/output); at the other, close substantive relations between every element of each (e.g., thoroughgoing interdependence, reciprocity). In general, we can distinguish cases of low and high degrees of connection. Where early-stage relations between explanatory models are compatible with (and plausibly often require) a high degree of connection between models, late-stage relations between models are more likely to hold up if their degree-of-connection is low. On this point, the social action framework and empirical studies of interdisciplinarity converge. If interdisciplinary explanation

(IDE) is conceived as a collaborative activity, the shared goal is the activity of connecting explanatory models across these disparate specializations. Participants are committed to different methods, values, and norms for constructing explanatory models. Because of those differences, features that tend to stabilize collective intention are lacking. The more background beliefs and values bearing on the shared goal are held in common across participants, the more stable the shared intention, and so the more useful for planning/practical reasoning. Conversely, participants in a shared intention with different reasons, values, and background rationales run the risk of their shared "planning intention" being shallow and unstable (Bratman 2014, Tollefsen and Gallagher 2017). For many social actions, negotiation and compromise are effective strategies when participants' interests conflict. But it is both unrealistic and unreasonable to expect scientists to forego important values from their own explanatory contexts. A more effective strategy, then, is to tightly circumscribe the connection between explanatory models. Empirical studies of interdisciplinary research corroborate this result, showing that such projects tend to succeed best when connections between fields are limited and leave most practices unchanged (e.g., Shrum et al. 2007, Gerson 2013, MacLeod and Nagatsu 2016, MacLeod 2018). Therefore, the activity of connecting explanations amounts to constructing a conceptual bridge that leaves each specialized contribution mostly unaltered; i.e., a low degree of connection between related models. Late-stage connections between explanatory models tend toward a low degree-of-connection.

The lowest possible degree of connection is purely formal. For example, different linguistic models of a target X can be united in a long conjunction that simply adds one to another: simple additivity between models.[39] Simple additivity is bare conjunction of statements from different models. This minimal connection corresponds to another category in interdisciplinary studies. Along with transdisciplinarity, it is now commonplace to distinguish multidisciplinary research (see, e.g., Mattila 2005, O'Donnell and Derry 2005, Leavy 2012, Frodeman et al. 2017). In multidisciplinary research, members of different specializations contribute to a common project without transcending disciplinary boundaries, "each discipline maintains its autonomy during the collaboration . . . Borders are left intact" (Leavy 2012, 18–19). Connections are made at the end of the project (i.e., late-stage) and fall short of integration between models. This corresponds closely with additive or merely formal degrees of connection between explanatory models. Multidisciplinary practice (and, it's often claimed, its output) is only a minor adjustment of ordinary research within disciplinary boundaries. The epistemic upshot of such a process is not IDE. Taking on board this insight from studies of interdisciplinarity, more than simple additivity is required to connect

[39] Chakravartty (2010) proposes a contrastive account of explanations from different perspectives along these lines.

models into an IDE. This makes sense, given the difficulty and challenges of interdisciplinary work. Simple additivity (or any bare formal connection) involves no substantive connection between models from different perspectives; their contributions are simply strung together in a conjunction. The connection is logical consistency, and nothing more. Many cases might fit this formal pattern; that's not a problem for EP's framework. But we now have a further requirement for IDE explanation: diverse models contributing to IDE need to be connected in a manner stronger than simple additivity. These connections, whatever they are, involve *coherence*. The next chapter examines this idea in detail.

Finally, what of mixed cases? These are IDR projects such that one specialized scientific community has a late-stage explanatory model, while another community's is early-stage. This situation involves an explanatory asymmetry that is prima facie incompatible with collaborative activity in the sense of Part I. The specialized community with an explanation largely in hand will, quite reasonably, consider this to be part of their contributory role in an interdisciplinary project. So an IDE constructed under these conditions is likely to be one-sided, its features and contribution to understanding largely determined by the community with the late-stage model. This is not to say that the framework of Part I cannot apply to such cases. But the ways in which it applies will depend on particular features of the interactions between members of the relevant specialized communities. The general model–model relation framework here does not provide additional guidance. That is, for mixed cases, full-blown explanatory particularism is the best philosophical approach. So this category of cases can be bracketed in what follows (although not from particularist studies more broadly).

To sum up: the two axes for representing possibilities for model–model relations suggest two clusters of relations and associated types of IDR (Figure 4.2a). Transdisciplinary projects are characterized by early-stage high-degree connections between explanatory models from different specialized communities, which effectively establish a new community and hybrid explanatory style. For these projects, the account set out in Part I applies straightforwardly, although with more challenges than explanation in scientific communities. (Early-stage low-degree connections are also possible, though these fall outside the category of transdisciplinarity.) The other cluster is of late-stage low-degree connections between models. These are cases in which different explanatory models are (more or less) finished constructs, according to the standards of different communities. Multidisciplinary research marks an extreme of this type: connections between explanatory models are purely formal or additive, and the different explanations stand on their own as achievements in their respective communities. (Mixed cases will only rarely yield IDEs. When they do, those will emerge from features of the particular communities involved, and their contingent interactions.) Between these two extremes sprawls the broad, as-yet-uncharacterized category of "interdisciplinary research" (Figure 4.2b). The term "interdisciplinary" here is in the

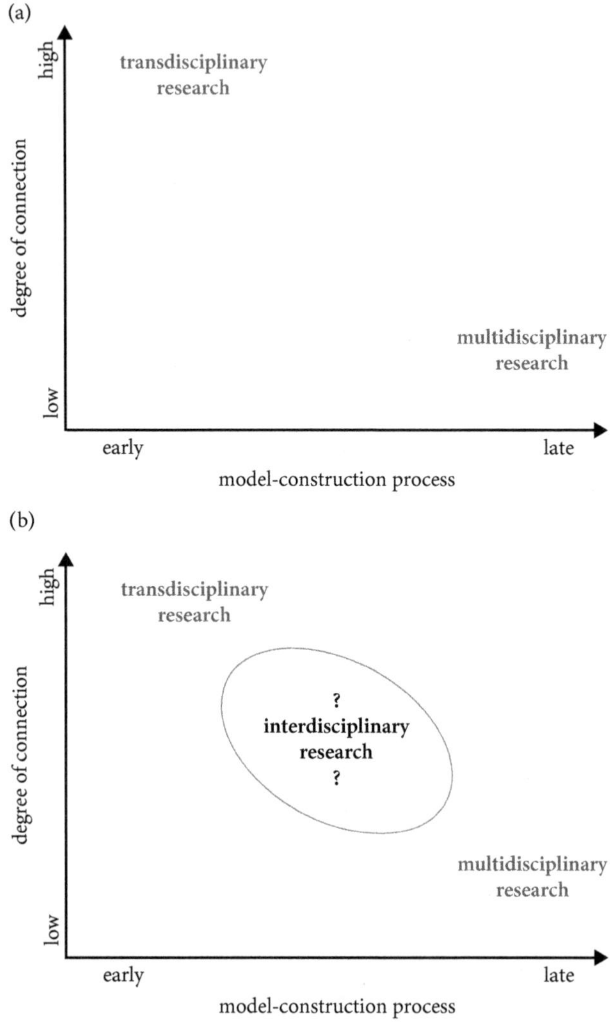

Figure 4.2 Two modes of interdisciplinary research. Types of interdisciplinary research mapped onto general modeling framework (stage of process, degree of connection): (a) early- and late-stage relations between explanatory models from different scientific specializations; (b) interdisciplinary research in the narrow sense an intermediate category.

narrow sense; i.e., excluding the other two categories. But that's still a very broad swathe of projects! Scholars of IDR characterize this narrower sense of the term as intermediate between trans- and multi-disciplinary projects. This kind of research does involve integration: contributing members adjust their methods and results to construct an integrative understanding of a problem and its solution. However, there is no deep transformation of discipline-specific concepts, methods, and results. Within this category, the vagueness and ambiguity of "interdisciplinary" remains in force.[40]

This offers a further way for EP's model–model framework to contribute to IDR studies. I've argued, building on earlier studies of IDR, that late-stage connections between explanatory models are more likely to succeed if they are of low-degree; i.e., involving one or a few elements of the models. Purely formal relations are not the only possibility for low degrees-of-connection. Many, many other kinds of relation are possible. "Low-degree" refers to the extent of connections between different explanatory models, not the nature of those relations as such. This brings us to the second part of EP_M's framework. We can use the two cross-cutting distinctions to analyze integrative (i.e., substantive, not simply additive) connections between distinct explanatory models at late or intermediate stages of construction. Relations between models are either direct or indirect; that is, models (or elements thereof) either relate to one another directly or via some further mediator. In the latter case, something more is needed than both models taken together. Orthogonally, models can relate to one another in virtue of similarities or differences. Similarity-vs.-difference concerns features of models in virtue of which they connect with one another. Models can relate to one another via having features common to both. Or they can relate to one another through difference: a contrast leveraged into a positive link. So, late-stage, low-degree relations between explanatory models can be classified as any one of four subtypes: direct similarity, indirect similarity, direct difference, indirect difference (see Table 4.1). Of these four, only one is traditionally associated with scientific explanation. For interdisciplinary projects, however, this is too narrow. I'll work through each category in turn, for the case of IDR.

Consider two explanatory models from different fields, related by direct similarity with respect to one or a few elements. This means that models from different particular communities share one or a few elements in common. They are related to one another by similarity, and this similarity relation is direct. So there are one or a few points of overlap or coincidence between explanatory models constructed from distinct perspectives. This point (or points) of coincidence serve as joints connecting different models. In interdisciplinary contexts, direct similarity relations are unlikely to hold between elements of distinct models, unless those elements are conceptualized in a rather shallow, thin, or abstract way. 'Thicker' notions will differ according to their communities of origin. Specialized

[40] Where models are involved, this broad category of research practice corresponds roughly to late-stage, low-degree interdisciplinary research—with two important differences. Most scholars using "interdisciplinary" in the narrow sense of this tripartite schema do not focus on explanation, and characterize research of this kind as requiring interaction among participants from the outset, and involving a high degree of connection. So the late-stage, low-degree type of explanatory connection could be seen as a hybrid of multi- and inter-disciplinary research—a distinctive form for explanation.

explanations at a late stage of construction will tend to have accrued considerable nuance, culminating in an explanatory achievement. So substantive elements of each model are suited to their respective contexts, and thus differentiated from one another in full detail. However, a common element may appear on a shallower assessment; e.g., the shared term "gene" in human behavioral genetics vs. developmental biology, or "molecule" in physical chemistry vs. molecular biology. The polysemy of "gene" across different fields of life science has led to extensive confusion and miscommunication, prompting proposals to give up the term entirely (e.g., Keller 2001). There is, however, a potentially positive explanatory role for 'thin' connecting concepts of this sort (see Part III).

The next type of model–model relation to consider is indirect similarity. In this case, we have two different explanatory models with an element (or a few elements) that are similar. But rather than coinciding or overlapping directly, the elements are connected—revealed as similar—via some additional mediating element. That is, something is added to both models; neither is as it would be if used in its community of origin. One indirect similarity relation is traditionally associated with scientific explanation. That is subsumption: two distinct models related by means of a more by a more "fundamental" (or abstract, or general) model of their common features (e.g., Thagard 1997, Morrison 2011; see Chapter 2). Both Kitcher's explanatory unification and the covering-law view define explanation in terms of subsumption of diverse facts/phenomena via an explanatory argument pattern or law. The association of explanation with subsumption is deeply entrenched in philosophy of science, to the point that a different view of explanation can seem contradictory or absurd. However, focusing on IDR reveals a subtle yet important difference from the traditional view. For the latter, the subsuming mediator—the abstract argument pattern or general law—is the sole source of explanatory power. That which is subsumed is thereby explained. And so the association of explanatory power with abstraction and generality is reinforced. But for IDR projects, explanatory power doesn't rest only in the mediating link. Rather, each model is explanatory on its own terms. The indirect mediating element shows a similarity relation between one or a few elements of each. Explanatory power is distributed throughout the integrated IDE. If the integrating relation is subsumption, the subsuming abstraction (or whatever) is not imposed by one field on the other. Instead, the subsuming mediating element is something added to both models—a new construct for an IDE. Many more relations than subsumption are possible, in such cases.

Models can also relate to one another via difference; some contrast or dissimilarity. Explanatory models constructed by different communities can be expected to differ in many ways. But for such differences to connect different models, they must involve some form of complementarity: direct or indirect. Models related directly by difference are LK-complementary—meaning that they fit together like pieces of a jigsaw puzzle, or a key into a lock, or the anatomical features of orchids and their pollinators. The result is a more inclusive whole formed from diverse elements; for example, covalent bonding between molecules with complementary shape and charge.[41] The disparate things, in the case at issue, are (elements of) models from different communities. In an interdisciplinary context, this would yield an inclusive explanation—a larger whole constituted by disparate parts. The different models complement one another; their contrasting features an advantage rather than cause for epistemic concern. Many scientific explanations fall into this category; e.g., multi-level explanations that relate phenomena at different levels of biological organization through the details of particular experimental results. However, because the complementarity is so-to-speak frictionless, there's no obstacle to IDE. Direct difference relations are epitomized by the seamless fitting-together of jigsaw puzzle pieces—complementarity in key properties. But that happy situation is very unlikely for elements of models constructed in different scientific communities. Complementarity of this type is unlikely to arise by chance. In the case at issue, low degree of connection, only one or a few elements of the models are involved in the connection. Even so, without some prior shaping of those elements, direct contrast/complementarity is unlikely. This kind of fitting together takes work. An explanatory model is shaped by the context of the particular epistemic community in which it's constructed. It's unlikely that elements of models shaped by two distinct, independent scientific communities would directly complement one another. So this kind of connection would mostly apply to explanations involving a single specialization or very closely related fields.

Finally, models can relate to one another via indirect difference relations. These involve some additional third mediating element that establishes complementarity between the different models. Examples include part–whole hierarchy relating different levels of description, causal schemas relating elements of each model as cause to effect, and many more.[42] There are many prospects here for IDE, explored in the next chapter.

Integrating models: other accounts

Before moving to that answer, I want to situate the results so far in the philosophy of science literature. The latter includes a number of studies on integrating models, especially in biology. One influential account is Sandra Mitchell's integrative pluralism (2002, 2009). Like EP, integrative pluralism is grounded on studies of scientific practice, notably the multiplicity of models and explanations

[41] Interestingly, this is close to the everyday meaning of "unification": "combining two or more things to or with one another so as to form one whole or unit" (paraphrase of *OED online*, "to unify," accessed August 2015).

[42] Mechanistic levels are an important example of this kind of integrative relation.

of complex biological phenomena, such as division of labor in social insect colonies.[43] In several key respects, Mitchell and I are fellow-travelers. Explanatory particularism, coming several decades after Mitchell's work, benefits from her and others' arguments against reductionism in biology and beyond. There are some important contrasts between our approaches, though. Mitchell is concerned with a coarser grain of social epistemic organization: "biology" broadly construed. (Her main examples are from evolutionary biology of social insect behavior; specifically, division of labor in colonies.) Mitchell is also committed to a substantive view of biological explanation. What is explained, on her view, is a complex real-world situation. What explains is an integrated collection of idealized causal models applied to that situation (one cause per model). The specifics of integrative relations between models are left open. But that's not a particularist tenet—it's because biologists have not yet worked out the correct causal history in the main case Mitchell considers. There is a particularist aspect to Mitchell's account, but it's very unlike my EP and its framework. EP, as presented in Part I, emphasizes social relations among scientists and scientific communities, framing inter-model relations in terms of collaborative social interaction. Models, targets, and model–target relations are left unrestricted, apart from EP_M's minimal schema. In addition, my conception of "model" is more inclusive (and less focused on idealization) than Mitchell's counterpart notion. The crux of my model–model framework is *connecting* relations between models constructed by and for different specialized scientific fields. Mitchell's examination of these relations does not emphasize social interactions among scientists and scientific communities, but rather complex causal mechanisms that afford multiple compatible explanations. There is a causal world-picture in the background of her view, making salient the distinction between idealized theoretical models and real-world causal processes (Mitchell et al. 1997). Explanatory particularism does not include any such world-picture, remaining neutral as to the nature of explanatory targets. Instead, the key background idea for EP is that science as a human activity has a complex social epistemic organization, involving many levels or grains. The worldly target of scientific inquiry is left open, to be filled in by studies of particular epistemic communities. So while our respective accounts proceed along similar lines, there are significant differences of scope, emphasis, and framing between Mitchell's account and that presented here.

Lindley Darden and Nancy Maull's (1977) account of "interfield theories" is another classic treatment that covers similar ground to explanatory particularism. On Darden and Maull's view, an interfield theory is constructed when different fields "share an interest in explaining different aspects of the same phenomenon," each needs to look beyond its borders for answers, and background knowledge

[43] Mitchell and Gronenborn (2017) also apply integrative pluralism to different experimental methods for predicting protein structure in biochemistry.

suggests that there is a relation between them (49–50). Darden and Maull's purpose is to rebut the then-prevalent philosophical view that relations between distinct scientific fields must be understood in terms of reduction, i.e., logical derivation of statements of one field from those of another. By a "field," they mean a central problem and a set of accepted facts relating to that problem, "general explanatory factors and goals," techniques, and methods—irrespective of institutional categories. What I term a "specialization" is very similar to Darden and Maull's view of a scientific field, although they explicitly eschew any sociological aspect. Ingo Brigandt (2010, 2013) uses Darden and Maull's account as a point of departure for studies of explanatory integration in biology. Brigandt's approach is broadly particularist, and his account has significant parallels with EP_M. The main contrast is with my social action framework and emphasis on collaboration, which to my knowledge has no counterpart in other accounts of model-integration.

4.4 Conclusion

This concludes the application of EP's framework in this chapter. Preceding sections showcase the philosophical and scientific significance of IDR, building on previous empirical and philosophical studies. I've argued that explanation presents an important challenge for interdisciplinary research, and set the stage for answering that challenge. This is accomplished through application of ideas from EP/EP_M and the framework developed in Part I. Here I briefly summarize the results so far.

The importance of IDR for solving practical problems, and for scientific progress more generally, is reflected in national funding initiatives, the rise of interdisciplinary departments, and proliferation of local institutes and centers. Yet IDR faces persistent obstacles, which are especially pronounced for projects involving disparate fields (e.g., physical and social sciences, computer science and the humanities) or aiming at transformational understanding. These obstacles are various: social, institutional, practical, and conceptual. Explanation, I've argued, can be a social epistemic obstacle. Many (if not all) interdisciplinary projects aim at or require for their success an explanation of some problem and/or solution. The preceding sections clarify this aspect of IDR. I began by positioning my account of interdisciplinary explanation (IDE) with respect to other philosophical work on interdisciplinarity (by, e.g., MacLeod, Mäki, and O'Rourke). One takeaway is that 'interdisciplinary research' is a broad and fuzzy category; I next identified the more precise, narrower aspect that is relevant for EP: relations between differently-specialized explanatory models.

The central question is: How, if at all, are genuinely interdisciplinary explanations possible? Answering it is at once a theoretical challenge for explanatory particularism, and a practical challenge for interdisciplinary researchers.

The challenge for particularists, as clarified above, is to account for explanatory achievements of this sort in interdisciplinary research. How can there be EP-explanations constructed by more than one specialized scientific community? I've argued that it's a point in favor of explanatory particularism that such achievements are not straightforward. Interdisciplinary research is challenging, and IDE arguably one of its most challenging aspects. The task is difficult, but not impossible. What remains is to show how this can work. The framework of model–model relations offers a starting point. The social action framework completes the task. So the next chapter applies that framework to the issue of 'bridging explanatory models.' This shows how such bridges can yield EP-explanations, answering the interdisciplinary challenge set out here.

5
Bridging Models

This chapter continues the application of EP, bringing in the social action framework. Constructing an explanatory model in a particular scientific community is a collaborative activity (see Part I). The Higgs mechanism, the double helix, density function theory in nanoscience, the elaborate network of embryo models, and myriads more, are each specialized achievements of a particular community. How, if at all, can distinct explanatory models (or families of models) be combined into an interdisciplinary explanation (IDE)? This chapter answers that question, using results from the previous chapter plus the social action framework (Chapter 3). An IDE too is constructed via collaborative social activity. But its participants are from multiple, different specialized scientific communities. The framework for model–model relations distinguishes two kinds of IDE case: early-stage high-degree, and late-stage low-degree. The former are constructed by a new, transdisciplinary scientific community, using newly-negotiated explanatory standards and values. Norms for this kind of IDE are the same as for an established scientific community (Chapter 3), although satisfying them is more challenging. But late-stage low-degree IDE is different. Each specialized community brings to the project an explanatory model that's already a significant achievement for that field, or at least well on the way. That's the class of case examined in this chapter. How is an IDE constructed in such cases?

For this class of cases, we should let go of the idea of a fully integrated overall model, with its own set of explanatory standards and values to guide the construction process. Some IDR projects do require negotiating and articulating a new, "bespoke" set of explanatory standards and values. But these tend to fall into the early-stage high-degree cluster, characteristic of transdisciplinary IDR. For late stage low degree IDE, the activity is to connect two (or more) distinct, divergent explanatory models, each constructed according to different values and standards—the style of explanation in those particular communities. The most efficient way to proceed is to leave each specialized contribution mostly unaltered. In cases of this kind, social action norms apply to the activity of constructing a narrow, sturdy bridge between explanatory models constructed in different specialized fields. That is, the collaborative activity is 'bridging models' instead of constructing a new, transdisciplinary explanatory model from scratch. The social action framework applies to this case as well, but somewhat differently from the 'one community' case. This application proceeds as follows. I first reprise the social action framework, then apply it to late-stage low-degree IDE (Section 5.1).

Explanatory Particularism in Scientific Practice. Melinda Bonnie Fagan, Oxford University Press.
 DOI: 10.1093/9780191998874.003.0006

Section 5.2 explicates the three-layer structure of collective intention for this case. The result is a set of coherence requirements for IDEs, analogous to norms for collaborating agents. Section 5.3 illustrates these results with a detailed example: the array of life science models concerning hemoglobin. This example extends the bridging account to more complex, realistic IDE projects. Section 5.4 begins to articulate the epistemic and social significance of bridging models, a task continued in Chapter 6. Section 5.5 concludes with a summary of the results so far: the nature and norms for bridging explanatory models constructed in different scientific communities.

5.1 Social action framework, reprise

Chapter 3's social action framework is adapted from philosophical accounts of social action, which aim to analyze collective intention. The latter attitude is thought to be distinctive and constitutive of social activity. Philosophical theories of social action share four consensus elements: multiple participants, interdependent participant intentions, a shared goal, and common awareness. Social action consists of multiple agents who form interdependent intentions to participate in the activity (the shared goal) under conditions of common awareness. This rough consensus account applies straightforwardly to the activity of constructing a particularist explanation within a specialized scientific community. The shared goal is constructing an explanatory model, which requires multiple participants. Interdependence among them is analyzed as a three-layer intercalated structure. Common awareness is a given: a significant epistemic product that advances understanding for members of a scientific community must be publicized within that community. Also, the process leading to that outcome requires critical assessment of explanatory models. Such criticism, according to familiar scientific norms, is "out in the open" for that community.[1] So all four consensus elements are present. Constructing an explanatory model is a collaborative activity for members of a particular scientific community; likewise for diverse members of a transdisciplinary IDR project. But what about late-stage low-degree IDE?

The social action framework applies to this case as well, but with somewhat different specifications of the basic elements (e.g., shared goal, common awareness). Late-stage low degree-of-connection IDE involves at least two explanatory models (call these M1 and M2), constructed in distinct specialized scientific communities (S1 and S2, respectively). Their dual construction involves multiple participants from each community.[2] Although more than one individual from each will be involved in practice, for simplicity we can consider two idealized agents, A and B, as

[1] See Longino (1990, 2002).

[2] What if a single member of one specialized community proposes a "solo" explanatory model to be combined with a model collaboratively constructed in another community? The former wouldn't,

avatars of each community. The challenge for IDE is not raised by multiple agents per se, but agents working in different specializations, committed to disparate styles of explanation. A and B represent different scientific specializations, committed to different norms, values and virtues that guide explanatory model-construction.[3] The main contrast with explanation in a single community concerns the shared goal. For many social activities, the shared goal is performance of the activity itself (see Chapter 3). But the activity of constructing an explanatory model is outcome-oriented; guided throughout by the purpose of advancing understanding of some target X for members of scientific community S. For IDE, the aimed-at advance in understanding is for members of both communities, S1 and S2 (for simplicity, A and B). Specific features of model M1, M2, target X, and model–target relations are left open and unrestricted, to be filled in piecemeal by philosophical study of particular scientific communities—that's explanatory particularism. The shared goal can be characterized further, however.

In late-stage low-degree IDE, explanatory models M1 and M2 are for the most part constructed independently in their respective communities. Each is an (incipient) significant achievement in its community of origin, advancing understanding of target X, which is of interest to both communities. The activity of constructing an IDE amounts to connecting two distinct explanatory models. Because the case at issue involves a low degree-of-connection between models, this connection is between one or a few elements of each model. So the interdisciplinary collaborative activity is quite restricted: constructing a narrow connecting bridge that leaves each specialized explanatory model largely unaltered.[4] The simplest scenario is two scientific communities, each with its own explanatory model to contribute to the project of understanding some target X. The task is to connect these different explanatory contributions. The purpose—the hoped-for outcome that guides the research process—is some form of unified understanding, constructed out of diverse components. This outcome is accomplished by connecting M1 and M2 via a narrow, sturdy bridge.

The narrow bridge connecting M1 and M2 must be sturdy—a stable focus for social activity. Within a particular scientific community there is usually enough shared background for consensus about major explanatory achievements. But not so for IDR. Distinct specializations, such as theoretical physics and molecular biology, have much less substantive overlap than those that share an intellectual lineage. So features that tend to stabilize collective intention are lacking for IDE. The more background beliefs and values bearing on the activity that are shared, or held in common across participants, the more stable the shared intention, and

I think, be a late-stage explanatory model, but a proposal from one community member that needs further engagement within that community (see Part I).

[3] A and B could also be interpreted as groups, each comprised of individuals from one specialization.
[4] This is not to say that more thoroughgoing integrations of explanatory styles are undesirable (see Chapter 6).

so the more useful for planning/practical reasoning. Conversely, participants in a shared intention with different reasons, values, and background rationales run the risk of their shared "planning intentions" being shallow and unstable (Bratman 2014, Tollefsen and Gallagher 2017). In the case of IDE, then, stability for collective intentions is prima facie undermined—precisely because so little pertaining to explanation is shared across disparate scientific specializations. Relatedly, in many social actions, negotiation and compromise are effective strategies when participants' interests conflict. But it's unrealistic to expect scientists to abandon or dilute important values from their own explanatory contexts. A more effective strategy, then, is to tightly circumscribe the connection between explanatory models. Explanatory gains, like other benefits, are more likely to be realized the less that each contributing discipline needs to adjust their practices of explanatory model-construction. Rather than wholesale merging of different styles of explanation, we should think of connection between disparate models by a narrow but sturdy bridge.

To sum up and further review: the social action framework from Chapter 3 applies in this case to the activity of constructing a narrow sturdy bridge to connect M1 and M2. The latter models are largely constructed separately in S1 and S2, respectively. A and B are the idealized multiple participants. The shared goal is to connect M1 and M2 via a narrow sturdy bridge. A's and B's attitudes toward that goal are intercalated via three layers of relations. First, each participant intends to perform the activity (shared goal). These intentions have two aspects: singular and plural. Singularly, each intends to do their part as a participant. Plurally, each intends that the group as a whole perform the activity. Second, each participant recognizes the other's analogous first-layer intention. That is, each regards the other as a participant in the group's activity, symmetric with their own self-regard. Third, each participant regards the other as other-regarding, with respect to the social activity. Not only do they recognize one another as participants, standing in the same relation to the shared goal, but each recognizes the other as recognizing this. Each agent is aware of the other's other-regarding attitude with respect to the shared goal. This three-layer structure intercalates agents' intentions into a collective intention, in a manner that ensures common awareness in several key respects. The three-layer interactive structure is symmetric, reciprocal, and interlaced. Normative constraints on participants' attitudes follow from this structure (see Chapter 6). Analogous constraints apply to the connecting bridge itself. Those are the main focus of this chapter.

The analogy is underwritten by several parallels. First, the basic elements of social action correspond to features of the problem-situation for IDE.[5] The shared

[5] Although it is conceivable that a single person might construct independent explanatory models of the same phenomenon from two or more specializations, and subsequently seek to connect them, to my knowledge this hasn't occurred in scientific practice.

goal is to connect different explanatory models. Participants are members of distinct scientific specializations, whose intentions to participate are interdependent.[6] Common awareness would seem to be a given for any interdisciplinary project, for reasons indicated above. So the elements of social action, as presented in philosophical theories of same, are a pretty good match for the case at issue. Second, the prevailing philosophical approach to analyzing social action also mirrors this situation. Major theories of social action focus on cases involving two agents who combine their efforts into a joint activity such as painting a house together or going for a walk. Philosophers focus on the question of what distinguishes these and other social actions from analogous individual actions. What makes the difference between two persons walking *together* and two persons who happen each to be walking in the same direction? Analyses of social action seek to identify the difference-making element, or set of elements, that when added to individual actions weaves them into social activity.[7] This parallels the situation for late-stage IDE. The basic question is the same: what makes the connection? So it's reasonable to look at social action theory for resources to analyze IDE. Third, both collective intentions and model–model bridges are constrained to be stable. Collective intentions need to be stable to fulfill their role in practical reasoning. This stability requirement underwrites the intercalated, looping structure of collective intention (Section 3.3). In late-stage IDE, the connecting bridge between distinct models is the crux of interdisciplinarity. In order for an IDE to be useful in any particular interdisciplinary project, the overall model must hold together—the bridge must be stable, and stabilize the IDE overall. The three-layer interactive structure of collective intention offers as an analogous guide for constructing an interdisciplinary bridge between explanatory models constructed in different specialized fields (i.e., late-stage low-degree connections). In light of these parallels, the analogy between an interactive structure of collective intention and a conceptual bridge linking disparate explanatory models is worth pursuing.

5.2 Bridging models

An IDE consists of two (or more) explanatory models contributed by different scientific communities (M1 and M2), and a connecting bridge C. To succeed as an explanation, an IDE must meet standards of explanatory adequacy in the relevant

[6] Participants could also be construed as the specialized epistemic communities themselves, but this commits one to a controversial theory of social action. The account presented here remains neutral, insofar as possible, on these matters.

[7] Cf. Roth (2017): "What is left over when we subtract what each of us did from what we did together?"

interdisciplinary context. The traditional way to conceptualize this requirement is in terms of robustness: an IDE should represent consensus across all contributing communities.[8] But that's a strong requirement, unlikely to be met by late-stage IDE. Models contributed by different specializations are prima facie very unlikely to conform to one another's standards for explanation. An overall IDE model satisfying all relevant specialized standards at once would be an extraordinary happy accident; explanatory standards aligning across disciplines. So requiring that an IDE model conform to explanatory standards of all contributing specializations at once is too strong.[9] A more promising approach is for constraints to apply not to an IDE model as a whole, but just to the connecting bridge. That is, constraints for IDE apply to that which makes the model–model connection indirect: elements added to, or altered within, M1 and M2 so as to integrate the two. Therefore, most of what is distinctive about IDE concerns the bridge itself, and its relation to the original specialized models. Those models will be largely unchanged. Bridge C is a conceptual link between specialized explanatory models, allowing passage from one to the other without collapsing the distinction between them.

More analytically:

> (IDE) An explanatory model M is an IDE, comprised of models M1 and M2 contributed by scientific specializations S1 and S2 (respectively), if and only if (i) M1 contributed by A is an adequate explanation according to the standards of S1, (ii) M2 contributed by B is an adequate explanation according to the standards of S2, and (iii) bridge C coheres with the standards of S1 and of S2.

Conditions (i) and (ii) are implied by the EP account already presented (Part I). Condition (iii) needs more unpacking. As stated, bridge C is constrained to *cohere* with the standards of all involved specializations. What does that mean? Here the interactive structure of collective intention offers guidance; a blueprint for normative constraints on the connecting bridge. The tripartite complex of relations that weaves together agents' attitudes is mirrored by three points of contact defining the bridge of an IDE: first, between the bridge and each original model (near-side); second, between the bridge and the other's model (far-side); and third, between the bridge as seen in both directions (meeting in the middle). In order to function as an explanation in an interdisciplinary context, the bridge must satisfy coherence

[8] Associated definition:

> (IDE) An explanatory model M is an interdisciplinary explanation in context C, where C includes contributing scientific specializations $s_1, \ldots s_n$, if and only if M is an adequate explanation according to the standards of all of $s_1, \ldots s_n$.

[9] Researchers involved in an ID project could of course commit to arriving at a robust IDE model. But this involves a lot of planning and effort, comparable to transdisciplinary interactions constituting a new epistemic community. Moreover, those efforts would remove (abstract away) distinctive explanatory features of each specialization's contributed model.

constraints at each of these points. These requirements entail norms for interdisciplinary explanation.

Coherence

Coherence has long been associated with explanation in general epistemology. Coherence theories of justification assert that "all beliefs may be justified by their relation to others by mutual support," in contrast to foundationalist theories that trace justification to a set of basic beliefs (Lehrer 1990, 13).[10] However, in what follows I set aside that association. This is for two reasons. First, theorists working in analytic epistemology do not typically clarify the notion of explanation that figures in coherence in any detail. Often, they presuppose one or another classic theory of explanation from philosophy of science; e.g., deductive-nomological or causal mechanical. As argued in Part I, explanatory particularism rejects any such monist account of scientific explanation. So the concept of explanation presumed by theorists of epistemic justification isn't applicable here. Second, defining coherence in terms of explanation raises circularity concerns for EP. My task is not to define coherence for a theory of epistemic justification, but for an account of IDE. In this account, coherence figures as a constraint on bridge C linking models from different specializations; relations internal to this form of explanation. To define coherence for bridging relations in terms of explanation thus at least flirts with vicious circularity. For both reasons, I eschew the notion of explanatory coherence that appears in general epistemology.

Instead, the sense of 'coherence' assumed here is minimal. In the philosophical literature, minimal coherence is associated with consistency; i.e., absence of contradiction.[11] Belief in propositions that directly conflict with one another—that cannot both be true—is incoherent. At minimum, then, coherence rules out belief in contradictory propositions. However, this is still a bit too restrictive for my purposes. The concept of coherence in general epistemology applies to propositional beliefs had by an agent. But the relations involved in the bridge between diverse explanatory models do not only connect propositions. One of the main insights of philosophical research on models in science is that scientific models are not solely (or even usually) propositional. Scientific models are extremely diverse in form and content; this holds for explanatory models too (Chapter 2). So a more liberalized concept of coherence as consistency is needed, to accommodate this wide and open-ended diversity. At this point, the three-layered structure of

[10] See also Quine and Ullian (1970), BonJour (1985), Poston (2014), Olsson (2017).

[11] Even Lehrer's (1990) notion of reasonableness presumes that beliefs can conflict or compete with one another, and builds the absence of such conflict into the theory of coherence as what is reasonable for an agent to accept.

collective intention is a useful guiding analogy. The task is to clarify the notion of coherence between a bridge between explanatory models and the explanatory standards of different specializations. I consider a simple, abstract case involving two explanatory models (Models 1 and 2) constructed by different specializations committed to different styles of explanation. Bridge C linking M1 and M2 should cohere (in a minimal sense) with the standards of contributing specializations. What this amounts to is indicated by analogy with the three-layer structure of collective intention.

Following this analogy, bridge C has a three-part structure: an extension from each contributed model, attachment to the other, and a middle allowing for crossing (Figure 5.1). Just as the structure of collective intention constrains attitudes of agents in social action, the structure of a bridge linking explanatory models is constrained by coherence requirements for features of specialized models. I'll work through these requirements in turn, with brief examples to illustrate. The next section considers a more detailed, complex example.

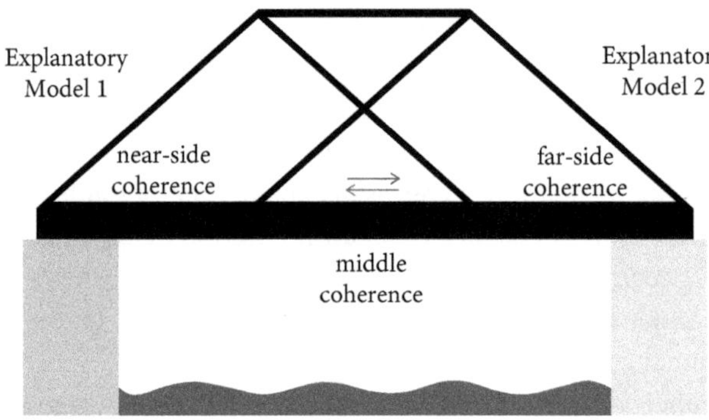

Figure 5.1 Schematic of coherence requirements for bridging models (from the perspective of Model 1). Requirements for near- and far-side coherence are reciprocally one-sided; middle coherence concerns these respective 'bridging efforts.'

Near-side

First, C must be securely anchored in each specialization—the "near-side," from each specialization's point of view. Some adjustment is made to each contributed model to enable connection with the other. The details, of course, will depend on the particular communities in question. This adjustment acknowledges the limits of that field's model—an added disclaimer or caveat to the effect that something of explanatory interest falls outside M1 or M2. Members of scientific communities routinely make such adjustments toward neighboring fields. Consider molecular genetics and molecular biology—very closely related fields, whose boundaries are fuzzy and porous. But there is a clear contrast in what is doing the explanatory work. Explanations in molecular genetics model how DNA sequence determines an organism's traits (see, e.g., Waters 2007). Differences in the nucleotide base sequence of a DNA molecule cause differences in the nucleotide base sequence of an RNA molecule, which in turn cause differences in the amino acid sequence of a protein molecule. Protein sequence differences are associated (causally or correlationally) with differences in organismal phenotypes. In contrast, explanations in molecular biology depict parts and processes within cells that contribute to an overall process, such as protein synthesis. These mechanistic models allows users to visualize the causal steps involved, "looking under the hood" at the inner workings of the cell (e.g., Craver and Darden 2013). Protein synthesis is modeled as the spatial arrangement and temporal order of activities (such as binding, breaking, diffusion and phosphorylation) involving diverse components (such as DNA strands, three kinds of RNA molecule, and two protein complexes). The result is a protein chain; a linear sequence of animo acids which then goes on to fold and participate in other activities.

The two explanatory models are obviously closely related, overlapping at several points. But there are important differences. The molecular genetic model locates explanatory power in DNA sequence, while in the molecular biology model explanatory power is distributed throughout the interacting parts and processes within a cell.[12] DNA figures in both models, but its explanatory role is different. The molecular genetics model depicts just one important causal role of DNA: determining protein sequence and 'downstream' organismal traits via coding. A connecting adjustment from that side would be to acknowledge other causal roles for DNA and other molecules; a caveat that other causal relations also influence protein sequence and 'downstream' organismal traits. The adjustment makes space for other explanatory contributions. This doesn't erase or undercut the DNA sequence model, as such. What's removed is DNA sequence's exclusivity as *the sole* determinant of protein sequence and organismal traits via its sequence-coding role. Although this example isn't strongly interdisciplinary, it does highlight an important conceptual change in biology, ongoing for about forty years. This is a broad shift in thinking about the explanatory role of DNA sequence, distributing explanatory power more broadly among molecular and environmental entities and processes. The shift plays out differently in different specialized communities, with some areas of genetics and biomedicine remaining committed to DNA's coding role as uniquely explanatory. Returning to the matter of bridging explanatory models, this requirement parallels the first-layer relation in collective intention. Each agent intends to contribute to the shared goal, recognizing that the activity isn't their solo project.

[12] This leads to ambiguity and tension within life and biomedical sciences, concerning DNA's explanatory role (see Section 5.3).

The new element of M1 (M2), whatever it is, should cohere with the explanatory style of that specialization. Ideally, the new element would be explanatory in its own right, by the standards of S1 (S2). But that's a high bar. Coherence is a minimal condition, ruling out only bridging elements that are inconsistent with M1 or M2. This rules out only bridge elements that present anomalies entirely outside a field's explanatory framework, explananda that immediately cry out for explanation, or direct refutations of M1 (M2). Recognizing a model's limits isn't likely to lead to the last, so I'll focus on the other two exclusions. An anomaly in Kuhn's sense (1962) is outside the epistemic reach of a paradigm, calling attention to its limits. Although my concept of an explanatory model is not the same as a Kuhnian paradigm, there is some overlap. Specialized norms for constructing and using explanatory models delineate the domain of what is explicable in a particular scientific field. That which is outside that domain will appear anomalous from the perspective of that specialization. It can't be explained in accordance with a specialization's style of explanation. The near-side of a bridge linking explanatory models across specializations can't be anomalous in this sense. An anomalous adjustment of M1 (M2) would be an explanatory non sequitur; say, that Schrödinger's equation or an experimenter's psychological state bears on the DNA coding relation. (There are probably scientists who agree with one or both—but they are outsiders to molecular genetics.)

Positing an explanandum is inconsistent with explanatory standards in another way. A bridge that posits something (event, pattern, etc.) that itself cries out for explanation is a non-starter or a reductio ad absurdum for an explanatory model. If an adjustment to M1 adds an explanandum for S1 as an element of the model, then M1 itself is a target of explanation for that specialization. The connecting bridge would raise a question for S1, rather than pointing outward toward another field. The bridge would then itself become a target for explanation. If the connecting relation itself poses a question of the kind the specialization aims to answer, then explanatory efforts will reasonably be directed to that connector. This distracts from the interdisciplinary project, effectively dismantling the bridge. Therefore, explananda for S1 cannot be included in the near-side adjustment to M1, and similarly for S2 and M2. The general requirement is that C's near-side not present an immediate explanatory challenge to S1 or S2; i.e., the bridge itself does not demand or confound explanation from either side. When considered from the near-side by each specialization (i.e., as an extension of that specialization's model), the bridge should be neither mysterious nor cry out for explanation according to that field's own standards. The molecular genetics example meets this requirement. Allowing for other causal roles had by DNA and other molecules, alongside their coding role, is neither anomalous nor crying out for explanation. It is not anomalous, because geneticists have long recognized that DNA sequence alone is insufficient to explain organismal traits; more than just DNA sequence differences underlies organismal traits. It does not include an explanandum, because additional causal roles for molecules (e.g., protein-producing cellular machinery) aren't organismal traits or protein sequences. The adjustment is to conceptualize DNA as one cause among others, even if it is still privileged as the most important (for geneticists).

Now consider the bridge extending from molecular biology. Only a modest adjustment to the mechanistic model of protein synthesis is needed, to connect it with the model of DNA as encoding protein sequence. The adjustment is to add sequence complementarity of DNA bases, RNA bases, and amino acids to the mechanistic model of protein synthesis. This is not merely coherent with the style of explanation in molecular biology, but improves the model by adding more molecular detail. Adding the details of sequence complementarity and nucleotide to amino acid coding makes for a more complete mechanistic description. From the perspective of molecular biology, the bridge amounts to wholesale addition of the DNA specificity model into the mechanistic description. This situation reflects the closeness of the specialized fields involved. Indeed, it may seem misguided to distinguish molecular genetics and molecular biology at all, given that bridging relations between the two are a seamless fit (especially from the molecular biology side). However, they do differ with respect to the concept of the gene and its explanatory role—and this tension motivates a distinction between the two explanatory models. Specialized fields that are more widely separated from one another are, as noted above, very unlikely to exhibit mutual improvements directly. In interdisciplinary projects, the near-side adjustment is on both sides. When the fields are not near neighbors, the point of connection is likely to be rather simple and "low-tech." This brings us to the far-side connection between an adjustment to M1 and M2.

Far-side

The new bridging element extending from one model should make contact with the other. That is, the adjustment to one model (M1) should also cohere with the other (M2), on the "far-side." This requirement parallels the second-layer relation of the interactive structure: each agent's recognition of the other's reciprocal intention to contribute. The new extension added to a specialized explanatory model should cohere with the other's specialized contribution. One might suppose that this coherence requirement would be the same as for the near-side: absence of anomalies or explananda in the bridge, relative to the standards of the other specialization. However, that's too demanding. The model-construction practices of each specialization are different, and cannot be assumed to be common knowledge across community boundaries. Members of a specialization involved in an interdisciplinary project are knowledgeable about the explanatory standards of their own field, but largely ignorant of the explanatory standards of others. The expertise that informs coherence at the bridge's near-side is unlikely to be available for the far-side. It is too much to require that M1's extension be assessed on its far-side by the standards of S2.

Instead, what is wanted is a less demanding condition: coherence for non-experts. That is, the bridge extending from one specialization's model should make contact with the other's model, without invoking any specialized standards. That rules out obvious incompatibility between one specialization's bridging element and the other's model (C1 and M2, C2 and M1), where 'obvious' means something like 'accessible to a non-expert.' The far-side connection will be vague and hand-wavy. It may even contradict something in the other specialization. For example, molecular geneticists may be committed to the idea that molecular mechanisms within a cell, environmental influences on an organism's development, etc., are less important causes of traits than DNA sequence. Cell and developmental biologists might reject that, lumping DNA sequence with other causal factors. That's fine! Each explanatory model is extended or altered, so as to cohere with the near-side specialization is a nuanced and sophisticated way, and with the far-side in a very blunt, commonsense way. In other words, from the perspective of each specialization, bridge C makes an intelligible connection with the other's model. That is, A can intelligibly (i.e., without contradiction) describe the connection between C1 (the extension of her own model) and M2, and similarly for B, C2, and M1. But those descriptions needn't agree or align. From the side of molecular genetics, the molecular biology account builds on their own explanatory model, tracing effects further 'downstream' to encompass more intra-cellular processes and (ideally) effects on organismal phenotype. From the molecular biology side, the genetic explanation highlights one component (DNA sequence) and its causal sequelae in their more inclusive model. All that's required is that the relation between C1 and M2 is intelligible from the perspective of M1, and the relation between C2 and M1 is intelligible from the perspective of M2. Together with near-side coherence, this requirement ensures that the bridge is seen as anchored on both sides, from both sides. In this sense, the bridge is balanced, a mutual relation between models contributed by different specializations.[13] New elements comprising the bridge, spanning near and far sides, are complements in the sense of linking models in opposite directions (M1 to M2 via C1; M2 to M1 via C2). Each is rooted in a distinct specialization, with which it coheres on near- and far-sides according to different standards (specialized and not).

[13] This is the counterpart of A and B having *equal standing* with respect to the shared goal of connecting their respective models.

Middle

The tripartite link between A and B constrains the two sides of the bridge to 'meet in the middle.' But as just discussed, reciprocal extensions will rarely if ever coincide. Instead, the analogy with collective intention suggests a third coherence requirement: the bridge from one side (C1) should not undermine the reciprocal connection in the other direction (C2), and vice versa. This ensures that the bridge allows traffic in both directions. In other words, efforts at connection from opposite sides must be mutually realizable, as parts of a more inclusive project. The different resources used by A and B to construct the bridge between their respective models must not interfere with one another. This corresponds, albeit roughly, to the third-layer of the interactive structure: each participant sees their own activity through other's eyes, so to speak. Such perspective-shifting isn't practically possible for most interdisciplinary projects. So this coherence requirement too is very weak. It rules out bridging elements premised on intolerance or explanatory imperialism: ideas about the other specialization that undercut the bridge coming from that direction.[14]

Summary

I've articulated three norms of coherence for bridging explanatory models in late-stage low-degree IDE. First, the bridge must attach to each contributed model, via some new or adjusted element. This corresponds to the first layer of the interactive structure, which relates each agent to the shared goal. Second, the bridge must attach to the other model; the new element of one model must lead to the other. This corresponds (albeit roughly) to the second layer of the interactive structure, which relates each agent to the other's participatory attitude. Finally, both models' new elements must allow for 'two-way traffic' across the bridge. This corresponds (also roughly) to the third layer of the interactive structure, which closes the loop of agents' attitudes and establishes them as equal participants in the activity. An IDE satisfying these minimal conditions coherently connects two different specialized explanatory models. The connection itself may differ, depending the specialization from which it is envisaged—just as the contributed models may differ. All that's required is that neither bridging relation undermines the other, while each is an acceptable extension of the appropriate contributed model and does not present any obvious incompatibility with the other. Putting all this together yields minimal necessary conditions for connecting distinct explanatory models into an IDE:

[14] See Chapters 7 and 8 for examples and further discussion.

(IDE) A model is an interdisciplinary explanation produced by specializations S1 and S2 only if (i) model M1 contributed by A is an adequate explanation according to the standards of S1; (ii) model M2 contributed by B is an adequate explanation according to the standards of S2; and (iii) bridge C comprised of C_1 and C_2 satisfies the following coherence requirements:

(a) C_1 does not include any anomalies according to S1; similarly for C_2 and S2;
(b) C_1 does not include any explananda for S1; similarly for C_2 and S2;
(c) C_1 does not obviously conflict with or contradict M2; similarly for C_2 and M1;
(d) C_1 does not undermine C_2, and C_2 does not undermine C_1.

Importantly, no element of a collaborative explanation must conform to standards of all the involved specializations at once—not even the bridge. No "neutral" stance outside scientific specializations from which to assess the full IDE is presumed.

A bridge C satisfying (a)–(d) above affords a path from each contributing model to the other, which all participants in an interdisciplinary project can traverse if they choose. This is not the same as requiring that each contributed model be acceptable to all involved perspectives, although such cases are not ruled out. There is no requirement for an overall unified model, except in the minimal sense that models of multiple perspectives are connected. The bridge must conform to standards of all specializations thereby connected—but not in all its parts at once, and not necessarily in the same way depending on the 'direction' of crossing. This is a much weaker requirement than traditional robustness. Rather, the requirement concerns the relation (direct or indirect, hinging on similarity or difference) that connects models from different specializations. Coherent bridges linking different kinds of explanatory model are common in some areas of biology, though few traverse wide disciplinary boundaries. Next, I consider an extended example of this sort: explanations of hemoglobin and related phenomena.

5.3 Example: array of hemoglobin models

Many IDR projects involve more than two different specializations. Pairwise analysis applies in these more complex cases, too. If there are, say, four specializations involved in a project and all contribute an explanatory model, then (exceptional cases aside) no single bridging relation connects each to the others. Indeed, not all contributed models need connect to all the others. The nature of pairwise connections will differ, according to the particulars of each field and their explanatory modeling practices. Bridging relations in these more complex cases form a connected graph, or network.[15] Network structure, as well as the specifics of bridging relations, vary across particular IDE projects. As with other features of IDE and EP models more generally, the details need to be worked out by studies of science in practice—particular interdisciplinary projects. To illustrate this idea, an example will be helpful. I'll consider explanations of hemoglobin's role in (human) health and disease. Scientific understanding of hemoglobin takes the form of an interrelated array of models. Different areas of life science converge on that target, coordinated with one another in rather straightforward ways.[16] I'll begin with a trio of explanatory models of hemoglobin from three distinct fields: biochemistry, animal physiology, and hematology (the study of blood cells). I'll introduce each specialization's explanatory model in isolation, and then describe pairwise bridge relations among them. Next, I'll widen the treatment to include more scientific and medical fields, although more briefly. For ease of exposition, the specialized models described here will be somewhat simplified; "textbook science" instead of cutting-edge research.[17]

Biochemistry

The biochemical model of hemoglobin represents that molecule's structure and shows how that structure contributes to hemoglobin's role in living organisms. The structural model depicts chemical bonds between atoms, larger-scale molecular shape, and changes to the former mediated by chemical interactions.[18] Hemoglobin in vertebrates is a protein composed of four polypeptide chains (linear sequences of amino acids encoded by genomic DNA). Each polypeptide chain is wound around another molecular structure: the heme group. The heme group is composed of four rings of protoporphyrin and an iron atom. Protoporphyrin is an arrangement of carbon, nitrogen, and hydrogen atoms that resembles a snowflake, with side-chains dangling from the corners. The iron atom slots into the center of the protoporphyrin rings, a configuration that allows it to form additional bonds on either side of the 'snowflake' plane (Figure 5.2). So each polypeptide chain is a carriage for the oxygen-binding heme group, by virtue of its molecular shape. The iron atom can take one of two oxidation states (+2 or +3). The +2 state can bind oxygen atoms, the +3 state cannot. Glossing over the stereochemical details, the

[15] Not a directed graph, because bridging relations are mutual, bidirectional.
[16] Bridging connections in IDR projects are rarely as simple and clear-cut (see Part III).
[17] See Fleck ([1935] 1979).
[18] See Tahko (2020) for more on chemical/structural and biological/functional perspectives on proteins such as hemoglobin. This section is based primarily on classic textbook treatments: Stryer (1988), Hazelwood (2001), Steinberg et al. (2009).

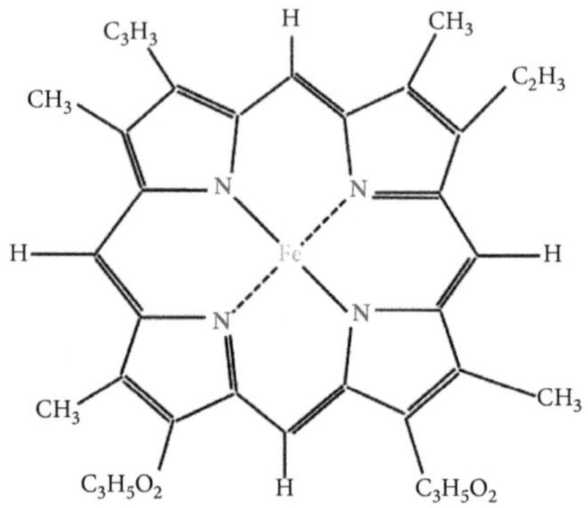

Figure 5.2 Biochemical model of the heme group

hemoglobin protein makes an environment for the heme group that allows reversible oxygen-binding. Hemoglobin is an allosteric protein; i.e., its different chains interact with one another at specific contact locations, resulting in conformational change of the whole molecule. So the shape of hemoglobin is determined not only by the amino acid sequence of its four constituent chains, but also by interactions between those chains. Those interactions, in turn, are regulated by the binding of small organic molecules that exist in varying concentrations at different locations within the vertebrate body.

The four polypeptide chains of vertebrate hemoglobin are of two types, termed ∝ and β. Two chains of each make up the most common form of the molecule (in human adults). The two pairs of polypeptide chains are packed together in a roughly spherical shape, with ∝ and β chains alternating and heme groups nestled in clefts near the exterior. Hemoglobin's allosteric interactions confer several properties, the most biologically significant of which are cooperative binding and dissociation of oxygen to heme and variable affinity for oxygen dependent on features of the chemical environment.[19] The chemical reactions and structural changes that affect oxygen binding rates of hemoglobin are well understood; I'll pass over those details. The key point is that hemoglobin's sensitivity to environmental conditions vis-à-vis oxygen binding and dissociation make it an effective transporter of oxygen within a vertebrate organism. This brings us to the model of hemoglobin in animal physiology.

[19] Cooperative binding means that one heme group facilitates binding to the other heme groups of the same molecule; similarly for dissociation. Affinity is a measure of binding strength. The affinity of hemoglobin for oxygen is mediated by pH and the presence/absence of organic phosphates.

Animal physiology

The explanatory model of hemoglobin-mediated oxygen transport is a major contribution to animal physiology.[20] Gas exchange in animal respiration involves transport of oxygen from lungs to tissues, with carbon dioxide moving in the opposite direction. The physiological explanation hinges on the fact that oxygen binding is enhanced under conditions in the lung, while dissociation is facilitated in the capillaries (the furthest reaches of the vertebrate circulatory system into the tissues). The model presumes a bounded system: oxygen (O_2) enters the lungs, is taken up by the blood, transported to tissues where it is off-loaded, with carbon dioxide (CO_2) returning to the lungs to be exhaled. The cyclical structure is typical of physiological explanations, which often presuppose the principle of homeostasis. That principle states, roughly, that an organism's body maintains a relatively constant state in a changing (often hostile) environment. So physiological explanations typically describe the return to a starting or set point after some change or disruption. For simplicity, only oxygen's journey is presented here; models for CO_2 exhibit the same features (and complete the cyclical form).[21] The model applies general gas laws to the particular conditions of O_2 in lungs and tissues. The gas laws, named after their discoverers (e.g., Boyle, Charles, Dalton) describe exact relations between quantitative properties of gases, such as temperature, pressure, and volume. Physiological conditions are represented abstractly, as assignments of value to the quantities figuring in the laws.

This model typically takes the form of a stylized spatial representation of O_2's journey, broken down into discrete steps. In the first step, O_2 enters the body in the lungs, diffusing across alveolar capillary membranes and dissolving into the blood. Oxygen uptake by blood is determined by the law of partial pressures (Dalton's law), which relates the partial pressure of a gas in a mixture of gases to its concentration and the mixture's total pressure.[22] Structural features of the lungs, alveoli, and capillaries determine the values of these quantities, such that the pressure gradient (from high to low) induces O_2 diffusion into the blood. The pressure gradient has this effect because the system operates so as to return to equilibrium. For O_2 uptake by blood, equilibration is rapid; by the time blood leaves the lungs via arteries en route to tissues, the pressure gradient is approximately balanced.

[20] Most generally, physiology aims to explain integration and control of the internal systems necessary for an organism to continue living. A number of different specializations fall under this broad banner, distinguished by focus on particular types of organisms (e.g., human, plant); organizational levels (e.g., cell, molecular); or systems within an organism (e.g., nervous, respiratory, digestive). To my knowledge, there are no philosophical accounts of explanatory models in physiology.

[21] For more detailed textbook treatments, see (e.g.) Eckert et al. (1988), Thibodeau and Patton (2007), and Hill et al. (2016).

[22] The partial pressure of a gas (O_2) in a liquid (blood) is in turn determined by the partial pressure of that gas in the environment of that liquid. Also essential is Fick's law, which specifies the quantities relevant for diffusion of O_2 across the capillary membrane.

Oxygen uptake is greatly enhanced by hemoglobin, which binds O_2 and so acts as a 'sink' for O_2. Hemoglobin lowers O_2's partial pressure in blood, thereby facilitating more uptake into the liquid. Once bound to hemoglobin, O_2 is carried by blood circulation to the tissue capillaries, where the pressure gradient is reversed.

The core of the explanation is the equilibrium equation for the reaction combining hemoglobin and O_2. The higher O_2's partial pressure, the more hemoglobin-O_2 is formed, and as that pressure decreases, the more O_2 and hemoglobin dissociate. This relationship is expressed as an oxygen–hemoglobin dissociation curve, representing the percentage of hemoglobin molecules combined with O_2 (saturation) as a function of partial pressure in liquid blood plasma (Figure 5.3). The partial pressure in the lungs is such that hemoglobin binds O_2, while this quantity in the tissue capillaries is such that O_2 and hemoglobin dissociate so as to maintain equilibrium. The result is delivery of O_2 to tissues, with a reciprocal process transporting CO_2 from tissues to the lungs to be exhaled. In describing several stages of changing spatial relations of components in a complex overall system (the respiring animal body), the physiological model resembles the kind of mechanistic explanation characteristic of molecular biology (a near neighbor of biochemistry). But, unlike explanations in those fields, general gas laws play an essential role in the physiological model: simple formulae and constants describing relations among quantitative properties of a gas; pressure, temperature, volume, concentration, and solubility. Gas laws are essential to the explanation; necessary for "true understanding of respiratory function" (Thibodeau and Patton 2007, 886).[23] It is also, as noted, an equilibrium model. The explanatory core is the effect of a pressure gradient on the rates of binding/dissociation reactions between O_2 and hemoglobin. Reactions between the latter two occur so as to move the system toward equilibrium; that's a fundamental assumption of the model. The behavior of O_2 in the lungs is governed by a pressure equilibrium; gas molecules in the environment "dissolve in the liquid until the partial pressure of the gas in solution becomes equal to its partial pressure in the environment of the liquid" (Thibodeau and Patton 2007, 899). The rate equations describing behaviors of gases under various conditions are premised on that equilibrium assumption. The full physiological model includes other relations between quantities assigned to parts of the respiring organism and its environment (e.g., respiratory membrane surface area, volume of air inspired). Dependence relations among these quantities enable prediction of O_2 uptake/dissociation under many different circumstances.

[23] The full physiological model includes other relations between quantities assigned to parts of the respiring organism and its environment. These are presented as dependence relations, which enable prediction of O_2 uptake/dissociation under many different circumstances. For example, the amount of O_2 that diffuses into liquid blood plasma depends primarily on four factors: the O_2 pressure gradient (described above), the surface area of the respiratory membrane, the rate of respiration and volume of air inspired, and the amount of air inspired that reaches lung alveoli and capillaries. Changes in any of these quantities make a difference to the diffusion of O_2 into blood plasma. So the full model embeds the central relations governed by gas laws within a broader set of quantitative difference-making relations.

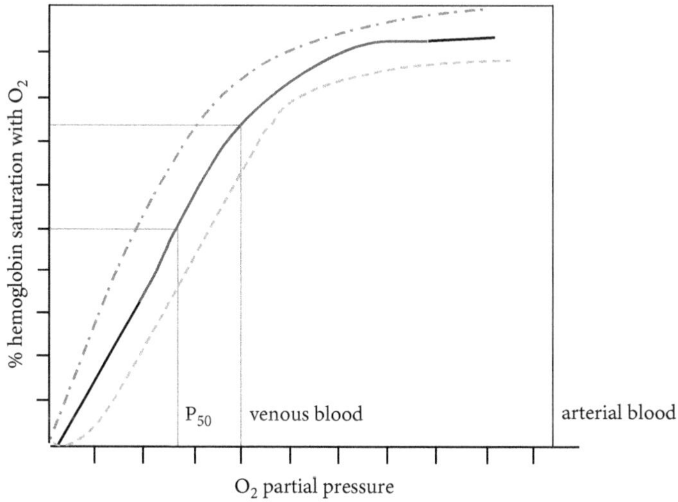

Figure 5.3 Oxygen–hemoglobin dissociation curve diagram, central to physiological explanation of gas exchange mediated by hemoglobin.

Coherence: biochemistry and animal physiology

Biochemical and physiological models are distinct from one another. The biochemical model represents hemoglobin's structure in terms of animo acid sequence, polypeptide shape, allosteric interactions, and free atoms in the chemical environment. The model includes chemical bonds between atoms, the structure of large molecules, and changes to that structure mediated by chemical interactions. Hemoglobin's molecular structure explains its oxygen-carrying (and unloading) function. In contrast, the physiological model represents the equilibrium-based cycle of O_2 transport from lungs to tissues in terms of gas laws. Hemoglobin is a factor influencing values in rate equations governing O_2 uptake and dissociation. Their many differences notwithstanding, the bridging relation between these models is simple on both sides. Consider first the bridge from biochemistry to physiology. For biochemists, hemoglobin is a protein made in animal cells. Those cells are paradigmatically parts of an animal body. An animal's body requires functioning physiological systems. That's enough to link biochemical models of hemoglobin to physiological models of same. In the model–model relation taxonomy, this relation is one of 'indirect difference.' It is indirect because something is added to link the two models: the idea that the hemoglobin molecule (in its differing functional forms) is part of a larger biological context. This is a part-whole relation, highlighting differences between the two models rather than similarity. Near-side coherence is satisfied, because conceiving hemoglobin as part of a more inclusive biological whole is neither anomalous nor an explanandum for biochemistry. It is not anomalous, because part-whole relations make up a standard view of biological levels of organization: molecules, cells, tissues and organs, organisms, populations, ecosystems.[24]

As an aside: most biologists encounter some version of this 'background "Big Picture" of the biotic world very early in their scientific education. This widely shared background assumption facilitates bridging connections across many different life science communities. Indeed, that may be its most important role in scientific practice. Philosophers of science have cogently criticized the concept of "levels in biology" for its unwieldy metaphysical baggage. As a biotic worldview or metaphysical foundation for life science, the part-whole hierarchy is problematic. However, as shown here, it can play a different role, affording coherent bridge relations between explanatory models in different biological fields. That role has important explanatory value, across and among life sciences.

To return to coherence requirements for bridging models: conceiving biochemical hemoglobin as part of a more inclusive biological whole doesn't present biochemists with an immediate explanandum. Part-whole relations at atomic and molecular scales are explananda for biochemists. But whole organisms and their physiological systems are at another scale entirely; not an explanatory concern for biochemists. Physiological processes are outside the scope of biochemical modeling. (For evolutionary theorists interested in major transitions in evolution, the part-whole hierarchy is an explanandum—but that field is quite distinct from biochemistry, and doesn't bear on its bridging relation with animal physiology.)

The far-side connection to animal physiology identifies the larger whole as a model of an animal's respiratory system—very roughly and loosely conceived. Hemoglobin-mediated oxygen transport is a key part of that system, and there's no obvious incompatibility in 'scaling up' from biochemical models to that larger physiological context. So far-side coherence is satisfied. The bridge's far-side embeds hemoglobin's biochemical structure/function in the wider context of an organism's oxygen transport mechanism. From the perspective of biochemistry, physiological models are a wider containing context of their own explanatory model, all the better for being based thereon. Biochemists may privilege their own explanatory models as more fundamental than models at higher levels. This isn't a problem for coherence. All that's required is that biochemists recognize physiologists as making an explanatory contribution connected to their own.

[24] There are different versions of this hierarchy in life science pedagogy, but the basic idea is fairly consistent.

The bridge from physiology to biochemistry is also simple. Hemoglobin is a direct link; an element common to both models, although represented differently in each. This is a direct similarity relation between models (with more subtle differences). The physiological explanation represents hemoglobin as a term in the key binding/dissociation equation, an essential difference-making factor in O_2 transport from lungs to tissues. The only adjustment needed to connect this to biochemical structure/function is to allow for intracellular factors that make a difference to the role of hemoglobin in gas exchange. The physiological model does not itself include any such factors, but recognition of hemoglobin as a protein implies that it has some molecular structure. Intracellular mechanisms can affect hemoglobin structure and thus its role in O_2 transport. Allowing for intracellular factors to affect the role of hemoglobin is neither anomalous nor an explanandum for physiology. It is not anomalous, because the physiological model includes difference-making relations among properties of a gas, including binding and dissociation with hemoglobin. The bridge to biochemistry does not alter that equation or the fundamental gas laws.

Rather, the new element is to allow factors outside the physiological system (i.e., cells and their internal components) to affect the concentration of hemoglobin in blood. This is just one of the variable quantities in the model, which makes a difference to the effective uptake of O_2 by hemoglobin. The added assumption is not an explanandum, as cells and their components are not targets of physiological explanation. The extension from respiratory physiology is through this narrowly circumscribed part of the model of O_2 transport. From the perspective of physiology, the connection is consistent with biochemistry, although none of hemoglobin's structural details are included. So the linking bridge assumptions above are coherent on near- and far-sides.

Finally, the bridge is stable in both directions. From the biochemical side, it scales up from molecular structure to a more inclusive physiological system, while from the physiological side it allows for variation in the role of hemoglobin in rate equations of the O_2 transport system. The two alterations each gesture toward the other's explanatory domain. They are different—a part-whole relation and an expanded causal picture—but don't undermine one another. If either community thought of part-whole relations as inconsistent with causal relations—i.e., causal processes don't cross scales of biological organization—then middle coherence wouldn't be satisfied. (To my knowledge, life scientists make no such claims, though some philosophers have.) With these straightforward, coherent bridging relations, the biochemical model of hemoglobin slots smoothly into physiological accounts of respiration like the heme group nestles within hemoglobin protein. For more disparate fields, without established explanatory 'scaffolding' to link explanatory models, bridging models is more of a challenge. Coherence requirements apply in both kinds of case.

Hematology (blood cell development)

Another field that further extends the hemoglobin explanation is hematology: the specialized study of blood. Not all animal cells make hemoglobin protein; only those in the red blood cell lineage. Iron atoms bound to the heme group give red blood cells their eponymous color. Red blood cells have very short lifespans, and are generated continuously via cell developmental pathways beginning with stem cells in bone marrow. Development, activities, and turnover of red blood cells are well understood in hematology, itself a diverse field with ties to pathology, immunology, developmental biology, cell physiology, and more. Explanatory models of red blood cell structure and function from hematology merge smoothly with biochemical models of hemoglobin, and models of vertebrate physiology with both. I'll focus here on just one kind of hematology model: lineage trees of blood cell development. (These models are more distinct from biochemistry and physiology models than others common in hematology.) Lineage tree models are important explanatory achievements for communities investigating cell development.[25] A lineage is a complex entity (or process) made up of ancestor–descendant relations between lower-level entities (Neto 2019). In hematology, the lineages of interest are of cells descended from hematopoietic (literally, "blood-making") stem cells. In the body of a multicellular organism, cells reproduce by binary division (mitosis): a parent cell divides to produce two offspring cells. Generations of cells linked by reproductive division form a lineage. Hematology models also depict differentiation; i.e., change in the properties (or state) of a cell such that the cell moves forward on some developmental pathway, toward a mature, fully developed state. Red blood cells are generated as one branch of the blood and immune cell lineage tree (Figure 5.4).

The model represents the process of cell development as initiated by a stem cell, which gives rise to a lineage tree with a particular structure, characterized by depth, breadth, and an arrangement of branch-points. Depth is the number of stages a cell passes through in its process of development; a convention based on available techniques for distinguishing between cell types. Breadth reflects a stem cell's developmental potential: the number of 'final' cell types it can produce. Some arrangement of branch-points links the stem to developmental termini. Working out the exact structure of blood cell lineage trees is a major task in hematology. The red blood cell branch is fairly simple, but fits into a more complex lineage model. This model represents the cellular scale of development. That is, cells are

[25] The phenomenon of biological development is, most broadly, directed changes of an organism and its parts, from the origin point of a single cell, to reproductive maturity (and perhaps beyond). However, there is no widely accepted general framework for conceptualizing the developmental process. Instead, scientists and philosophers alike propose a number of different schemas for understanding development (see Fagan and Maienschein 2022). Lineage tree models are also central in phylogenetics, although there is very little contact between these and studies of cell development.

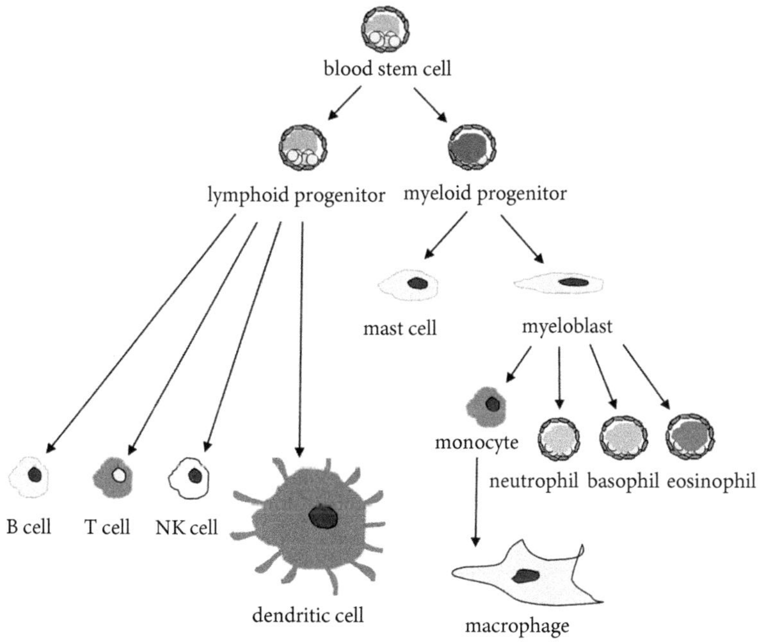

Figure 5.4 Blood cell lineage tree (partial), an explanatory model from hematology (study of blood cells)

the basic entities represented, with their reproductive abilities and the stages of differentiation comprising a lineage hierarchy. The model is quite abstract, covering and grounded by a wide variety of experiments and results. Rather than unifying through laws, this model identifies three variables specified by hundreds of experiments. The main norm involved in constructing the lineage model is accurate representation of a wide range of experimental methods and results. In this the developmental model resembles the mechanistic models of molecular biology. But in other respects the lineage model is unlike the latter. Although both cell reproduction and differentiation are causal processes, 'thick description' of causal relations is not emphasized in the lineage model. Instead, this developmental model is topological, displaying systematic relations among diverse cell types in development. The context, or environment, of cell development is omitted entirely—abstracted away. In practice, blood cell development is understood as contextualized within an organism's body, but the lineage tree deemphasizes that context.

Coherence: biochemistry and blood cell development

Cell lineage models' abstraction is such that hemoglobin molecules don't appear directly. So there's no direct similarity between these two models. From the biochemical side, again the background framework of biological parts and wholes comes into play. Hemoglobin protein is made by red blood cells; the part-whole relation is between cells and molecules. Coherence relations are much as described above. But there's another link to cell lineage models also. Cell development is a causal process antecedent to hemoglobin's biochemical structure and function. So there's a dual connection between these models from the biochemical side; two additional elements of the hemoglobin model. First, the cellular scale of biological organization is an inclusive whole of which hemoglobin protein is a part. Second, cell development is a causal process that gives rise to cells that synthesize hemoglobin protein. Neither presents an explanatory anomaly or open question for biochemists studying hemoglobin, and both make coherent links to studies of blood cells and their developmental pathways. From the hematology side, hemoglobin is a protein synthesized by cells in the red blood cell lineage (at end-stages of that process). The near-side adjustment is the uncontroversial idea that cells at those stages make proteins that give those cells their distinctive traits and abilities—specifically, oxygen transport in blood. (That assumption is central for hematology, though not included in the cell lineage model per se.) Absent any molecular or biochemical details, this idea makes for a straightforward connection between the blood cell lineage model and biochemical models of hemoglobin protein. The bridging connections from either side are coherent, dovetailing with respect to part-whole, temporal order, and structure/function relations. Models of red blood cell physiology segue smoothly into biochemical ones. Notably, the three-scale hierarchy of molecules, cells, and tissues serves as an indirect difference relation for models in both directions.

Coherence: animal physiology and hematology

These two fields' respective models also share very little. Both are abstract, but in very different ways. The physiological model applies general gas laws in rate equations, while the cell lineage model indicates the structure of developmental pathways and relations between blood cell types. One point of overlap between them is the concept of the organism. This is unlike the situation of hemoglobin, a central element of both biochemical and physiological models. 'The organism' is a crucial part of the background for both physiological and developmental models, conceptualized somewhat differently in each. Its role is as a bounding concept. Physiology conceptualizes the organism as a working system constrained by homeostasis principles, cell development as the background environment of cell lineages. To bridge the gap between these models, interpretive changes to the organism concept are needed on both sides. Consider first the bridge extending from the physiological model of gas exchange (O_2 transport). To connect with the model of cell lineage development, the most straightforward addition is to conceptualize the organism as the product of developmental processes. That is, the overall physiological system (explained in terms of applied gas laws) is not a static, timeless entity, but results from developmental processes. More specifically, the anatomical context in which an animal's respiratory cycle takes place—lungs, blood vessels, tissues, circulating blood—can be conceptualized as a product of developmental processes. Those processes needn't be specified in terms of cells or lineages; just the idea of a causally antecedent developmental process is enough. This is not anomalous, as causal processes figure in physiological models also; the extension is just to allow for a causal origin of the system as it is understood in physiology. Nor does this extension presume an explanandum for physiology. The idea of development is left unspecified, with only the general idea of a complex causal process as the origin. There's no obvious conflict here with cell lineage models.

In the cell lineage model of development, the concept of the organism has a similar bounding role. The organism is coextensive with a single cell at the very begin ning of its developmental process. More abstractly, the organism is the 'upper limit' and boundary for developmental cell pathways. To connect with the physiological model, another aspect is added to this concept of the organism: that the organism is an actively maintained living system. As with its counterpart, this is a minimal assumption. The idea is just that life-sustaining processes are necessary for the organism throughout the process of development. This extension of the organism concept in the cell lineage model is neither anomalous nor an explanandum—although it is peripheral to the explanatory aims of that model. The bridge between specialized models is thus characterized in two directions, each an extension of one model that makes coherent contact with the other. Finally, neither extension undermines the other. On the one side, it is assumed that the physiological system has a causal origin in developmental processes. On the other, it is assumed that the developing organism is maintained by ongoing life-sustaining processes. In both specialized models, the concept of the organism plays a bounding rather than central role, making these adjustments less fraught than those engaging core concepts of either model.

"The first molecular disease"

So far, I've illustrated the IDE account of bridging relations with three models involving hemoglobin: biochemical structure/function, O_2 transport in animal physiology, and pathways of blood cell development. Each model connects with the others via a bridge that satisfies minimal coherence requirements. But the interrelated array of hemoglobin models spans more fields than those three. This is especially noticeable when sickle cell hemoglobin comes into focus as a target phenomenon. Relations between models from different areas of life science bearing on this variety of hemoglobin cast a wide net. Stryer's classic introduction to proteins indicates this range:

> The transition from anaerobic to aerobic life was a major step in evolution because it uncovered a rich reservoir of energy. Eighteen times as much energy is extracted from glucose in the presence of oxygen as in its absence. Vertebrates have evolved two principal mechanisms for supplying their cells with a continuous and adequate flow of oxygen. The first is a circulatory system that actively delivers oxygen to cells. The second is the use of oxygen-carrying molecules to overcome the limitation imposed by the low solubility of oxygen in water. The oxygen carriers in vertebrates are the proteins hemoglobin and myoglobin. Hemoglobin, which is contained in red blood cells, serves as the oxygen carrier in blood and also plays a vital role in the transport of carbon dioxide and hydrogen ion. Myoglobin, which is located in muscle, serves as a reserve supply of oxygen and facilitates the movement of oxygen within muscle.
>
> (Stryer 1988, 143)[26]

These few short sentences encompass major transitions in evolution, organic chemical reactions and energetics, organismal physiology (the latter involving cells, tissues, and organ systems), and the interactions of molecules large and small. The breadth of topics and scales is not just an introductory flourish. Hemoglobin and myoglobin are among the best-understood proteins. For hemoglobin, in particular, our understanding extends from molecular structure to evolution and ecosystems, spanning many fields of life science.[27] Explanations with this reach and inclusive extent are aspirational for IDE. I'll sketch the models and relations of this extended array more briefly.

Stryer's is a work of biochemistry, and that field is a good starting-point to articulate the wider network of interrelated models. Sickle cell hemoglobin is a mutant form of hemoglobin protein with one amino acid substitution in the β chain (glutamate to valine). That single substitution changes the chemical properties of hemoglobin, creating a surface 'patch' that can bind to other, similarly valenced hemoglobin molecules. In the de-oxygenated state, mutant hemoglobin molecules stick to one another to form complexes: long helical fibers that wind around one another into stable braids. These hemoglobin braids can precipitate out of solution within red blood cells, creating long fibrous filaments. Red blood cells packed with these fibrous filaments become distorted in shape. That is the eponymous diagnostic feature of sickle cell anemia: a "large number of thin, elongated, sickle-shaped and crescent-shaped forms"—'sickled' red blood cells, unlike the normal

[26] Stryer's *Biochemistry* is a classic textbook, in its ninth edition at the time of writing. The third edition's (1988) section on proteins begins with a full chapter on myoglobin and hemoglobin, the opening paragraph of which is reproduced here.

[27] For more detail, see Hazelwood (2001), Steinberg et al. (2009) and, of course, Stryer (1988).

round corpuscular form (Herrick 1910, 518). Sickled red blood cells do not function as well as their smoothly rounded counterparts in the circulatory system. The deformed cells have shorter life spans and, worse, can become stuck in narrow blood vessels far into the tissues. Because these tissues have lower oxygen concentrations, this leads to a vicious cycle of more sickling, blockage of blood vessels, and eventual organ damage. The proportion of sickled cells fluctuates in a patient, and symptoms at the organism-level are complicated and diverse. This is partly because sickling of mutant hemoglobin in the body depends on the kinetics of blood circulation between oxygen-poor and oxygen-rich environments. If cells shuttle quickly back into higher oxygen concentrations, the ratchet of fiber-formation, precipitation and sickling is deferred.

This brief sketch of sickle cell hemoglobin and its impact on organisms draws on explanatory achievements of a number of scientific fields: biochemistry, hematology, pathology, clinical medicine, animal physiology. The biochemical structure of sickle cell hemoglobin is modeled much as for the normal molecule, highlighting contrasts between the two. A single amino acid difference leads to different chemical properties for the molecule. These are in turn related to hemoglobin's oxygen transport function, as oxygen-binding prevents sickle cell proteins from braiding together in a red blood cell. Hematology models show how hemoglobin braids affect red blood cell shape and function, in the context of animal respiration. Sickled red blood cells can become stuck in narrow blood vessels within the tissues—precisely where O_2 concentrations are lower, initiating a vicious cycle of more sickling, blockage of blood vessels, and a suite of organism-level symptoms including severe anemia (low red blood cell count). These two models link smoothly together via part-whole and causal relations, as described above.[28] The explanation of sickling tracks structure/function effects through a series of increasingly inclusive contexts: from a single molecule to collections of molecules, to molecules in an intracellular environment, to a whole red blood cell, to red blood cells in an organism's circulatory system. Physiological models of animal respiration explain the normal homeostatic process, adding to overall understanding of sickling's impact on an animal. These models in turn connect to medical models of sickle cell anemia. Explanations in clinical medicine target the disease, identifying patterns of symptoms in widely variable patient responses. The explanation shows how severe anemia and organ damage, as well as more mild symptoms, can result from physiological processes. In this way, the clinical explanation itself links physiological and medical models. It is also geared toward treatments in ways that basic science explanations are not. Another medical field—pathology—is primarily responsible for diagnosing sickled cells via microscopy and other tests. Pathology's

[28] Cell lineage models are more distinct from biochemistry than hematological models of red blood cell structure and function. So links between sickle cell hemoglobin models are even more straightforward: part-whole framing in both directions.

visual cellular model of sickled red blood cells further extends the array, and makes for another tie between medical and bench science. Each field contributes a different model of sickle cell hemoglobin, which connects with one or more others. The result is an interconnected (but not maximally connected) network of models (Figure 5.5).[29]

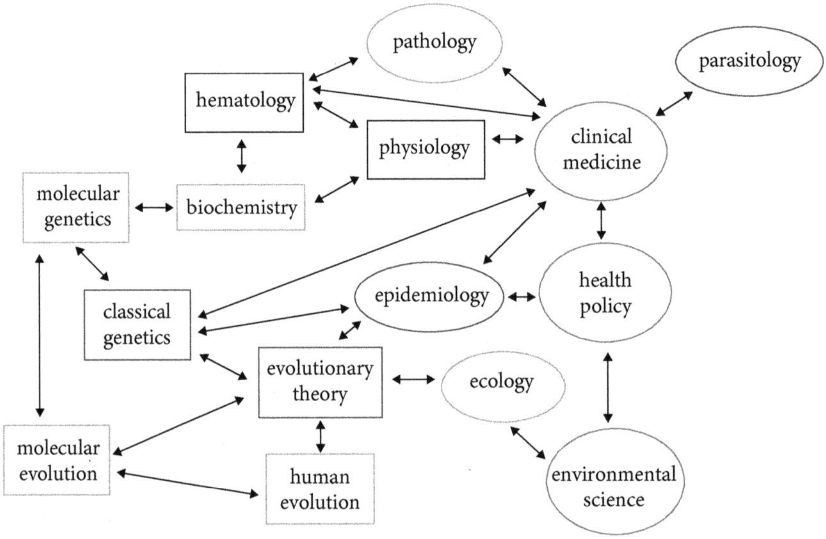

Figure 5.5 Array of explanatory hemoglobin models, variously linked by bridging connections. This array, although incomplete, spans many areas of life science.

The array of models dealing with sickle cell hemoglobin goes beyond those linked by part-whole relations or shared focus on the organism. Notably, several sub-fields of genetics contribute to the extended biomedical understanding of sickle cell anemia. Indeed, explaining sickle cell anemia as a "molecular disease" is arguably the apotheosis of molecular genetics. Sickle cell hemoglobin, recall, has one animo acid difference from normal vertebrate hemoglobin: valine replaces glutamate at one spot in the β chain. That contrast is central to the biochemical model. Molecular genetics highlights protein coding via DNA sequence within a cell. The animo acid substitution is encoded by a single nucleotide base substitution (thymine to adenine; a point mutation) in the genomic DNA sequence identified as the vertebrate β chain gene. On these grounds, sickle cell anemia was termed "a molecular disease," vindicating the causal chain associated with molecular biology's Central Dogma (Pauling et al. 1949, Crick 1970).[30] But sickle-cell

[29] The array of stem cell-based embryo models (Chapter 2) prefigures this network form.

[30] Over ensuing decades, many other molecular (or genetic) diseases have been characterized, in myriad other proteins (and a number in hemoglobin itself; e.g., the thalassemias). Our current practices of genetic screening and diagnosis stem from these early studies of hemoglobin.

anemia is an exceptional case: a neatly linear causal chain of difference-making from DNA sequence to physiological outcomes. Most genetic causal pathways are more complex, involving many more genes with diverse impacts on phenotype. (Indeed, sickle cell anemia symptoms vary considerably across patients; gene-to-phenotype mapping in this case appears simple only if coarse-grained phenotypic categories are used.) As noted, explanation in molecular genetics appeals to DNA sequence as the primary causal determinant of organismal traits. Differences in the nucleotide base sequence of a DNA molecule localized to a chromosome produce differences in the linear nucleotide base sequence of an RNA molecule and, through the RNA intermediate, differences in the linear amino acid sequence in a protein molecule. Sickle cell hemoglobin fits this explanatory pattern perfectly.

Aspects of animal physiology and blood cell function are also implicated in the phenotype of sickle cell anemia. Bridging relations between models in molecular genetics and those of molecular biology, described above, help flesh out this more complex causal story (see Section 5.2). Models of intracellular molec ular mechanisms are used to explain a wide variety of biological phenomena, including aspects of health and disease. This style of explanation is paradigmatic of molecular biology. Some of its key features have spread to other fields, with associated uptake of molecular biology's concepts and experimental methods: i.e., "molecularization" of those fields. Models of hemoglobin and sickle cell anemia in other fields straightforwardly link into these molecular explanations via the idea that hemoglobin is a protein made in cells of the red blood cell lineage via the molecular mechanism of protein synthesis.[31] That model is a signal achievement of molecular biology; indeed it helped launch the field as a distinct specialization. The molecular genetics model is tightly integrated with molecular biology explanations. However, there is a key difference: molecular genetics singles out the DNA sequence mutation as the primary or sole cause of sickle cell anemia, while in molecular mechanisms causal power is more distributed. Primacy of genes as causal agents is what distinguishes genetic explanation from explanatory models in other life science fields. Bridging relations between molecular genetics and molecular biology finesse this contrast without eliding it (see Section 5.2).

Because the concept of a gene has changed considerably since the term's introduction, there are also diverse genetic models within the extended array.[32] Where the molecular genetic model traces the origin of sickle cell anemia in an organism to a single DNA sequence change, classical Mendelian genetics explains the disease's pattern of inheritance across generations. The latter begins with classification of organismal traits into sharply delineated alternative phenotypes: sickle

[31] The process of hemoglobin synthesis begins before red blood cells reach the end of their developmental pathway. It has to, because fully differentiated red blood cells have no nuclear DNA. Genomic changes are part of blood and immune cell development; jettisoning the genome is a drastic one.

[32] See Johannsen (1911) for the original definition; Keller (2001), Griffiths and Stotz (2007).

cell disease, sickle cell trait, and no sickle cell ("normal"). Each category is associated with a different combination of sickle cell or normal hemoglobin genes. Because human genomes have two copies of each chromosome, there are three possibilities: both sickle cell, both normal, and one of each (homozygous mutant, homozygous normal, and heterozygous, respectively). Across human generations, the sickle cell mutation exhibits classical Mendelian inheritance. That is, persons homozygous for the mutation developing chronic anemia and a cluster of related symptoms such as enlarged heart and kidney damage. (The physiological, biochemical, and molecular underpinnings of these symptoms are well understood via the tidily linked models discussed above.) All hemoglobin is mutated in homozygous carriers, and about 50 percent in heterozygous carriers. But the phenotype is largely recessive. About 50 percent of red blood cells are sickled in homozygotes, compared to only about 1 percent in heterozygotes, who are diagnosed as having "sickle cell trait." The latter are usually asymptomatic, although they experience impairment in some circumstances (e.g., exercise at high altitude).

The classical genetics model dovetails with that of molecular genetics, so smoothly that the bridging connection is all but invisible. In classical genetics, genes are assigned to an organism's chromosomes, inherited by the next generation according to Mendel's laws. These patterns are classically visualized using Punnett squares. Because sickle cell anemia is traced to a single DNA sequence change, there's a clear point of overlap in the two genetic models: organisms carry different combinations of DNA sequence at the β chain gene on their chromosomes. Part-whole-connected molecular, cellular, and physiological models account for an individual organism's phenotype, while Mendelian models extend this picture across generations. Proportions of normal vs. sickled hemoglobin in different organisms become a focus, related to the 'dosage' of genes on chromosomes. The concept of "gene" anchors the bridge between classical and molecular genetic models, on both sides.

There are other ties between genetic explanatory models as well. The difference-making explanatory pattern originated in classical genetics, which aims to explain patterns of inheritance in breeding experiments. Explanations in classical genetics describe how differences in inherited factors (genes in the classical sense, localized first to cells, then chromosomes, then to particular regions of a chromosome) produce differences in organismal traits. This basic structure, explanation of patterns of variation in a population of organisms, is retained by molecular genetics, although the entities that figure as cause and effect are not the same and inheritance across generations is not part of the basic model. These points have been made before by a number of philosophers of biology, who distinguish various gene concepts associated with different aspects of biological thought and experimental practice (see, e.g., Keller 2001, Griffiths and Stotz 2007, Waters 2007). However, the 'bridging models' account synthesizes this body of work in fruitful ways. Pluralism about the gene concept within biology makes sense when this diversity is contextualized, located in the ideas and commitments of different, though closely integrated, scientific communities with subtly different explanatory practices.

The classical genetics model of sickle cell hemoglobin is relatively simple, so its basic ideas (with some molecular gloss) are easily incorporated into other fields. One such is epidemiology, itself a complex and pluralistic community with ties to health policy as well as biomedicine. Epidemiological models of sickle cell disease characterize its incidence in geographic regions and human population subgroups.[33] Intriguingly, sickle cell anemia is more common than one would expect, for an inherited disease with serious health impacts. The epidemiology of sickle cell disease is well characterized. In the US, the incidence is about 4 per 1000 in African American populations. In parts of Africa, the frequency of the sickle cell gene is much higher, despite the homozygous form being lethal without prompt medical care. The epidemiological pattern motivates an explanation. Because the etiology of sickle cell anemia is traced to genetics (integrated classical and molecular models), that explanation is in terms of evolution by natural selection—the province of neo-Darwinian evolutionary theory. This interfield connection is via an indirect difference relation – but unlike others considered so far, the connecting relation is itself explanatory. Human genetics and epidemiology reveal the distribution of sickle cell genes in human populations (the explanandum), evolutionary biology explains that distribution (explanans). Evolutionary biology doesn't bear this same relation to explanatory models in other fields, though. The nature of sickle cell as a molecular disease is such that those causes fall within the scope of evolutionary explanation.

The neo-Darwinian explanation of sickle cell anemia is a model quite distinct from those discussed so far, invoking natural selection on populations over long time-scales. Briefly: there is very strong selection against homozygotes for the sickle cell gene, yet the gene persists in human populations. Hundreds of other mutant forms of hemoglobin have been characterized, but all are many orders of magnitude less frequent in humans than the sickle cell mutation. Why is this deleterious gene so common? The answer is that the sickle cell mutation is maintained in human populations via balanced polymorphism: a combination of positive and negative selection resulting in overall maintenance of the mutant gene in human populations. Crucial to this explanation is that the mutant gene confers some selective advantage—uncharacterized in the sickle cell models discussed so far. The advantage is as follows. Sickle cell trait, while having few physiological consequences, confers protection against the blood-borne, mosquito-conveyed parasite that causes malaria, *Plasmodium falciparum*.[34] So negative

[33] Epidemiological explanations of a disease usually seek to identify its causes, and to find ways to offset them or otherwise reduce disease incidence in human populations (see Chapter 7). In the sickle cell case, the primary cause is inheritance—and further causes are sought in evolutionary models.

[34] Ferreira et al. (2011).

selection against homozygotes is balanced by positive selection for heterozygotes in environments that include certain lethal strains of malaria. The sickle cell gene is subject to both positive and negative selection, the balance of which maintains the gene at a low but persistent frequency in populations at risk from malaria. Those populations are presently mainly in Africa, with relatives sharing their lines of descent also carrying the gene at higher rates. This neo-Darwinian model connects with explanations in a number of other fields: ecology, environmental policy, parasitology, tropical medicine, critical study of race in medicine, and more. Ecological models clarify relationships between the malaria parasite, human hosts, and environmental factors like the distribution of swamps and standing water. These in turn influence organismal traits and fit into models of evolutionary change. Environmental science and health policy models in turn link to these. The array of sickle cell hemoglobin models extends far beyond biochemical structure and function, to encompass studies of host–parasite interactions, complex ecological relationships involving humans, mosquitos, and wetlands habitats, and health policy questions about (for example) the use of racial categories in medicine.

Detailed accounts of all these connections are beyond the scope of this chapter. Here I'll indicate only some of the main features of the neo-Darwinian explanation of the distribution of the sickle cell gene in human populations. That model is an application of evolutionary theory, the eponymous (and continuously developing) explanatory achievement of evolutionary biology. The latter field has significant ties to theoretical population genetics, which in turn connects with classical genetics (as well as other fields, such as quantitative genetics). Neo-Darwinian evolutionary explanations use a variety of abstract, formal models to analyze variation in populations, genetic variation being the priority and focus. Such models represent the state of a population at a given time in terms of distribution of different types of genes; i.e., alternative alleles on chromosomes at one, two, three, or (rarely) more loci.[35] For sickle cell anemia, one locus suffices. The core of the model is a set of equations for calculating gene frequency in a population of organisms across generations, omitting nearly all features of organisms' reproductive and genetic systems. Solving the equations describes a sequence (evolution) of states of a population over time. Such models are highly idealized, assuming 'atomism' for fitness values assigned to genes; i.e., no interactions between genes, no effects on a gene's fitness via other genes. These assumptions cohere well with those of molecular and classical genetic models of the sickle cell gene and its inheritance. More realistic (less idealized, more loci with effects) population genetic models are less tractable mathematically and theoretically. So population genetics tends to trade in very simple (one-locus) models. Models in population genetics are highly idealized, abstract, and formal. Bridging relations to other models in

[35] See Lloyd (1994, 107).

the array are via the concept of the gene. 'The gene' in population genetic models is an abstraction that follows rules of inheritance in the population as modeled. Often, these rules incorporate aspects of organismal phenotype, via the parameter of mean fitness (fitness being a function of phenotype)—genotype and phenotype aspects of a population both combined in the calculation of gene frequency across generations. This is quite distinct from the molecular gene (DNA sequence) and more subtly different from chromosome-localized factors of classical genetics. In these evolutionary models, organisms appear as containers for different genes and bearers of fitness. There is no direct link to biochemical, cellular, or physiological models. But other evolutionary models do link more directly with that region of the model network.

Closely related to the biochemical model of hemoglobin is that of molecular evolution. As discussed, vertebrate hemoglobin is made up of four polypeptide chains. In adult humans, as noted, the most common form of the molecule has two α chains and two β chains. Embryonic and fetal hemoglobin, however, are made up of different kinds and combinations of polypeptide chains; there are also minor variant forms of the molecule in human adults. Comparison of amino acid sequence across developmental stages, species, and with myoglobin indicates a common evolutionary origin for all globin protein chains. Different forms of the protein, encoded by genomic DNA, are thought to have diverged from their common ancestor via gene duplication and selection in different cellular and physiological environments. The basic shape of the "globin fold" is conserved in all vertebrate myoglobin and hemoglobin, and so thought to be necessary for oxygen transport. This conformation is maintained by the size, shape, and charge of particular amino acids, also conserved (though the amino acid sequence overall is not). The molecular evolutionary model takes the form of a phylogenetic tree displaying inheritance and mutation (reproduction and divergence) relations among different genomic DNA sequences and associated protein structures.

5.4 Takeaways and significance

The array of models depicted in Figure 5.5 could be further extended and enriched with additional ties. But I'll stop here and indicate the main takeaways from this example. One reiterates a disclaimer. The hemoglobin example illustrates connections for the most part between models constructed in very closely allied fields. So there is more common ground for bridge-building than one would expect to find between more widely separated communities (e.g., life and computer sciences; psychology and applied mathematics). Shared background ideas such as levels of biological organization defined by part-whole relations, or diverging yet related gene concepts, or focus on an organism's health, present opportunities for direct or indirect relations between shared elements of different models. Many bridging connections are possible. Indeed, some readers might think of others than those presented here. If so, great! The exact network configuration of hemoglobin models across the life sciences is not the main takeaway from this example.

The key takeaway is that explanations of hemoglobin and its biological role form a network of models, linked by bridging relations that satisfy coherence requirements for IDE (sketched in Figure 5.5). The network's structure is dynamic, changing over time as scientific communities putter along. Because all the models considered are from life science fields, there are many possible connections for bridging. Models with disparate targets afford fewer opportunities for connection, but with sufficient ingenuity they can be found. Opportunities for IDE are extensive and various. It follows that bridging models is a relatively low-effort interdisciplinary strategy. Bridges can take a variety of forms: part-whole relations, causal ordering, shared terms with conceptual differences, and more. Relations between any two models are constrained only by coherence requirements: near-side, far-side, and middle. The basic epistemic form is of diverse explanatory models of a single target, related to one another in ways that bear on understanding of that target. In practice, any late-stage low-degree IDE forms a network of related models, each constructed by a different scientific field, linked via a particular structure of bridging relations. An IDE's target is whatever participants in the project seek to understand (X, in EP_M's schema). In the hemoglobin example, the target is, broadly speaking, hemoglobin's role in health and disease. But that is a rough characterization only. Indeed, the example begins with the hemoglobin molecule, then shifts to sickle cell anemia. So there is some slippage or ambiguity in the explanatory target.

Here one might raise an objection. On close inspection, each model in the array has a somewhat different target. Biochemical models of hemoglobin don't address the same questions as evolutionary models of the sickle cell gene—and so for each explanatory model in the array. If each model in the array has a different target, then they cannot be 'diverse models of the same target.' Thus, the idea of an array of models advancing understanding of some target falls apart. That's a serious objection! But the IDE account cannot be dismissed so quickly. The objection relies on a premise inconsistent with EP_M. The premise is that an explanatory model's target is what fixes an explanation's boundaries; i.e., what individuates an explanation as such. Granting that assumption, if targets differ then there is no overall IDE. There is no question of integrating the various explanatory models, because each model targets a distinct explanandum. So IDE falls apart. This objection effectively rules out the possibility of connecting different explanatory models at all, and a fortiori those connections having any epistemic significance. Because explanatory models constructed in different scientific communities reflect many features of that community context, it's nearly always possible to discern differences across targets, even if those targets are related.

In EP_M's schema, however, a model's target X is one variable among others. There's no ground for singling out target X as what individuates an explanatory model M. Indeed, EP_M leaves the features of M and X entirely open; only community S is specified precisely (using sociological criteria). Purpose P is specified only insofar as understanding target X; what that amounts to is left open, to vary with community context. For explanatory particularists, there's no tidy formula for individuating explanatory models (or models generally) at all. Identifying and individuating scientific explanations is a task for particularist studies of science in practice, responding to communications by members of the epistemic community at issue. Targets of an explanatory model in practice are often multiple, malleable, vaguely defined, or ambiguous. The key feature indicative of a scientific explanation, on the particularist account, is recognition by a community of a major achievement that advances understanding of—something. Variable X is largely a placeholder; that which a scientific community's explanatory model advances understanding of. Often, X is not fully determinate within a community, or has multiple interpretations. All the more so, in interdisciplinary research. In such projects, a broad, vague characterization of the target to be understood is often all that can be articulated. The general ideas of hemoglobin and related phenomenon of sickle cell anemia, as these figure in the example above, are a good match for notions of a shared target in interdisciplinary research. Discernable differences in X as conceived within each particular community don't exclude an overall IDE, any more than such differences halt IDR projects.

Another takeaway from the hemoglobin example is that, although many different communities contribute, bridging connections are pairwise. So the two-model case (Section 5.2) straightforwardly extends to more complex cases. Bridging relations among models in the array are various—no single relation links them all. (Some relations appear more than once in the network—part-whole, cause-effect—but no single bridging relation covers the whole network.) This is in keeping with the tenets of explanatory particularism: minimal generic requirements for bridging models, applied according to features of the particular epistemic communities involved. Relatedly, a network of models connected via bridging relations is non hierarchical. This non-hierarchical character holds for bridging relations between any two models, and for an IDE as a whole. Each bridging relation is mutual; a two-way street, so to speak. Although bridging relations in an array or network may be diverse, no pre-set order among them structures the network. (Different models can be ranked according to degree of connectedness to others. But the epistemic significance of this feature is not obvious.) And, as noted, the exact structure of an explanatory network is malleable, changing over time as scientific inquiries progress and communities engage in new interdisciplinary efforts.

Relatedly, bridging connections are not reductive in any traditional sense. The overall hemoglobin/sickle cell explanation doesn't reduce to one or a few models in which explanatory power is concentrated. Particular scientific communities are liable to privilege their own explanatory contribution; this doesn't rule out coherent bridging relations. We can see this by considering the hemoglobin network from the standpoint of biochemistry. Stryer's reason for devoting an entire introductory chapter to hemoglobin and myoglobin is because the two proteins "illustrate many important principles of protein conformation, dynamics, and function" (1988, 143). Models of genetic encoding, molecular structure, and biochemical regulation for hemoglobin can be extrapolated, *mutatis mutandis*, to many other proteins. These models function as general principles for protein synthesis, structure, and function. But the network of models extends beyond protein fundamentals to involve cells, tissues, organs, and physiological systems; organismal genetics and development; human health and disease; evolution by natural selection, and more. To be sure, chemical properties of atoms and small molecules have effects on molecular structure and function, with ensuing causal consequences for organisms and populations. The physiological role of hemoglobin, however, is explained in terms of the 'geography' and mechanics of the vertebrate body; red blood cell sickling in terms of cell shape, development, and location within that body. Key aspects of sickle cell hemoglobin are explained not in terms of molecular structure and function, but principles of genetics and evolutionary change, alongside complex ecological interactions.

Not everything comes down to atoms, molecules, and their arrangement—even in Stryer's own treatment. Models of hemoglobin that center on cells, tissues, organisms, etc., are not derived from principles about proteins or their molecular components. Indeed, Stryer's chapter states only one biochemical principle explicitly, that the three-dimensional structure of the globin fold "is nature's fundamental design for an oxygen carrier: it places the heme in an environment that enables it to carry oxygen reversibly" (1988, 153). Biochemical and molecular models are not expressed as generalizations, nor does Stryer derive their consequences. The array of hemoglobin models does more than illustrate principles of protein biochemistry. It also shows how those principles, exemplified by hemoglobin, relate to concepts and principles in other areas of biology. These 'extra-protein' concepts and principles concern vertebrate circulation and respiration, the organs and tissues involved in these physiological processes, cell shape and integrity, Mendelian inheritance, genetics of human populations, the clinical presentation of sickle cell anemia and sickle cell trait, the distribution and infectiousness of malaria, balancing selection, and more. The array of models—the overall hemoglobin explanation—shows how all of these diverse concepts and principles are *connected*. But the connection among them is not a relation of reduction, derivation, or subsumption. It isn't a single relation at all. Bridging connections among diverse explanatory models in an IDE are not univocal.

This view of IDE, as networks of models linked by non-hierarchical bridging relations, departs from traditional philosophy of science. The latter defines an explanation as a simple, subsuming formula (or set of formulae) that covers a wide range of facts. Explanation, on this view, is limited to principles, laws, or other generalizations of broad scope. The concept implies a hierarchy of general over particular, abstract over concrete, simple over complex. Explanatory particularism rejects all these. Yet the hemoglobin network does conform to the traditional view in some respects. The array of models covers a broad range of diverse facts. And some field-specific contributed models in the array include elements that can fairly be called laws or principles, which are central to those particular explanations. For example, principles of protein biochemistry explain the molecular structure and function of hemoglobin. Principles of Mendelian genetics explain the phenotypes of sickle cell anemia and sickle cell trait in human populations. Principles of organismal physiology explain how sickled red blood cells produce pathological symptoms in human patients. Principles of molecular genetics and functional genomics explain how normal or mutant hemoglobin is expressed in vertebrate red blood cells. Principles of population genetics explain the distribution of sickle cell gene in human populations. And so on. However, if the overall IDE is reduced to such principles only, crucial features are lost. If the hemoglobin network is reduced to principles only, then we end up with the same fragmented situation that ensues from carving up targets of explanation. Each life science community traversed by the hemoglobin explanation can use its principles to assert its boundaries, laying claim to its domain of explanation and fencing it off against all others. But this way of dividing up scientific territory is exactly what IDR seeks to overcome. An overall IDE, if successful, doesn't collapse into a set of tidy, more limited, and more philosophically familiar explanations.

Comprehension

What is gained by IDE, over and above the various field-specific explanations taken separately? The hemoglobin example offers some suggestions. What's distinctive about IDE is its span or reach, across the explanatory terrain of different life science communities. There is a sense in which the array of models covers a large domain of real-world facts and targets, ranging from bioactive molecules to ecological webs and evolutionary processes. But those targets are not just a mass or heap. The network of models includes and connects many different aspects of the living world. The overall IDE traverses many different fields of biology, linking molecules, cells, organ systems, organisms, ecological communities, and evolutionary lineages into a coherent, inclusive representation. I'll label this explanatory virtue *comprehension*. In everyday language, the term means both 'inclusion' and 'understanding' (literally, "to grasp with the intellect").[36] The overall hemoglobin

[36] The *Oxford English Dictionary* identifies two main meanings for the noun "comprehension." The first is *inclusion*—the action, fact, quality, or faculty of. The second is *understanding*—the action, fact, ability, mental state, or faculty of. For example, "[t]he faculty of grasping with the mind, power

explanation connects diverse explanatory models; in this sense, it *comprehends* them all. The explanation as a whole shows how these different models are related to one another. Comprehensiveness is, to be sure, just one explanatory virtue among others. Some epistemic communities may not place much if any value on it, and those that do will interpret it in different ways. Those are just basic tenets of explanatory particularism. However, IDEs are comprehensive by definition, due to bridging connections linking models from different epistemic communities into a network.[37] Comprehension via IDE suggests a non-traditional ideal of *unity of science.*

Unity of science for particularists

Traditional explanatory unification is defined as subsumption of many diverse facts by a single general principle, covering law, or argument pattern.[38] The phrase "unity of science" often refers to a very extreme view proposed by Oppenheim and Putnam as a "working hypothesis" (1958). They hypothesize that scientific knowledge is structured as a hierarchy of laws and hypotheses ordered by logical derivation, such that the reductive hierarchy of scientific laws maps neatly onto an ontological hierarchy of parts and wholes. This neat reductive picture has been rather decisively disconfirmed by studies of scientific practice (and critiques of its core assumptions). But its background ideas remain influential. In a recent review, Woodward (2019) distinguishes three kinds of unificatory achievement in science: a common classification scheme or descriptive vocabulary for a previously pluralistic area of science (e.g., Linnaeus' system of species classification), a common mathematical framework that applies to many different phenomena, and showing that phenomena previously thought to have diverse causes are results of a common cause or mechanism. All three involve reduction or subsumption of diversity to a single classification scheme, mathematical formula, or causal basis. Concepts of unification traditionally associated with explanation reduce diversity in phenomena (or models thereof) via some unifying achievement. A hierarchy is implied, with the unifying element 'above'—prioritized or privileged over—diversity. On the other side is thoroughgoing 'disunity of science.' But hierarchical unification and disunity are not the only possibilities. IDE offers another vision

of receiving and containing ideas, mental grasp." The adjective "comprehensive," relatedly, is defined as "complete; including all or nearly all elements or aspects of something … of large content or scope; wide-ranging." (The term also applies to examinations that test a student's command of a special field of knowledge—a particularist sense.) Combining the two everyday meanings yields the idea of 'understanding by way of inclusion.' There is a connotation of inclusiveness, completeness, encompassing a lot.

[37] See Yilmaz (2017) for a compatible view of explanation across areas of life science, from a very different theoretical approach.

[38] Friedman (1974), Kitcher (1981, 1989); see Chapter 1.

of the unity of science. The array of hemoglobin models doesn't unify via a single classification system, formula, or common cause. Rather than covering many diverse facts with one simple formula, the overall array shows how different aspects of hemoglobin biology are related. This points toward a conception of unity of science consistent with explanatory particularism.

An IDE is unified via bridging connections. Those connections are pairwise, mutual, and informed by features of the particular communities in which contributed models originate. No contributed model or element thereof is elevated over others. Bridging relations connect explanatory models without loss of diversity. Differences among specialized explanations are not elided or subsumed or reduced to a common denominator. Instead, minimal coherence requirements allow (but don't guarantee or require) connection between even very disparate explanations. There is no 'overall explanation' satisfying all the norms of contributing specializations (or elevating one over the others). Indeed, different contributed models might make incompatible predictions about aspects of the target—that isn't necessarily a problem. (Whether it is depends on the particular interdisciplinary project.) Bridging connections are circumscribed, various, and constrained only to be minimally coherent. Because the coherence conditions are so permissive, limited by our own ingenuity in adjusting explanatory models and seeing connections with others, all of science can be included in this potential 'tapestry.' So it's fair to call this 'unity of science,' even though the unity is potential and preserves diversity of specialized scientific explanations. This (potential) unity of science takes the form of a patchwork of lightly connected explanatory achievements.

'Patchwork unity' keeps different explanatory models connected yet distinct; neither assimilated nor confounded with one another. This is helpful for explicating situations where closely related fields use similar concepts in subtly different explanatory models, such as concepts of 'gene' and 'cause' in molecular genetics, molecular biology, and medicine. The account is also a corrective to oversimplified explanations that conflate distinct models, thereby distorting our understanding. So bridging relations in an IDE are a tool for disambiguation, and a way of contextualizing pluralism in, for example, gene concepts. Moreover, as my brief discussion of comprehension suggests, there are further advantages to explicating bridging relations in an array of models or IDE. These are explored in the next chapter.

5.5 Conclusion

This chapter has articulated coherence constraints on bridging models. Building on the model–model framework from Part I, I've focused on late-stage low-degree IDE, a product of "low-effort interdisciplinarity." In this kind of interdisciplinary research, participants bring to the project two or more explanatory models, each constructed in a different scientific specialization. The shared goal is to construct a narrow, sturdy bridge between two (or more) distinct explanatory models. Some new element (or adjustment of an existing element) is added to each of the specialized models, which affords connection with the other. Extending the social action framework of Chapter 3, I explicate these ideas in terms of coherence constraints. The main result is a set of minimal requirements for connecting two or more disparate explanatory models ('bridging models'). For collaborative activity to succeed, connecting efforts from each side of the bridge (i.e., from each specialization) must be coherent with one another. This implies constraints on the bridge itself, analogous to constraints on participants' attitudes. The constraints are very minimal, however. No element of an IDE need conform to standards of all involved specializations at once—not even the bridge. There's no general requirement for an overall IDE to function univocally. Connections between explanatory models are as diverse, or more so, as what they connect. This type of IDR is, accordingly, "low-effort" relative to more resource-intensive transdisciplinary efforts.

This result answers the challenge for EP, the central problem of Part II: how to combine, integrate, or connect explanatory models from different specialized fields. Bridging relations show how this is possible, given EP/EP_M. So explanatory particularism doesn't imply unacceptable (or at least very unwelcome) elitism and insularity about scientific explanation. Explanatory particularism begins with explanations in scientific practice—in all their context-dependent particularity—and *then* considers how different explanations so conceived might relate to one another. Most real-world problems for which scientific solutions are sought today—such as environmental sustainability, global health, and managing impacts of anthropogenic climate change—demand solutions that transcend the boundaries of scientific specializations. It is therefore important to discover and articulate ways of connecting different explanations. But those links require effort; they don't follow ineluctably from within-community explanation. The conditions I propose are relatively undemanding, amounting to a narrow 'conceptual bridge' that leaves mostly unaltered each specialized explanation thereby connected. The extended hemoglobin example shows how bridging relations apply in more complex cases. The idea of a linked array of models points toward a new conception of unity of science, compatible with tenets of explanatory particularism. The next chapter examines the significance of these results, for interdisciplinary research and philosophy of science.

The traditional view among philosophers of science is that explanation is sparse and abstract; covering much with little. The new view propounded here allows for richness of explanations; a patchwork of explanations, understanding from diverse combinations. This is what we see in the hemoglobin case—not reduction, but profusion of explanatory models; not competing, but connecting. An IDE of this sort can be rich and diverse. Not only do we have a good explanation of hemoglobin's biochemical structure and function, but that explanation connects to many others, encompassing biological phenomena at other scales, understood in diverse ways: energy transformations, atoms, molecules, molecular complexes, cells, tissues, physiological systems, whole organisms, populations, ecosystems, evolutionary change. Comprehensive explanations of this kind have distinctive epistemic value. The next chapter further articulates that value, fleshing out the idea of understanding in the framework of EP/EP_M.

6
Social Understanding

This chapter continues to examine the implications of IDE. I first consider some practical implications. The IDE account offers tools for facilitating interdisciplinary research (IDR). It does so by identifying attitudes conducive (or not) to constructing minimal bridging connections with explanatory models from other fields. This completes my answer to the challenge for particularists: if explanations are localized to particular epistemic communities, how can IDR spanning multiple communities yield explanations? But what looks at first like a problem for my version of explanatory particularism turns out to be an advantage. Interdisciplinary explanations *are* challenging to construct, especially among disparate sciences. This is so even though coherence requirements for bridging models are minimal, satisfiable in so many possible ways. It still takes work to connect different explanatory models so as to satisfy coherence constraints. But the main obstacle is not in the different specialized models themselves. Even stark contrasts can be interpreted and articulated so as to satisfy the minimal coherence constraints set out in Chapter 5. The main obstacle to IDE is scientists' own attitudes. Section 6.1 addresses this obstacle. I then turn to more theoretical matters. Section 6.2 applies IDE's pro-social norms to articulate a nuanced account of scientific understanding, building on the earlier discussion of 'particularist objectivity.' Section 6.3 puts my new account of understanding into conversation with other philosophical accounts. Section 6.4 concludes, summarizing the key results of Part II.

6.1 Norms for interdisciplinary explanation

That scientists' attitudes are a major obstacle to IDR is not news. Their attitudes are scaffolded and reinforced by social/institutional structures, values, and incentives.[1] A number of these discourage IDR. One indication of this is the prevalence of "boundary-work" in scientific practice; i.e., talk by group members that draws a line between "us and them" (Gieryn 1999).[2] Such talk is often

[1] Havstad (in preparation).

[2] Some sociologists use other terms to characterize the same phenomenon; e.g., Campbell (1969) on disciplinary "ethnocentrism." Campbell's proposed solution is the "the fish-scale model of omniscience ... collective comprehension through overlapping patterns of unique narrownesses" (1969, 329). There are obvious similarities with IDE. The main contrast is that Campbell's solution is social-organizational, aimed at university administrators and department heads. He advocates encouraging

Explanatory Particularism in Scientific Practice. Melinda Bonnie Fagan, Oxford University Press.
 DOI: 10.1093/9780191998874.003.0007

deployed around explanation. Consider the contrast between "explanation" and "mere description." This familiar trope means different things to different epistemic communities. There's no consensus on what explanation is, among scientists or philosophers. Thus, the contrast doesn't simply indicate two kinds of epistemic product, severed by a bright dichotomous line. It draws different distinctions, depending on who is holding the pen. Often, the point of the contrast is boundary-work: to contrast one's own practice (explanation) with the epistemically less significant work of other groups. On a particularist approach, this form of boundary-work makes perfect sense. An explanation is by definition a major achievement for a scientific field. Favorably comparing one's own group to outsiders is an effective way to build group identity and solidarity. The latter facilitate within-community collaboration that advances understanding, and is thus epistemically beneficial. The same kind of boundary-work, however, indicates attitudes that block IDE.

A famous example is the aphorism that "All science is either physics or stamp collecting." The source is revealing. The remark is attributed to Ernest Rutherford, but doesn't appear in his scientific publications. It first appears in print (to my knowledge) in a report about the Rutherford Jubilee International Conference in 1961. That event took place more than twenty years after the eponymous scientist's death, so it's obviously not a remark delivered by Rutherford at that meeting. But the aphorism fits there, as an influential piece of scientific boundary-work attributed to a giant of that epistemic community. The purpose of the Jubilee conference was to commemorate Rutherford's conception of the nuclear atom and Bohr's quantum theory of radiation. It was an important community-building event. During World War II nuclear physics as a scientific community was riven apart, transformed, and rebounded that transformation onto the entire world with the development of the atomic bomb.[3] The 1961 conference was an event helping to renew and rebuild physics as an international scientific community in the post–WWII era. "Physics or stamp collecting" expresses the epistemic importance and superiority of the field—a positive rallying cry to move the field forward, out of the Bomb's shadow.

Explanatory boundary-work is more than team-building rhetoric (although it is that). Researchers committed to one specialized community's view of explanation quite reasonably find the explanations of other fields wanting. The latter are deemed poor explanations, or even unrecognizable as such. And this too follows

narrow specialization in the areas that remain as gaps in the existing landscape of disciplinary expertise. My proposal is aimed at scientists and philosophers of science, focusing on explanatory models in EP's sense. Another, related contrast is that I focus on very narrow, limited zones of overlap between explanatory models, rather than large or substantial interfaces.

[3] There is a large historical literature on this topic; Rhodes (2012) is one of many excellent accounts.

from EP's basic tenets. Scientists working in a specialized epistemic community take up the values and standards of that community. Many of those values concern explanation and understanding. If members of a particular specialized epistemic community attend to explanations from other scientific communities at all, they're not likely to recognize those other epistemic products as explanations. That's because the latter don't look like the explanations those community members construct and use. The contrast between explanation and mere description looks obvious to those who make it; a fact of scientific method. However, what falls on either side of that line depends on the speaker's epistemic context—the community in which they work to construct explanations, among other things. Putting down the explanatory models of other specializations is an effective way to strengthen community ties underpinning a field's own style of explanation. But that strategy hamstrings interdisciplinary efforts. One of EP's philosophical contributions is highlighting this tradeoff.

So, against IDE, we have the human tendency toward boundary-work and prioritizing one's own group over outsiders.[4] Complicating and exacerbating this tendency is the entrenched idea of a *hierarchy of sciences*. Theoretical physics has arrogated a great deal of epistemic authority, which some sciences seek to emulate and thus borrow. Interdisciplinary efforts do not take place under conditions of social epistemic neutrality. Many scientists have attitudes reflecting outdated hierarchies and blind spots inculcated via boundary-work. Philosophers of science have had a role in creating and sustaining this situation. Explanatory models from fields of theoretical physics have long dominated studies of explanation in philosophy of science (see Chapter 1). Philosophy of science has tended to exacerbate this situation, valorizing particular styles of explanation (traditionally those of theoretical physics, now causal models of complex systems) over others. In philosophy of science, this hierarchy has outlived its usefulness. Our field's traditions around explanation have impeded philosophical studies of interdisciplinarity and social epistemic aspects of science more generally.[5] For scientific practice—for IDR, specifically—these entrenched ideas of hierarchy are outright obstructive.

Explanatory particularism deliberately eschews traditional views of scientific hierarchy. Modeling and social action frameworks developed in Part I are part of this tack. Conceptualizing a field's major explanatory achievements as models (rather than theories or laws in the traditional sense) helps lower barriers to interdisciplinary work. A model is, by its nature, incomplete. It can be a massively significant achievement of understanding—but even so, a model necessarily allows for other, different efforts at understanding. The social action framework offers further resources along these lines. Norms for IDE can help counteract the effects

[4] There are of course many other incentives against IDR across the sciences; see Chapter 4.

[5] See, however, works by Hanne Andersen on philosophy of physics that emphasize interdisciplinarity (e.g., 2010, 2016). Andersen's ideas are an important influence on this work.

of explanatory boundary-work within particular fields, thereby lowering an important obstacle to IDR. This section shows how. I first reprise the social action norms and apply them to IDE. I then operationalize these norms further, for use in IDE projects. The result is a set of guidelines for counteracting explanatory boundary-work and entrenched ideas about epistemic hierarchy.

Social action framework: recap

The social action framework (Chapter 3) implies norms for collaborating agents. These norms derive from the three-layer structure of collective intention, which interweaves attitudes of participants in collaborative social activity. The norms are, briefly:

- The activity is not one's own exclusively; others are involved.
- One's own contribution to the activity is incomplete; exercise humility.
- Participation is public; no one acts behind the scenes.
- Other participants have equal standing in the activity; respect them as full partners.

These norms also apply to the activity of constructing an IDE, just as for explanations constructed within a particular scientific community. But satisfying them in the former context is more challenging, for reasons discussed previously. Applied to IDE construction, social action norms yield a checklist for whether an interdisciplinary project can succeed. I'll first work through the three-layer structure of collective intention as applied to IDE. Then I'll use the results to formulate practical guidelines.

The first layer of the interactive structure of collective intention relates each agent separately to the shared goal. I'll keep the case artificially simple: agent A from specialized field S1 and agent B from S2, each bring an explanatory model (M1 and M2, respectively) to an interdisciplinary project. The shared goal is to construct an IDE by connecting M1 and M2; i.e., bridging the two models. Agent A intends to bridge the models, as does B. Each intends this as an individual participant and also as the group; as both 'I' and 'we.' For the former, each intends to do their part: to connect their field's explanatory model with the other's. A intends to connect M1 with another model, and B intends to connect M2 with another model. In order for that task to make sense, both A and B should recognize that IDE isn't a task for their own specialization exclusively. That is, explanation is not the sole province of S1 nor of S2. A and B should each recognize that the bridging activity engages another field's explanatory model, acknowledging a role for members of other specializations in the activity. This is the norm of openness: acknowledge other approaches to explanation as legitimate, at least for the purposes of the interdisciplinary project. In their intention to connect, participating agents must be open to contributions from outside their own epistemic community. Otherwise, their intention is incoherent and unstable for practical purposes. It follows that participating in IDE requires some tolerance of diversity and pluralism about explanation—at least, within the limited sphere of the interdisciplinary project. Relatedly, A and B should see their respective models as contributing to, but not wholly accomplishing, the overall IDE. In a collaborative activity, an individual agent's intended contribution to a shared goal isn't sufficient to accomplish that goal. The part they intend to play is just that—a part, which combines with others in overall performance of the activity. Each participating agent should recognize that their individual contribution is incomplete on its own. Recognizing this incompleteness requires some humility. Acknowledgment that one's own specialization's explanation is not sufficient for IDE on its own is a key requirement for such activity to succeed.

There is also the group aspect of the first-layer relation to consider. Both A and B intend that their interdisciplinary group connect M1 and M2 into an IDE. Each participant, no matter their specialization, should identify to some extent with the interdisciplinary group. However, excepting labor-intensive transdisciplinary projects that create a new community from diverse participants, group-level identification in many IDR projects remains shallow and vague. A and B have little sense of 'we,' beyond the bare idea of both specializations making a contribution. Participants' notions of the overall IDE are also perforce vague and hand-wavy. What is shared across divergent scientific fields is at most a very thin and abstract concept of the understanding sought, and of the explanatory model that can provide it. Beyond this vague shared notion (which may be largely inchoate), each field brings to an interdisciplinary project the view of explanation appropriate to their field. Participants will have a much clearer sense of their own contribution: their specialization's explanatory model. Each participant contributes an explanatory model (M1 and M2) and resources to construct a bridge to another. The key role of each participant in this case is to find a point of connection linking their explanatory model with the other's: one or a few elements to make the bridge's near-side. Because each participant approaches the bridging task with different epistemic resources, the bridge is comprised of distinct but parallel efforts: C1 and C2. Call these reciprocal bridging efforts C1 and C2 (for bridge C). Because the group-level concept is so thin, there is little scope for interplay in each agent's individual- and group-level identification. This source of stability for collective intention is also undercut, in late-stage low-degree IDE. It is therefore all the more important that the bridge connecting M1 and M2 be stable and coherent as such.

The second layer relates each agent to the other's first-layer relation; i.e., the intention to connect their own specialization's model with the other's. Each recognizes the reciprocal intention of the other, which mirrors their own intention to contribute to the activity. A regards B as intending to connect M2 with another model, and B regards A as intending to connect M1 with another model. This amounts to recognition that bridge-building is initiated from both sides of the specialization divide; C1 (C2) isn't all there is to the bridge. This adds another requirement (alongside non-exclusivity): that the other's participation in the activity is 'on par' with one's own. Both A and B should each see the other as contributing to constructing the bridge linking the models. This is the equality norm. Participants are equal in this regard, although the specifics of both explanation and connecting bridge will perforce differ. Because there is little sense of 'we' in this case, the boundary between A and B remains obvious. Rather than combining their agency, the activity of connecting the explanatory models is conceptualized as constructing some connection between them—a bridge. The key point here is that efforts on both sides are recognized; the bridge between models is constructed from both ends. In this sense, A and B have equal standing toward the shared goal of connecting their respective models. It's not necessary for A or B to see M1 and M2 as having the same explanatory status (although that attitude is an aid to interdisciplinarity). All they need do, to satisfy minimal norms for this collaborative activity, is recognize that other specialization's model connects to theirs, and that across that divide there is also explanatory activity going on relevant to the interdisciplinary project. It does require some humility, to accept that C1 (C2) connects with another explanatory model. This implies that explanation extends beyond the boundaries of S1 (S2).

Relatedly, there's an important sense in which M1 and M2 are on par. Neither is a model (or a target) for the other; neither explains the other.[6] Both are models that contribute to the same interdisciplinary project of constructing an IDE. The target of explanation may be the same for M1 and M2, or they may bear on different targets (though all relevant to the interdisciplinary project). Either way, there's no pre-existing explanatory priority ordering the relation between M1 and M2. Metaphorically, the bridge connecting the models is well-balanced. In accordance with equal standing in relation to the shared goal, A and B should recognize symmetry of their own and other's attitudes toward the bridging activity. Each should recognize that the other's intention complements their own. This amounts to conceiving the bridging activity as proceeding bi directionally. Each participant seeks to connect their specialization's model with the other, and sees the other as doing the same. So the two sides of the bridge must meet in the middle, somehow. Participants in interdisciplinary explanation are also equal in the sense that their attitudes about constructing a connection are public. Neither A nor B is aloof or hidden from the other. Yet common awareness across specialized fields is limited. Within a scientific specialization, there is a fair amount of common

[6] Predecessors of this idea include Longino's norm of equality of intellectual authority (1990); Andersen's (2016) interlocking models, and Nersessian's ethnographic and cognitive studies of scientific modeling (2022).

(shared) knowledge. But very little of this specialized knowledge is shared across fields, unless those fields are closely related to begin with. In such cases, IDR faces fewer obstacles. For harder cases with more disparate collaborating fields, knowledge common within each is 'uncommon' to the other; e.g., ways of acquiring and interpreting data, of constructing models, of crafting explanations.[7] That's why a low degree-of-connection is more likely to succeed in such cases (see Chapter 4).

The third layer of the interactive structure closes the intentional loop, relating each participant to the other's other-regard. In IDE, A recognizes that B recognizes A as intending to connect M1 with M2, and B regards A as recognizing that B intends to connect M2 with M1. This completes the 'back and forth' interlacing structure of relations that weaves individual intentions into a single collective intention. Each sees their own intention to participate in connecting the two models through the other's eyes, so to speak. Because the other has little knowledge of the specifics of that effort, the most important feature of this third-layer relation is that each agent recognizes the other as a full partner in bridge-building. The associated norm is respect: each participant should see the other as making a contribution analogous to, and on par with, their own.[8] Participants in the bridging activity should respect one another as having equal standing in that project. This has further implications for understanding, explored below.

To summarize, the social action framework implies norms for bridging models to construct a late-stage low-degree IDE:

- Acknowledge a role for other fields in IDE (non-exclusivity, openness).
- Recognize other fields as contributing to the bridge (equality).
- Recognize the other field's bridging efforts as complementary to one's own, even if one's understanding of those efforts and the models from which they emanate is very limited (respect).

One may notice that IDE norms concern individual participants' attitudes, rather than focusing on the community-grain that is primary for explanatory particularism (see Part I). This shift of scale is appropriate for practical applications to IDR projects, as these tend to be planned and funded at the scale of research teams. The guidelines that follow are designed with that sociological grain in mind. It's worth emphasizing here that although my version of explanatory particularism focuses on specialized communities at the grain of sub-fields, this doesn't exhaust science's social organization. For particular IDR projects that include IDE

[7] Of course, a new field of inquiry that develops from the juncture of earlier ones will, as part of that process, add to a store of common knowledge of the nascent field. But the problem at issue here is how such "interfield" perspectives can become established in the first place—or how explanatory models can be constructed without presupposing an interfield perspective (cf. Darden and Maull 1977).

[8] Wylie (2015, 194) "At a minimum, what is at stake in demands for accountability is that archaeologists *respect* the interests and sensibilities of descendant communities, even if they don't credit them with epistemically compelling norms of justification" (italics mine).

aspirations, focus on attitudes of individual participants is helpful. That fine-grained perspective is the main focus of this section—with the caveat that this is not the whole story. More 'macrosocial' perspectives are needed too. (Section 6.2 returns to the community grain.)

Guidelines for low-effort interdisciplinarity

These norms can be further articulated into guidelines for low-effort IDR. They take the form of a set of ordered questions posed by an 'interdisciplinary counselor.' Although anyone in principle could play this role, I envisage a philosopher of science.[9] Philosophical theories of explanation, causality, and other concepts associated with explanation are a resource for brainstorming potential bridging relations between explanatory models. No one directly participating in the IDE project should play this mediating role, nor anyone from one of the specialized fields involved. The counselor role should be taken by someone who doesn't have an immediate stake in any of those explanatory models. It's difficult for anyone to put aside hard-won and valued identification with a particular scientific community—hence the need for an interdisciplinary counselor at all. Their role is to facilitate early stages of bridge-building, by encouraging participants to take on attitudes that satisfy coherence constraints. This begins the process of bridge-construction, which participants should complete on their own. Alternatively, if participants are unwilling to adopt those attitudes, the counselor can detect the situation quickly. In such cases, the interdisciplinary project should be modified so as not to require an IDE.

Set-up: I'll assume an IDR project combining explanatory contributions from two distinct scientific communities. The overall explanatory aim is to advance understanding of some target X, where substantive explanatory contributions from both fields are needed. IDR projects are more challenging when communities are very widely separated; e.g., computer scientists and sociologists, physicists and anthropologists.[10] Multiple members of each community are involved in the project. Nothing else about the context is assumed. The following steps should be applicable to many different interdisciplinary projects aimed at or requiring explanation.

Preliminaries:

(i) Begin by setting aside any "Big Picture" questions. Participants will of course have ideas about the project's goals and how different tasks are allocated among participants. But, especially when those ideas vary widely across fields, it'll be more productive to focus on a narrower issue. Here the focal

[9] In this respect I am following the lead of the Toolbox Dialogue Initiative (see Chapter 3).
[10] As opposed to, say, mechanical and electrical engineering, or structural biology and biochemistry.

task is bridging distinct explanatory models. Start with this and build up to the Big Picture as needed.

(ii) Meet with members of each scientific community one-on-one to discuss their own field's explanatory model. Of what does this model advance understanding? What qualities of the model afford that advance? What specific benefits does the advance in understanding confer; i.e., how does the model bear on project objectives? These questions are just those that EP prioritizes in studying scientific explanation. A philosopher asking these questions is ipso facto carrying out a particularist study of explanation in those fields, albeit with a small number of individual scientists. These one-on-one discussions will indicate each participant's conception of their own field's contribution to the overall IDE. There's likely to be some variation within a field (this is why meeting one-on-one is advised). If variation within a field is very pronounced, however, this undercuts prospects for a successful IDE, due to "internal" debate about explanation. (That kind of debate is normal, indeed endemic for some fields, but would be destabilizing within a single IDE project.)

(iii) After each participant has characterized their field's explanatory model, ask: What are its limits? What about the target (if anything) is put aside or left out? Some scientists may not acknowledge any limits of their model. But it's helpful to encourage them to think about it, even so. Scientists that reject the idea that anything of epistemic importance falls outside the scope of their field's explanatory model should probably step aside from further bridge-building.

(iv) Wait awhile after one-on-one discussions about each field's own model. Bridging questions are about how it relates to another, quite different explanatory model. Immediately shifting to that issue after focusing on one's own model's virtues sets the stage for infelicitous comparisons. Participants will be thinking about their own explanatory approach, and less inclined to focus on others'. Instead, let participants sit with their thoughts about their models' limits. Bridge-building should commence a day or two later.

Bridge-building: This could be one-on-one or grouped by specialization; whichever is best for a particular IDE project. The questions are designed for members of a single specialized community, as a precursor to more inclusive discussions. Figure 6.1 shows the process as a flow-chart, terminating at the stage just before such discussions (after the counselor's work is done).

(i) Ask each participant if there's an element of their explanatory model (a term, method, object, etc.) that overlaps with the others'. A shared term, like "gene" or "organism" or "noise," doesn't imply a shared concept, but is likely to offer some point of connection. If the answer is "no," skip to (iii).

(ii) If the answer to (i) is "yes," ask each participant if the other's use of that concept (or method, object, etc.) is the same as theirs. If the answer is "yes," then there's a direct similarity relation between models; skip to (v). If the answer is "no," then ask: How are their uses different? The idea is to get participants thinking that others are doing something similar yet different; e.g., focusing on another aspect of an object, another use of a method, a related meaning of a term. Ask them to try to articulate what the difference is, exactly. This amounts to specifying an indirect difference relation between models. Then proceed to (v).

(iii) If there's no obvious point of connection, philosophical accounts are a valuable resource. Causal theories of explanation may be especially helpful. Is some element of their model an effect of some outside (antecedent) cause? Or a causal factor for some 'downstream' effect? Part-whole relations? Changes of scale? Many potential 'connectors' could specify an indirect difference relation between models.[11] Both Darden and Maull (1977) and Bechtel (1986) offer lists of possible relations to consider, including: localization of an entity or process, physical description, structure underlying function, new levels of organization, model-construction techniques, and extensions of a theory. Offer suggestions as to points of contact with the other field's explanatory model. If the participant agrees with any, skip to (v). The task is then the same as for earlier points—making explicit the bridging relation. If participants don't accept any of these suggestions, go to (iv).

(iv) If the participant doesn't notice or agree with any suggestions above, bring up the Big Picture for IDE in the project. What's the primary goal of the project? How is that goal served by the "other side" contribution? Is IDE really the goal, or necessary to it? If so, there must be some explanatory contribution from the other field. Can the participant articulate how the far-side bridge element fits into that contribution? What might that be? If the participant has no answers, stop. That participant isn't interested in doing ID research. That's fine; they may have other reasons to participate in this project. Such participants should work on their own field's explanatory model; the IDR project can go on around them. As long as some members of each field do continue, IDE can go ahead. If not, the project should be reframed so as to not to require an IDE. If the participant does have some Big Picture ideas, then ask them to articulate these. What's the IDE's purpose? How is that goal served by the "other side's" contribution? These questions should encourage participants to see members of the other field as equal (yet different) members of the same project. (Not, for example, the downstream users, or people supplying some data-points, or the like.)

(v) Build on the connection, direct or indirect, similar or different. What does the common or contrastive element look like on the other side; i.e., to

[11] The only relations that won't work are those that arrogate all explanatory power to that participant's model.

participants working in the other field? Participants should think about the bridge's far-side, a counterpart to their own field's use of that bridging element. You want participants to have a sense of others' working "over there" on a different yet connected element of the explanatory model. This gives a rough picture of the bridge's far-side. It's unlikely to match what participants in the other specialization are actually doing, but that's not a problem.

(vi) What would participants on the "other side" say about the connecting element and relation (change of scale, cause-effect, etc.)? That's the 'seeing oneself through others' eyes'—not really, but it's making the effort that counts. Get each participant thinking about what their model's element looks like from the other side. They won't actually be able to do this, but it's the effort that matters. Participants who are dismissive or who can't do this at all shouldn't be central to the ID parts of the project (though of course they can contribute a lot).

(vii) Build on the minimal bridge connection. Participants who respond and are interested should sit down regularly and find other points of contact. This can develop more intricate and varied ties between explanatory models and other features of different specialized fields. Each bridging connection that's made explicit affords opportunities for further connections. However, if the discussion is one-sided—all those able to articulate connections are from one field, none from the other field—then the team should abandon the IDE project, or make the connection purely additive ("multidisciplinary"). If participants' attitudes aren't such as to satisfy norms for collaborative activity, coherent IDE won't ensue. (The project could still be interdisciplinary, it just won't involve IDE.)

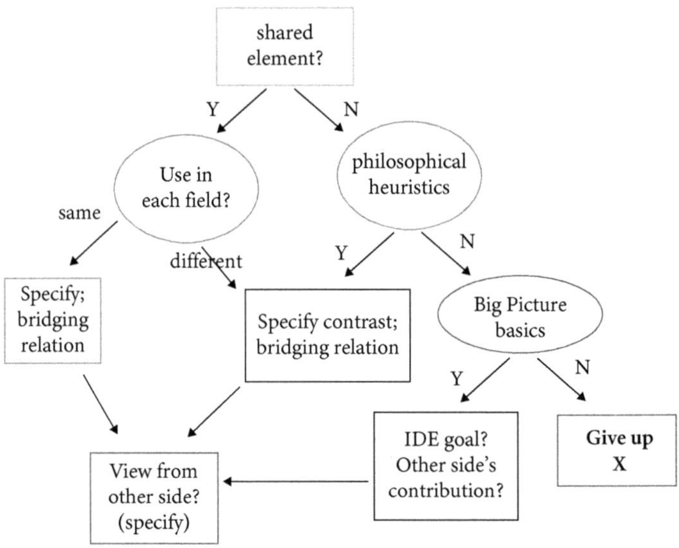

Figure 6.1 Bridge-building facilitation diagram, in flow-chart form

The above guidelines provide a recipe for explicating bridging relations, and for encouraging scientists engaged in IDR to adopt collaborative attitudes. A scientific community's commitment to a particular view of explanation (i.e., the ideas and values associated with explanation in that community) can be a major barrier to IDR. These guidelines are intended to lower that barrier, without compromising community-based commitments. Practicing scientists should conceive of explanations in their own specialization as one partial viewpoint among others, with no single specialization prima facie explanatorily privileged over others. Humility about one's own perspective is crucial, alongside a sense of accomplishment for one's own field's achievements. Perceived status differences across disciplines can affect patterns of interaction and thus group effectiveness. IDE norms let each field prioritize their own part—within limits.

As noted earlier, these guidelines target a finer-grain of science's social organization than EP/EP_M. The units of analysis for particularist studies of explanation and for particularist studies of bridging models are not the same. An EP-explanation is a major scientific achievement that advances understanding for members of a scientific community, while IDE projects involve at most a few members of multiple such communities. After IDE connections have been established, they may 'solidify' into achievements that resemble EP-explanations. IDE connections between hemoglobin models across the life sciences are of this type. But in practical terms, the EP account is most fruitfully applied to work-in-progress; projects aimed at IDE that might someday yield achievements for entire scientific communities. Hence, a shift in the grain of social analysis. The primary shift, though, is from studying established explanatory achievements to studying work-in-progress. The shift in social unit or grain follows because any explanatory achievement has a historical origin story in one or a few groups—scientific teams. If an IDE project at this finer-grain is successful, the achievement should eventually emerge at the community-grain.

There are other epistemic benefits to explicating bridging relations for an IDE. Participants gain a clearer sense of differences between their contributed models. When faced with complex phenomena, it's common to run together distinct yet related ideas. An IDE's bridges maintain different explanatory models as connected without conflation. This is a corrective to oversimplifying explanations or conflating distinct models. This is especially advantageous in contexts where closely-related fields use similar concepts in different explanatory styles, such as concepts of 'gene' and 'cause' in molecular genetics, molecular biology, and medicine (see Section 5.3). So IDE guidelines can aid disambiguation in these situations. Making explicit the bridging relations for IDE is also a corrective for top-down initiatives that aim to spur IDR without attending to the particulars of the fields involved. The latter include not only diverse explanatory styles and achievements, but also explanatory boundary-work and assumptions of scientific hierarchy. Many top-down calls for interdisciplinarity ignore these complexities and challenges. These advantages notwithstanding, a central question remains.

Why should scientists seek IDE? Why should philosophers facilitate? Answering these questions amounts to a particularist account of understanding, grounded in the social action framework. This yields another key result: an EP/EP_M view of understanding.

6.2 Social understanding

According to EP, both explanation and understanding are context-dependent. An explanatory model advances understanding in accordance with features of the relevant epistemic community, including (perhaps especially) its values. For an explanatory particularist, there is no short-cut around the work of characterizing these contextual features on a case-by-case basis. This holds for interdisciplinary projects involving IDE, just as for within-community explanations. However, the IDE account also offers some general insights about understanding in interdisciplinary practice. Consider first early-stage high-degree IDE. For this kind of IDR, the prospect of conceptual innovation is a major incentive. Successful innovative results are achievements of a new scientific community. Insofar as those achievements advance understanding, they do so by combining explanatory insights of different specializations. Establishing a new scientific community from diverse roots takes considerable effort—social and epistemic work, enabled by institutional resources. This kind of IDR, as discussed, maps roughly onto "transdisciplinary research"—high-effort, resource-intensive, and requiring extensive ongoing interaction among participants from different specialized fields. Such interactions can, with a lot of effort, yield a novel explanatory style, grounded in values and standards negotiated within the new transdisciplinary community formed by project members. This hard-won explanatory style implies a new, particular idea of understanding, honed and suited to the newly-formed community. But this is a tall order. Transdisciplinary research is demanding, high-risk, resource-intensive.

In contrast, late-stage low-degree IDE is comparatively low-effort. Explanatory models from different scientific fields can be integrated with much less effort, satisfying minimal coherence requirements for bridging models. But by that token, it doesn't offer the same innovative promise. Late-stage IDE does not constitute a new community with its own distinctive explanatory style. Connections between models are very limited. Participants from different specializations needn't understand much about one another, nor interact regularly. Any community formed around an IDE of this sort is likely to be thinly constituted. And yet, 'low effort' isn't 'no effort.' As shown in the previous section, research involving late-stage low-degree IDE does demand some efforts from (some) participants, to explicate bridging relations and adopt collaborative attitudes. Why go to the trouble? What epistemic benefit does IDE provide? What advance in understanding is achieved by bridging models?

Relational understanding

In one sense, the answer is obvious. Two explanatory models are better than one, so combining two or more models of a target must advance understanding of that target. But understanding for whom, of what? The interdisciplinary projects at issue involve late-stage low-degree connections between distinct explanatory models. Explanatory models at a late stage of construction are already (or incipiently) significant achievements of understanding in their respective communities, prior to any interdisciplinary efforts. Bridging relations between models involve one or at most a few elements of each. It follows that bridging leaves each explanatory model contributed to an IDE mostly unaltered. Indeed, that's the distin guishing feature of this kind of IDE: different explanatory achievements are largely the same before bridging and after. By the same token, there's no (or very little) loss of specialized explanatory advantages for each community. But that just makes IDE no worse than its contributing models taken separately. This sharpens the question at issue. How is a low-effort IDE better than its contributed models taken separately? What kind of advance in understanding does it confer?

From a group cognition perspective, the advantage is obvious. Participants in an interdisciplinary project comprise a group. That group gains an integrated explanation of the target of interest, where before there was no such explanation. This follows directly from the social action framework, applied to the shared goal of connecting distinct explanatory models. The achievement here is analogous to the parable of blind men studying an elephant. One is next to the tail, another to the trunk, another to a leg, and so on. From his experience with one part of the elephant, primarily via touch, each man conceives an idea of the whole animal: a rope, a snake, a tree, etc. While each man has a model that is correct, as far as it goes (i.e., to the part of the elephant that is similar to that representation), the parable depends on the intuition that the clash of incompatible models is unsatisfactory. Each man's model of the animal is different, and, taken as an image of the whole, contradicts the others'. Furthermore, each man's model taken separately is a poor fit for the real elephant. Importantly, both these inadequacies—mutual inconsistency, poor fit to target require that each man's model is interpreted as a representation of the whole animal. Each individual model of the elephant captures a key feature of one part of the whole animal. In some versions of the parable, the men are prompted to integrate their various models, yielding the familiar image of the animal we know. The prompt in these versions is external—a sighted king or a sighted pregnant woman tells the men how to integrate their different models. (These figures have interesting resonances to Western ideals of scientific objectivity, at least prima facie.) Lacking such external guides, IDR participants have to integrate their models some other way. The attitudes requisite for an IDE accomplish that. Jettisoning the assumption that each model on its own represents the whole animal—the entire target of explanation—is key. When each model is conceived as part of a more complex whole, then the men can integrate their individual perspectives into the bigger picture. That is to say, there is an epistemically-significant distinction between a set of connected models and the same (or similar) models in a heap.[12] Together, the men's models offer a more comprehensive representation of the target, capturing more of the elephant's distinctive traits. In this way, the men's models are better together than individually. The way their different models are connected is crucial to this advantage. The analogy indicates the importance of specific connections between individual models: IDE bridges. Of course, no model is perfectly or ideally comprehensive (e.g., Teller 2001). But as a comparative notion, comprehensiveness is a useful criterion for evaluating models of a complex target—individual models, collections of models, an integrated network of models. An integrated network of models is more comprehensive—allowing users to *at once* engage with more aspects of the target of interest—than the same models taken in isolation. Therefore, IDE offers one clear epistemic advantage: increased comprehension of a target.

However, for late-stage low-degree IDE, this advantage is rather diluted. An interdisciplinary project of this kind, recall, doesn't comprise a scientific community in EP's sense. The latter are more inclusive, more coarse-grained entities with established social infrastructure (see Chapter 1). An IDR project includes members of different scientific communities, who bring their various explanatory resources (among other skills and cognitive materials) to the task. Successful construction of an IDE within such a project is an epistemic advance for the group constituted by that project. But that group is a fragile, epiphenomenal entity within the dynamic social organization of science. Although the epistemic benefit for such a group is clear, the latter's status as an important epistemic agent is not. And the obvious epistemic agents, individual participants in an IDE project, do not thereby understand the combined model as a whole, comprised of different specialized elements. Participating in this mode of IDR doesn't confer expertise in another specialization. Rare exceptions aside, individual participants in IDR will belong to just one of the fields contributing to an overall IDE. Some might acquire specialized understanding of other fields in the course of the project. But the activity of bridging models doesn't require this of any participant, and for many it's infeasible. Individual researchers need not (and in most cases will not) have specialized understanding of the overall IDE. Nor will they have comprehensive understanding of the IDE's complex target. Comprehensive understanding of a complex target

[12] Cf. Morrison's (2015) contrast between models of turbulence (connected) and models of the atomic nucleus (unconnected).

might eventually ensue, however. An IDE may, through processes of criticism and community discussion, develop into a full-fledged EP-explanation in one or more of the contributing specialized communities. But that's a long-term outcome, and a gamble for interdisciplinary researchers. Is there any gain in the short-term, from an IDE constructed via bridging relations?

To get at the answer, it's first important to note that an individual participant's understanding of an IDE (not the target) is heterogeneous. There are at least three aspects to consider: (1) specialized understanding of the contributed model from one's own field, (2) non-expert understanding of the other field's contribution, and (3) specialized understanding of the bridging relation linking those contributions.[13] An overall IDE is understood differently from different sides of the bridge. Each participant perforce acts within their own specialized perspective to construct an IDE. They do so by connecting a specialized model that they understand in all its specialized nuance (which advances understanding of the target), to a different model that they (in most cases) understand as a non expert (affording little if any understanding of the target). The interdisciplinary effort is narrowly focused on one or a few elements of the models. The result is a conjoined model, expertly understood by the IDR group but not (most) individuals within it. Yet the epistemic gain conferred by an IDE on individual researchers is not trivial. The key advance is (3). What is gained via IDE is understanding of the bridging connection and its relational implications.

For individual scientists ensconced in a single specialization, the gain in understanding via IDE is *relational*. An individual researcher gains understanding of how their specialized explanation relates to those of other fields—one other field, per bridging connection. They do not thereby understand the explanation as a researcher from another discipline would, but do see "how to get there from here." One doesn't become an expert in another field's explanatory ways, but one sees how one could become such an expert. Understanding the bridging connection gives a researcher a path into a different field; a way to "travel" from one modeling perspective to others. So IDE gives individual researchers *directions* for gaining more comprehensive understanding of a complex target. In counterfactual terms: If A were to become proficient in the S2 resources used to construct M2, then A would understand the IDE as B does—from S2's specialized standpoint. Combining all the specialized expertise needed would afford specialized understanding of the overall IDE. More generally:

[13] This treatment assumes the simple case of two fields and two models, as elsewhere in this chapter. More complex cases are, as illustrated in the hemoglobin example, analyzed in terms of pairwise bridging relations between models that collectively comprise a network.

> For an explanatory model IDE constructed using resources from specialized fields S1, S2, ... Sn, if A (from S1) were to become proficient in the resources of disciplines S2 ... Sn, then A would understand the IDE in its specialized entirety.

Actually acquiring all this additional specialized expertise isn't practical for most researchers, even if it's in principle possible. Knowledge of the counterfactual (suitably filled in for a particular IDE) is a less costly advance. An IDE charts a path from each contributing model to another, which all participating agents can follow if they choose. The connecting bridge directs individuals toward a different field's way of understanding a target.

This epistemic benefit is not just a promissory note for possible future understanding. From an IDE, individuals working in a scientific specialization gain an expanded view of science's epistemic landscape. They can see how their model relates to others, and how explanatory insights go beyond the borders of their own scientific community. The epistemic advance for individual researchers is *social epistemic understanding* of how their own explanatory achievements are related to those of researchers working in other specializations. This notion of relational understanding can be explicated further in terms of the aspects of understanding discussed in Chapter 3, implications of conceiving understanding in the EP/ EP_M framework. To review: at least three aspects or modes of understanding together contribute to a particular scientific community's explanatory purpose of advancing understanding of some target X. These are: understanding of oneself, of others in the community, and of an explanatory model constructed in that community. These coordinated changes follow the tripartite structure of collective intention applied to the bridging relation.

Just as within a particular scientific community, these three aspects of understanding are involved in IDE. An advance in relational understanding via IDE involves changed understanding of oneself, others, and one's field's explanatory model. In the collaborative activity of bridging models, a participant's self-understanding is transformed. Rather than working entirely within their specialized field, they act as a member of a group extending beyond that community. Interplay between group and individual identification affects a participating agent's self-understanding. Concomitantly, a participant's understanding of their counterparts in another field is transformed by participating in IDE construction. They understand those others as having equal standing with respect to the connecting bridge, and thus as engaged in explanatory activity relevant to one's own. The most sophisticated expression of this advance is assessment of one's own model in terms of its relation to another. Users of a model in one specialized community can see how their model connects with those of other specializations in IDR. Members of a different specialized field are understood as collaborative partners in this explanatory activity. Lastly, understanding of an IDE includes the connecting relation between different contributed models. Relational understanding afforded by an IDE is a kind of "self-other" location; that is, locating one's own model in relation to a model constructed from another perspective. This affords a changed understanding of one's own model—from an isolated advance in understanding, to one related to other such advances.

All three aspects of understanding are integrated in the process of bridging models to construct an IDE. These coordinated changes are epistemically significant for members of a scientific community. An individual scientist attains a more comprehensive view of their own field's explanatory model as related to others, 'thicker' and more substantive identification with a wider scientific community than their own specialization, and a correspondingly richer and more nuanced view of the IDE's target. In this way, an IDE expands the 'community of understanding.' The collaborative activity of bridging models creates an IDE in a new, more inclusive community which benefits from both contributing models' advances in understanding of the target (whatever it is). This richer and more inclusive sense of scientific community has further epistemic implications. I'll survey some of these, thereby further explicating relational understanding and its significance.

Most obviously, IDE's minimal ties scaffold further connections. Bridges mark interstices that can be sites for additional connections between explanatory models. In this way, an IDE furnishes a platform for building up to richer modes of interdisciplinarity. Developing additional ties among different specializations is often advantageous for IDR. More extensive connections may in turn lead to changes in explanatory standards. So the distinction between late-stage low-degree IDE and transdisciplinary communities does not mark a dichotomy. The idea is better visualized as "interlace," connections spiraling and proliferating into a tessellation. The modest, relatively low-effort activity of bridging models can in turn generate further common ground, shared values, and so on. Epistemically, insofar as scientists' self-understanding includes identification with a more inclusive group, advances in collective understanding, of scientific knowledge had by "us," become more significant. Participation in collaborative activity affords a richer sense of identification with a more inclusive group—the "we" aspect of participation takes on more significance for individuals. Coherent bridges between different explanatory models accompanied by stable collective intentions to construct such bridges, feed back to enhance group identification. Where initially the sense of "we" involved in constructing an IDE is thin and minimal, a more inclusive scientific community gradually accretes through interdisciplinary activities.

Relational understanding reconciles traditional scientific goals with our epistemic limitations as individuals. No individual scientist today can achieve comprehensive understanding of all or even most explanatory scientific models and, thereby, those models' complex targets. Collectively, however, the scientific community can achieve such comprehensive understanding. Via IDE, individual scientists can both identify with more inclusive groups, expanding their shared understanding, and also, as individuals, understand how the explanatory achievements of their own specialized fields relate to those of other fields. In both ways, IDE advances scientists' understanding, placing their specialized knowledge within a wider epistemic landscape. This corresponds to one lexically important sense of "understanding": fitting that which is understood into a larger whole. In order to make this epistemic gain, participating researchers must let go of the idea of themselves as individuals grasping the entire IDE in its full subtlety, depth, or detail. A full IDE has 'shared explanatory power'—the explanation is distributed among the community. In its use, an IDE's range is likely to extend more widely still. For individuals, the epistemic benefits of IDE are inseparable from attitudes of epistemic (explanatory) humility and a pro-social stance of shared explanatory power. This insight exposes the normative ground of explanatory particularism.

Patchwork unity of science

Comprehensiveness is the explanatory virtue characteristic of IDE. A networked structure of connections between different explanatory models offers an epistemic advance over isolated counterparts. Connected models are, in some sense, better. The idea of relational understanding explicates the advance. Unconnected models cannot be located with respect to one another; they don't 'map together' in an epistemic landscape of science. Of course, a scientific explanation unconnected to any other explanatory models (in its own field or others) is a not thereby a failed explanation. Within a particular scientific community, there is much to say about an explanatory model's contribution to scientific understanding of a target (see Part I). But within-community advances do not exhaust scientific understanding. The comprehensive, relational aspect is exemplified in the sprawling array of hemoglobin models and smaller-scale IDE projects (see Chapter 5). Alongside context-dependent details of explanation and understanding within particular scientific communities, there is a general epistemic advantage to forging connections beyond the boundaries of one's specialization. The notion of relational understanding captures this advantage.

IDE advances relational understanding for members of participating scientific communities. These forms of understanding transcend the boundaries of a single scientific community, offering a more comprehensive view of models, users, and targets alike. One effect of these changes is to generate a more inclusive sense of the group engaged in explanatory activity. Relational understanding strengthens individuals' identification with a group extending beyond the boundaries of their own specialized field. This sense of identification and shared understanding can extend, in principle, to the entire scientific community. Potential IDE connections are rife throughout the sciences. Because bridging connections are potentially available for a far wider array of scientific models than actively sought at any given time through IDR, and the network pattern of bridging relations is unpredictable in practice, this view of IDE amounts to a general (potential) unity of science. This conception of unity presents important contrasts with more traditional views of understanding, explanation, and unity of science.

The IDE account is traditional in one respect, however. It explicates the idea that "understanding" means "fitting into a larger pattern." The latter is a familiar, colloquial meaning for "understanding," which has filtered into philosophical usage. Understanding in this sense is associated with unification. If that which is understood fits into some larger pattern, then that which makes the pattern unifies the target of understanding and the rest of the 'fabric.' It's a short step from these intuitive ideas to identifying the source of 'pattern-formation' with the essence of understanding and/or the locus of explanatory power. That idea is central to Kitcher's theory of explanatory unification.[14] Explanatory particularism makes this move too. But for EP/IDE, the pattern of connections between explanatory scientific models is not derived from a unifying core set of argument patterns. Instead, the pattern emerges from collaborative interactions among scientists. On this view, the locus of explanatory power and understanding is distributed, diversified, and decentralized. In this way, comprehension as an explanatory virtue extrapolates to a non-traditional ideal of unity of science. That ideal is realized via patchwork unity of scientific communities connected via bridging (and more substantive) relations in a fabric of explanatory models. Individual scientists locate themselves in this complex epistemic landscape via relational understanding; 'knowing where you are' as a scientist in a wider array of epistemic communities.

Patchwork unity of science, interconnected epistemic communities linked via bridging relations, replaces the old idea of a hierarchy of scientific knowledge. That earlier view was most strikingly articulated in Hempel and Oppenheim's "working hypothesis of the unity of science" (see Chapter 5). Although their hypothesis is now largely abandoned, entrenched background assumptions about scientific hierarchy continue to influence philosophical studies of explanation. The background character tends to conceal the fact that conceptions of unity of science positing hierarchical epistemic relations among the sciences are by virtue of that assumption *social as well as epistemic* theories of science. They are social in that they imply asymmetries of power and authority that impact social interactions across scientific community boundaries. Philosophical accounts of unity of science and systematization of knowledge are not socially neutral. They spring from considerations similar to those that motivate EP. Even in the mid twentieth century heyday of physics' primacy, philosophers of science were well aware of differences in explanatory practice across fields (Nagel 1961, Hempel 1965). Insofar as the classic accounts engage this issue, each offers a 'blueprint' of relations

[14] See Chapter 1.

among the sciences. Systematic, fixed relations among scientific disciplines became an intellectual preoccupation around the time scientific knowledge across multiple disciplines outstripped any individual person's cognitive capacities. If one can't grasp all of scientific knowledge, in all its specialized detail, then at least one can grasp its essentials and principles of organization. Commitment to such individualistic epistemic ambition in the face of rapidly proliferating scientific knowledge and specialization has profoundly influenced philosophy of science. Philosophers want to epistemically have it all.

While EP/IDE doesn't entirely jettison that ambition, my account does dispense with an individualistic orientation and any lingering assumptions of scientific hierarchy. Explanatory particularism asserts that there is no single, simple principle of organization for scientific explanation and understanding. Instead, we have a dynamic array of splitting and merging lines of inquiry, guided by commitment to different styles of explanation. Resulting explanatory models are a motley, but they can and sometimes do connect with one another. Those connections too are diverse. The model–model framework offers a way to classify them, and to identify general types of interdisciplinary projects aimed at explanation (Chapter 4). Late-stage low-degree connections are analyzed as bridges between models. More extensive, effortful connections are established in transdisciplinary projects. The complex, dynamic pattern of connections between explanatory models *is* the explanatory organization of the sciences. That organization may not be systematic, orderly, or univocal. But doesn't mean there's nothing to understand about it. On the contrary: mapping the various connections between different explanatory models is an important aspect of scientific understanding. Relational understanding of this sort helps consolidate explanatory understanding and generate a more inclusive view of scientific practice. Philosophers of science have long sought to map the epistemic organization of science, and in this if little else explanatory particularism follows tradition. However, EP does not impose 'external' ideals on different sciences as the price of this understanding.

Instead of any fixed hierarchy among sciences, the particularist account posits (potential) unity via a patchwork of lightly connected explanatory achievements. The most inclusive view of science is as a dynamic arrangement of diverse specialized epistemic communities. Abstracting away all but those communities' explanatory achievements and relations among them, yields an image of the patchwork unity of scientific understanding. Diverse achievements of understanding are variously connected via bridging relations into a tapestry of interconnected IDEs. The pattern of relations continuously changes as inquiry progresses. Some specialized explanations grow richer connections, while other IDEs fall by the wayside. This patchwork unity doesn't elide, reduce, or ignore diversity; differences among specialized explanations are not elided, subsumed, or reduced. Instead, coherence requirements for bridging models allow (but don't guarantee or require) connection between even very disparate explanations. For any pair of models, coherence conditions are permissive, limited only by researchers' own ingenuity in adjusting explanatory models and seeing connections with others. Unity in this sense is *possible*, through collaborative effort. So it's fair to call this 'unity of science,' even though the unity is potential and compatible with a wide variety of specialized scientific explanations.[15]

Contrast with traditional philosophical theories of explanatory unification and epistemic organization among the sciences is stark. Kitcher's explanatory unification is something of a limit case here, treating the scientific community as a monolith. On his account, to explain a phenomenon is to subsume it under a general argument pattern belonging to a core set of such patterns that together provide the best systematization of the entirety of scientific knowledge at a particular time. Although Kitcher does not explicitly define "the scientific community," he evidently treats it as unstructured and undifferentiated. All explanatory argument patterns accepted within a scientific community at a particular time are lumped together in the explanatory store. Furthermore, the concept of the best overall systematization of scientific knowledge, characterization of "explanation [as] a matter of instantiating patterns that are applied *globally*" to represent the order of nature (1989, 468), and requirement of completeness in determining membership in the explanatory store, indicate that Kitcher conceives the scientific community in a broad, inclusive sense.[16] But this monolithic approach to the social aspect of science elides differences across fields and disciplines in explanatory models' construction, evaluation, and use. Kitcher's unificationism sets aside questions of conflict, tension, and tradeoff among explanatory virtues that determine what argument patterns are included in a community's explanatory store at a particular time. And although Kitcher takes pains to show that his account allows for conceptual/explanatory change and progress, he does not address contemporaneous relations between different scientific fields.

In contrast, Hempel's classic D-N account is associated with a more variegated image of relations among the sciences. The overall, idealized image of science is as a hierarchical system of laws ordered by logical derivation, stemming from the most general covering laws to the myriad termini of particular observations. This image

[15] This recalls Neurath's pioneering "encyclopedic unity of science": "unity by integration whereby disciplines are connected to one another at the edges, guided by ideals of coherence and without requiring any sort of hierarchical subordination of other disciplines by one super discipline, for example" (Grüne-Yanoff and Mäki 2014, 53n.2; see also Cat and Tuboly 2019).

[16] See also: "We may think of the sciences not as providing us with many unrelated individual arguments which can be used in individual acts of explanation, but as offering a reserve of explanatory arguments, which we may tap as need arises" (Kitcher 1981, 512); "Intuitively, the explanatory store associated with science at a particular time contains those derivations which collectively provide the best systematization of our beliefs. Science supplies us with explanations whose worth cannot be appreciated by considering them one-by-one but only by seeing how they form part of a systematic picture of the order of nature" (1989, 430); "Conceive of science as a sequence of practices, each of which is distinguished by a language, a body of belief, and a store of explanatory derivations" (1989, 497).

of science as "unity of laws" was famously articulated by Oppenheim and Putnam as a working hypothesis (1958). They explicate this sense of unification in terms of theoretical reduction ("micro-reduction"), taking "theory" to refer broadly to any hypothesis, generalization, law, or conjunction of these.[17] Theory T2 reduces to theory T1 just in case (1) T2 contains vocabulary not in T1, (2) any observations that can be explained by T2 can also be explained (predicted) by T1, and (3) T1 is at least as well systematized as T2. Systematization, though not defined, refers to orderly logical relations among theories within a field of science, and is related to simplicity and explanation. The latter term is defined as for D-N; i.e., logical derivation given initial conditions. According to Oppenheim and Putnam, then, theories are unified by reduction, conceived and displayed as logical derivation of less from more general laws of nature (i.e., laws covering fewer or more empirical observations).

Oppenheim and Putnam go beyond D-N to consider reductive explanatory relations among "branches" of science. One branch is reduced to another by treating all the accepted theories of each as T2 and T1, respectively. This idealized account presumes that each branch of science has clearly distinguishable observational and theoretical vocabularies, and moreover that the observational vocabulary of reduced and reducing sciences coincides. Reduction relations are transitive (as well as irreflexive and asymmetric), so reduction among branches of science is cumulative. Oppenheim and Putnam's working hypothesis therefore entails that all explanatory power derives (literally) from one ultimate "reducing" branch of science, which has as its subject matter the most fundamental parts of all objects of scientific study. They further assume that D-N explanation tracks part-whole relations between the objects studied by distinct branches of science: "The essential feature of a micro-reduction is that the branch B1 [reducing] deals with the parts of the objects dealt with by B2 [reduced]" (1958, 6).[18] So Oppenheim and Putnam take an ontic view of relations among sciences that (they assume) parallel the derivational hierarchy of covering law (D-N) explanations. Our presently-recognized scientific disciplines, they propose, approximate idealized branches with distinct theoretical vocabularies, neatly stratified according to the part-whole hierarchy of elementary particles, atoms, molecules, cells, multicellular organisms, and social groups (9). These six "reductive levels" order the sciences according to their universes of discourse. Oppenheim and Putman stipulate that this part-whole

[17] Oppenheim and Putnam (1958, 29 note 2). Note that this sense of unity (of language and of laws) is distinct from unity of methods in scientific practice—including methods of constructing and evaluating explanations (1958, 5).

[18] Oppenheim and Putnam do not provide a detailed definition of the Pt. relation (part–whole), but indicate that "wholes" should be understood in a wide, permissive sense, and "parts" as anything spatially or temporally contained in a whole (11).

hierarchy parallels the logical hierarchy of D-N explanation; laws that predict attributes of parts are more general than laws that predict attributes of wholes.[19] This is a flagrantly ontic conception of relations among different sciences. Rather than engage explanatory practices of various scientific fields, Oppenheim and Putnam propose an objective hierarchy of targets of scientific explanation, which they take to coincide with D-N's hierarchy of laws.

Oppenheim and Putnam's classic reductionist account of unity of science is no longer defensible. Few if any philosophers today would defend this reductionist vision of unified science, with its six-layer hierarchy of objects, assignment of distinctive predicates for each level, a shared observation vocabulary across all sciences, and "over-all physicalistic reduction" (9). Reductionism of this sort asserts that the fundamental parts of all objects are the objects studied by fundamental physics, that those fundamental parts are the ultimate reducing level, and theories about those particles the ultimate reducing theories. Yet, although philosophy of science has moved beyond Oppenheim and Putnam's hypothesis, their view of relations among different scientific fields remains influential. Existing disciplines of physics, chemistry, biology, psychology, and social science approximate the hierarchically organized branches of ideally unified science. Assumptions about explanatory order and hierarchy among these different sciences, which conform to Oppenheim and Putnam's explicit proposals, often lurk in the background of contemporary debates about scientific explanation. The idea that more general laws governing more fundamental entities and processes explain the descriptions proposed as explanatory in disciplines concerned with less general/fundamental phenomena remains appealing to many philosophers.[20] Explanatory particularism jettisons the reductive presumption, as well as the associated idea that asymmetry of explanation is paralleled by unequal social relations among sciences. Patchwork unity of science via diverse bridging relations replaces traditional hierarchies of reducer and reduced, subsumer and subsumed, dominant and subordinate.

Explanatory values are pro-social values

Extrapolated to an ideal of scientific practice, the view of understanding emanating from IDE amounts to a conception of scientific objectivity grounded in pro-social values. We can now pick up a final thread from the end of Part I. There, I hinted at an ideal of scientific objectivity compatible with explanatory particularism. Chapter 3 proposed shared understanding of significant targets as a constitutive

[19] A crucial assumption is that the laws of one level include the objects of higher levels within their scope; "the objects in the universe of discourse of B2 are whole which possess a decomposition into proper parts all of which belong to the universe of discourse of B1" (6).

[20] The blueprint of relations among the sciences is grounded on, and shaped by, explanatory practices (including epistemic virtues) of theoretical physics.

aim of science; aka, a final end. This ideal holds straightforwardly within particular scientific communities. The IDE account shows how and why understanding can extend beyond particular communities. The argument for shared understanding as a constitutive goal of science holds for IDE connections as well. Briefly: explanatory practices must, on pain of incoherence, be directed towards achievable ends; i.e., goals that aren't impossible for the activity as conceptualized. This applies to social activity in general. Means-end coherence demands that participants in a social activity believe or accept that the shared goal of that activity is possible to achieve. Although explanatory models are deeply and significantly specialized, many scientists aim to transcend such specialization, producing understanding beyond the boundaries of their own community. Interdisciplinary projects that construct bridging connections arise unpredictably in practice. Members of scientific communities can't specify in advance the others they might collaborate with in future. It follows that bridging connections between models can't be specified in advance. Nor can the boundaries of scientific communities and specialized features associated with understanding. All these change over time, as inquiry proceeds. Therefore, scientific understanding sought over the long-term cannot be community-specific; to conceive of it that way is practically (means-end) incoherent.[21]

The ideal of shared understanding captures scientific aspirations to go beyond ephemeral, community-specific features of explanatory models. Science doesn't aim solely at isolated islands of understanding localized to particular communities. Relational understanding, and the ideal of patchwork unity of science, cash out this broader aspiration in terms stemming from tenets of explanatory particularism. Although not a requirement for all scientists all the time, comprehensive relational understanding is one important value for science. Commitment to forging connections between explanatory models is one way that scientific understanding escapes the boundaries of particular specialized communities. Not only do scientists seek explanations that fit the epistemic values and other contextual features of their own community, but they also strive to connect those achievements to explanatory models in other fields. Insofar as scientists identify as members of a more inclusive epistemic community, this reinforces the ideal of shared understanding and the associated norm of forging connections with other specialized fields. In this way, IDR involving IDE feeds back to justify and ground an ideal of ever-more inclusive shared understanding. Collaborative attitudes and associated norms ground a wider conception of objectivity; of understanding as shared. This ideal of scientific objectivity is important though not exhaustive of scientists' epistemic aims. Objectivity in this sense is a positive achievement,

[21] This argument has some parallels with Peirce's pragmatic ideal of scientific knowledge (1878), as independent of the opinions of any particular community.

overcoming countervailing parochial tendencies. Bridge-building to construct an IDE both conforms to and helps to realize that ideal in scientific practice.

This concept of objectivity is not ontic, veritistic, or hierarchical. But it articulates an important scientific goal even so. The ideal is grounded on epistemic benefits of overcoming fragmentary understanding and narrowness of specialized knowledge restricted to members of a particular community. The IDE account shows how those benefits can be secured by means of pro-social, collaborative attitudes had by members of scientific communities. Pro-social attitudes required for bridging explanatory models (tolerance, inclusion, and respect) are thus means to generate shared understanding and a richer sense of community across different sciences. In this aspect of scientific practice, explanatory and social virtues are one and the same.[22] An explanatory virtue, comprehensiveness, is entangled with social virtues. The minimal conditions for bridging different explanatory models are satisfied by attitudes that facilitate cooperation among agents: humility, tolerance, respect. Shared understanding—objectivity in a particularist sense—is achieved by means of pro-social virtues; norms of tolerance, inclusion, and respect. If scientists' attitudes are in accordance with these norms, they engage with members of scientific communities outside their own as equal partners in constructing an explanatory model—an IDE.

It follows that the IDE account has broader implications for science–society relations. My account rejects any strict, absolute, or univocal demarcation between 'science and society.' The normative ground of IDE is at once epistemic and social. In this sense, then, EP/IDE is *anti-demarcationist*. According to explanatory particularism, science is the furthest thing from a monolith. So there cannot be a single sharp dividing line separating scientific from other groups, at least as regards explanation. Furthermore, the IDE account extends straightforwardly to projects involving scientific communities and groups in other areas of society, provided there's a shared aim of understanding—some shared target. There are many, many opportunities to build bridging connections with other social groups.[23] A related implication of this view is that *non-expert understanding* plays an important positive role in fulfilling scientific aims. Non-expert understanding is featured in bridging relations: far-side and middle coherence. Everyday, non-technical notions of, for example, cause-effect or part-whole relations mediate bridging connections between specialized models. These non-expert aspects of a bridging connection blunt the edges of what might otherwise be incompatible specialized elements of explanatory models. In bridging relations, naïve, commonsense, generally accessible concepts and interpretations aren't a weakness, but a strength. It

[22] Others have put forward similar ideas; e.g., Rolin (2015, 173) "Some moral and social values should be permitted to play a role in acceptance because they are woven into the epistemic fabric of scientific collaboration." Dang (2019) also mingles social and epistemic values in science.

[23] For example, Alison Wylie's work (2015) on collaborations between archaeologists and indigenous communities.

follows that in the IDE account non-expert aspects of understanding are not only unavoidable in IDR, but also play an important epistemic (and pro-social) role. So non-expert understanding isn't merely watered-down or simplified scientific knowledge, nor is it inherently inferior to specialized scientific understanding.

This is a sharp break from philosophical tradition. The value and significance of non-expert understanding—which extends beyond scientific communities into the wider public—has generally been underrated by philosophers. A classic assumption shared by general epistemology and philosophy of science is that deep theoretical knowledge (understanding) is possible only within a narrow specialization. As expressed by Kvanvig:

> the common assumption is that understanding of the theoretical sort is a species of knowledge . . . the kind of understanding at issue when regarding our cognitive successes and achievements is some type of deep and comprehensive knowledge concerning a particular subject, topic, or issue (2003, 188).

I do not dispute that deep understanding (or knowledge) can result from working within a narrowly constrained specialization. Indeed, that's a core tenet of EP. But the IDE account does belie that this is the only scientifically important variety of understanding. On my view, non-expert understanding is not only crucial to science–society relations, but helps to constitute science as an inclusive community. An inclusive community of shared understanding is stitched together out of a tessellation of deep and shallow understanding.

The above considerations underwrite a *norm of connection* for scientific practice; i.e., a directive for practicing scientists. The directive is that scientists should aim at connections outside their particular epistemic community, so as to produce relational understanding and progress toward the ideals of patchwork unity and objectivity as shared understanding. There are, to be sure, many norms and values in play concerning explanation in any particular scientific community. The norm of connection is one among many. Other things being equal, it's better for an explanatory model to connect with those of other sciences than not. Connections between explanatory models are an important epistemic achievement for science. Failing to make these connections, conversely, is grounds for criticism. So explanatory particularism has some normative bite, as regards IDE. This sets my version of explanatory particularism apart from full-blown relativism about explanation and understanding. EP, as developed in this and preceding chapters, has grounds to criticize the explanatory practices of particular scientific communities. An explanation that satisfies standards of one epistemic community may be an achievement in its own right. But it should also connect with achievements of understanding in other epistemic communities, to provide relational understanding for members of that community. Failure to do so indicates an epistemic weakness in that scientific community. A field that touts its explanatory achievements yet lacks *any* bridging connection to explanatory models in *any* other specializations fails in an important aspect of science. Such a community might well show other kinds of interdisciplinary connections and practices. But if there are no mutual, non-hierarchical, collaborative connections with explanatory models in other specializations, then that community is isolated from the epistemic benefits generated thereby, whatever its other strengths.

What is the failure in such cases, exactly? For one thing, explanatory achievements that lack bridging connections are indicative of epistemic hubris. Absence of relational understanding goes hand-in-hand with attitudes that hinder collaboration. It's likely that such a field is in the grip of explanatory practices and values that inhibit IDE and encourage anti-collaborative attitudes, as well as epistemic hubris. Of particular concern are claims of explanatory privilege incompatible with clear, explicit bridging relations. Such claims are likely to arise in specialized fields committed to "explanatory imperialism" and/or dependent on institutions that have interests other than advancing scientific knowledge.[24] Cases of this sort (e.g., evolutionary psychology, certain areas of economics) are ripe for philosophical critique. Such critique is the other side of coin, so to speak, of the normative claim made here: establishing connections with other specialized fields, through IDE and other interdisciplinary ties, is an important norm for science. The coherence constraints for bridging relations are a minimal standard for such connection, directions for an interconnected fabric of scientific understanding.

Relatedly, the failure can be construed as lack of objectivity, in that a field's explanatory models remain isolated from others. Understanding of both model and target are then sequestered within the community. On the other hand, satisfying norms for IDE involves a shift in individual researchers' perspective, to a more inclusive view of science. A participating agent locates herself and others within a scientific community that is more inclusive than her own community. Disregard for such relational understanding implies disregard of ideals of shared understanding and failure to engage with a wider scientific community. Social, relational understanding counter-balances increased specialization and fragmentation of scientific knowledge. Such counter-balance is valuable for practical reasons as well as epistemic ones. Notably, fragmentation makes the sciences vulnerable to other interests: economic, political, and other social forces. Making more of a 'society of the sciences' via myriad IDE connections is therefore valuable. The upshot is that it's worthwhile, for a number of reasons, to view science as more than a set of distinct lines of inquiry vying for intellectual supremacy. On the account presented here, different explanatory models in a network are not competitors. Rather, each field contributes something to the overall fabric of explanatory models, with no

[24] The term "imperialism" here is from Mäki (2013).

single model dominating. None overcomes others. Instead, they form an interconnected, collaborative array.

6.3 Other philosophical accounts: understanding and beyond

How does the EP/IDE account of understanding compare with other philosophical accounts? Classic studies of explanation relegated understanding to psychology or history; a subjective byproduct of explanation.[25] More recently, understanding has emerged as a distinct topic for philosophy of science. But so far, there is little consensus on how to theorize understanding. Approaches range from traditional analytic epistemology to naturalistic science studies.[26] A few key debates have coalesced. Is understanding a requisite for successful explanation, or can the two come apart? Is understanding "factive"; i.e., must an agent in some sense "grasp the truth" to have understanding, or not? How is understanding implicated in the role of idealized models in science? There is currently no consensus on these matters. But a few features of understanding are generally accepted:

- Understanding applies to (is had by) a psychological subject or epistemic agent. (So there is a subjective aspect; whether any objective aspect is combined with this is debated.)
- Understanding is a cognitive achievement for a subject or agent. (The nature of this achievement, and its relation to explanation, is debated.)
- There are degrees of understanding; epistemic agents can have more or less understanding of something, or their state of understanding can vary with respect to qualities that make it more or less good.
- Understanding involves "grasping" something so as to see relations among things; how things hang together. (The nature of grasping is debated, but connotations of unification and coherence are generally accepted.)

The EP/IDE view of understanding concurs with all these—to a point. Some are implied by explanatory particularism's basic thesis:

> (EP) A scientific explanation is a product of a particular scientific community that is significant, specialized, and advances understanding for that community.

[25] Michael Friedman's pioneering unificationist account was an exception—but defined understanding reductively, shorn of subjective aspects. Friedman's objective constraint: "what counts as an explanation should not depend on the idiosyncrasies and changing tastes of scientists and historical periods. It should not depend on such nonrational factors as which phenomena one finds somehow more natural, intelligible, or self-explanatory than others" (Friedman 1974, 14). See Bangu (2017) for more discussion.

[26] See, for example, de Regt et al. (2009), Strevens (2013), Faye (2014), Le Bihan (2016), de Regt (2017), Khalifa (2017), Rice (2021).

Epistemic agents and their activities are obviously central for EP/IDE. Understanding as a cognitive achievement fits with EP also. Explanation is the primary achievement, cashed out as an advance in understanding of some target X for members of a scientific community (or interdisciplinary project, in the case of IDE). Degrees of understanding are implied as well; an EP-explanation *advances* understanding—that is, increases the degree of understanding for members of an epistemic community. The IDE account is an articulation of the idea that understanding consists in an epistemic agent's grasping relations among things, associated with unification and coherence. So the view of understanding presented here is not a complete departure from ongoing philosophical discussions. Nevertheless, the EP/IDE account contrasts with other philosophical views of understanding in important ways.

First and foremost, my account of understanding is particularist, eschewing any general substantive definition of 'understanding' for all of science. Beyond EP/EP_M's very thin general characterization, the concept of understanding is left open and unspecified. Specification is via study of particular scientific communities, notably scientists' own views about the features of good explanations and how those features advance understanding of some target. Clarification of a scientific community's notion of understanding arises in particularist study of its explanations. EP-explanations are suited to and reflect their community context: technology, other resources (funding, access, etc.), historical ties to earlier communities, epistemic traditions instilled by pedagogy, and more. Individual researchers within a scientific community are likely to have somewhat varying ideas as to the key features of an explanation and an advance in understanding. But for a model to be recognized as a major achievement in a scientific community, there must be some consensus on a cluster of its virtues and attendant advantages. The latter amount to a characterization of what 'understanding' means for that scientific community.

Most philosophical accounts of understanding take their cue from philosophical theories of explanation. So monism about scientific understanding is the default approach (see Chapter 1). The main accounts propose substantive conditions for an agent's understanding across all scientific fields, sometimes for all epistemic agents. (De Regt's contextual account is an exception here: more on his view shortly.) Even so, monism about understanding is less entrenched than monism about explanation. Most philosophers of science who concern themselves with that topic allow for varieties of understanding, individuated by object and/or mode: understanding-why, understanding-how, understanding-what; understanding of phenomena, theories, or objects; understanding of words or concepts vs. understanding of the natural world, etc. So pluralism about understanding is a common view. But this pluralism is akin to limited pluralism about explanation: there are several kinds of understanding, individuated by broad categories largely borrowed from general epistemology. Different philosophers offer somewhat different classifications. For example, Strevens (2013) distinguishes understanding-that, understanding-why, and understanding-how, while de Regt and colleagues (2009) distinguish understanding of theories, phenomena, and objects. These coarse-grained distinctions lump together the modes of understanding in different scientific communities. Particularism about understanding, in contrast, opens the door to a much greater variety of conceptions of understanding, each tied to and arising out of the complex combinations of values, technologies, methods, and other constraints of a particular scientific community (at a given time). So the gap between explanatory pluralism and particularism, argued in Chapter 1, holds for understanding as well.

However, my EP/IDE account reveals one kind of understanding as especially salient, because heretofore underappreciated: understanding advanced by IDE. Relational understanding doesn't supplant the particular conceptions of understanding attributed to different scientific communities, but flourishes alongside them. Contrasting this kind of understanding with other philosophical accounts of understanding will clarify the idea further. Yet the primary contrast between (most of) those accounts and my own is that between particularism and monism, just as for explanation.

Contextual understanding (de Regt)

Of the philosophical accounts of understanding currently on offer, the most similar to mine is the contextualist theory articulated by Henk de Regt and his collaborators (2005, 2009, 2015, 2017, 2019).[27] Although some details have changed since its initial presentation (de Regt and Dieks 2005), core ideas and themes remain consistent. For de Regt, understanding is a matter of agents' skills and abilities, not an attitude of grasping the truth. (Many other philosophers do endorse a "factive" conception of understanding, however.) As noted, de Regt distinguishes understanding phenomena from understanding theories. The former he takes to be a central aim of science, accomplished via explanation. On that point, de Regt's theory fits current philosophical consensus. Where he departs from other accounts is in the nature and role of understanding a theory. De Regt interprets the latter as pragmatic, a necessary condition for understanding a phenomenon. Pragmatic aspects of explanation are central, on his view: activities and skills of individual agents working in scientific contexts. Specifically, the activities and skills used to construct a model of some phenomenon on the basis

[27] See also Wilkenfeld's view (2013, 2017), which elaborates on de Regt's ability theory to define scientific understanding in terms of scientists' mental models and their (internal) manipulation, together with a requirement of accurate representation of the target object.

of a theory: "(CUP) A phenomenon P is understood scientifically if and only if there is an explanation of P that is based on an intelligible theory T and conforms to the basic epistemic values of empirical adequacy and internal consistency." (2017, 92)[28]

On de Regt's view, explanations that afford understanding of phenomena are based on—constructed in accordance with—*intelligible* theories. Intelligibility depends on not only on a theory's qualities (e.g., visualizability, simplicity, unifying power) but also on the skills scientists bring to applying it. The same theory can be made intelligible to different agents in different ways; intelligibility is subject-relative and context-dependent. De Regt defines "intelligibility" as the value attributed by scientists to qualities of a theory that contribute to understanding a phenomenon.[29] This is not unlike EP's idea that features of a scientific model advance understanding for a scientific community, in accordance with that community's explanatory values. As with EP, de Regt argues that subjective, pragmatic aspects of intelligibility do not thereby yield an unacceptably relativist theory of understanding. However, his argument strategy is unlike mine. For de Regt, subjective aspects of intelligibility are constrained by criteria for intelligibility that "objectively test" for that quality in a particular context. He discusses one such criterion in detail: scientists "can recognize qualitatively characteristic consequences of [theory] T without performing exact calculations" (2019, 336). This is in line with de Regt's near-exclusive focus on history of theoretical physics as a source of examples and case studies for his account. His contextual account is, in practice, restricted to a very narrow swathe of scientific contexts.

De Regt's account of understanding has significant parallels with the view put forward here. Most obviously, his view acknowledges that diverse epistemic values influence conceptions of understanding across scientific communities. The value-laden epistemic context in which scientific agents construct and use models of a phenomenon is an important determinant of scientific understanding, alongside features of theories and basic epistemic values related to scientific success. In a co-authored 2009 statement, de Regt explicitly endorses particularism about understanding:

> Whether a particular explanation makes scientists understand a phenomenon may depend on the context (both social and material) in which those scientists find themselves ... the historical and disciplinary context determines at least in

[28] De Regt and Gijsbers (2016) replace CUP with the "scientific effectiveness condition for understanding," which combines the intelligibility requirement with "reliable scientific success" cashed out as correct predictions, practical applications, and fruitful lines of research.

[29] More precisely, de Regt defines "intelligibility" as "the value that scientists attribute to the cluster of qualities of a theory T (in one or more of its representations) that facilitate the use of T (2017, 40); and as "the value that scientists attribute to the theoretical virtues that facilitate the construction of models of the phenomena" (2015, 3781).

> part when and how understanding is achieved ... This contextuality implies, in turn, that it is difficult, if not impossible, to give a universal definition of (the essence of) scientific understanding.
>
> (3–4)[30]

Relatedly, de Regt's account includes the dynamic social organization of science, casting understanding as an aim for science as a whole, for scientific communities, and for individual scientists (2017, 90).[31] That is, he allows for understanding at the grain of scientific communities, although individual agents are his main focus. Also like EP/IDE, de Regt's theory eschews ontic and realist requirements for scientific understanding. Instead, he approaches scientific understanding as an activity unfolding in historical context. His contextual account explicates understanding in terms of model-construction and use. This parallels EP_M and my distinction between aspects of understanding (self, others, model) that bear on understanding some target.[32] So there are significant similarities with my account.

There are also major differences. Although de Regt allows for understanding at the grain of scientific communities, social epistemic organization figures in his account only as background for individual scientists and their epistemic values. Individual scientists' values and their associated modeling practices take center stage. Indeed, two scientists' epistemic contexts may differ even if a phenomenon and its explanation are the same. So de Regt's primary grain of analysis is much finer than EP's, even IDE's. In my account, epistemic values, modeling practices, and resultant explanations are conceptualized primarily at the grain of specialized communities. Individual scientists participate in collaborative activities that produce EP-explanations and, for some interdisciplinary projects, IDEs. Another significant contrast is that de Regt's view has no counterpart to EP's social action framework. (Nor does any other philosophical account of understanding of which I am aware.) That framework grounds norms for collaborative activity, coherence requirements for IDE, the value of patchwork unity and relational understanding, and the associated ideal of shared understanding (an EP-compatible conception of scientific objectivity). So all these ideas are distinctive to IDE. The notion of fitting one's explanatory model into a broader epistemic landscape of connected models that together advance understanding of targets, is central to my account. Changed understanding of an agent's own activities aimed at explanation, and those of scientists in other fields, is implicated in IDE alongside relational understanding

30 Corroborated by a more recent statement: "the demand that theories be intelligible does not pose universal restrictions on theories or theory choice, but it does entail that certain theories are preferable over others in particular local contexts" (2019, 341–342).

31 See also Humphreys (2000).

32 Most pointedly, De Regt's distinction between understanding a theory and a phenomenon parallels my distinction between understanding a model and a target.

of scientific models. The norm of connection for explanatory models, objectivity as shared understanding, and bridging connections secured by pro-social attitudes among scientists, are all further implications. Although de Regt and a few others (Elgin, Potochnik) also associate understanding with models/modeling, their accounts don't focus on community structure of science and social action norms for epistemically significant connections within that structure.

Yet another contrast with de Regt is that his theory is based almost exclusively on examples from history of theoretical physics.[33] No other scientific fields, or practices of explanatory modeling, are considered directly. So although de Regt's account is open to contextual differences in criteria for intelligibility of theories, no other criteria are discussed. His account of understanding presumes that scientists achieve understanding of phenomena by using theories to construct explanatory models. But that practice is not universal among scientific fields—as the modeling examples in Chapters 2 and 5 show. Indeed, if "theory" is interpreted as a set of laws in the form of equations yielding exact predictions under ceteris paribus conditions (or something along those lines), then the modeling practices de Regt casts as fundamental for understanding are restricted to theoretical physics, neo-Darwinian evolutionary biology, and some areas of engineering. Of course, physics-centrism and accompanying myopia about scientific practice are long-entrenched biases in philosophy of science; this isn't a criticism of de Regt. A suitably relaxed notion of 'theory,' with more focus on scientific communities rather than individual scientists, would align de Regt's view quite closely with EP_M's framework. To sum up: de Regt's account of understanding is particularist and also resembles EP/EP_M in emphasizing activities and processes, diverse values in science, and the role of models in explanation and understanding. But his view has no counterpart to IDE and collaborative norms and ideals, nor the blending of social and explanatory values defended in this chapter.

Group understanding

There is, at the time of writing, very little philosophical literature on understanding for groups (though de Regt 2017 and Humphreys 2000 endorse it). Kenneth Boyd's (2021) treatment is the only work of which I'm aware that fills this gap. Boyd proposes a minimal conception of understanding, comprised of three necessary conditions for an agent S (individual or group) to understand (why/how/that) a proposition p: S must possess a representation of p and reasons that support p, S's representation of p accurately mirrors reality, and S grasps the relationship between p and its supporting reasons (6842).

[33] De Regt does discuss Chang's (2012) reevaluation of the Chemical Revolution, but this treatment remains within the history of physical sciences.

The resemblance to traditional analyses of knowledge in general epistemology is not coincidental. (More on that shortly.) Boyd's main focus is the third condition. "Grasping" for an individual agent, while not theoretically limpid, has been analyzed by a number of philosophers. But how to conceptualize grasping distributed among members of a group? Boyd proposes that a group G grasps some proposition p just in case "(i) G represents p and reasons for p, and (ii) the members of G are mutually p-reliant" (6852). That is, there is some representation of G within the group (distributed somehow, unspecified), and members rely on one another to use that representation. Relations that constitute mutual reliance involve the elements of social action: a shared goal, common awareness, and attitudes of trust toward other group members via trust. All this is compatible with my social action framework, although Boyd uses somewhat different terms and doesn't focus on bridging relations between models. The parallels make us fellow-travelers, though.

Jan Faye's pragmatic approach also engages social aspects of science (2014). His "organizational" theory exhibits several parallels with my account. Most distinctively, Faye argues that local epistemic norms are socially constructed (59). This is very much in line with EP/EP_M, although as noted the social action framework is unique to my account (and so the specific collaborative form that construction takes, is not the same in our two views). But the idea that 'explanatory power is social power' is one I share with Faye. Explanatory models yield an advance in understanding because and in accordance with ideas and values pertaining to explanation/understanding that are prevalent in a scientific community. A further parallel is that Faye explicates understanding in terms of coherence, as "grasping how pieces of information relate to one another" (2014, 59). The latter view of understanding is shared with many philosophers, including those working in general analytic epistemology. (See next section.) However, Faye's analysis of "reflective understanding," the variety associated with science, contrasts with my account in important ways. For Faye, reflective understanding is of states of affairs, in terms of a theory that is correct in accordance with an agent's background. As with explanatory particularism, understanding is context-dependent. But Faye explicates the relevant background context for an agent as a "worldview" (54). A worldview, or all-encompassing theory, is an obstacle to IDE, incompatible with bridging relations that demand recognition of explanatory contributions from outside one's own specialized field, via another kind of explanatory model.

Understanding in general epistemology

In general epistemology, understanding is theorized as an attitude, either distinct from or a special case of knowledge. The attitude of understanding is thought to involve *seeing or grasping relations between diverse items.* For example, Kvanvig

(2003) posits that "[u]nderstanding requires the grasping of explanatory and other coherence-making relationships in a large and comprehensive body of information ... understanding is achieved only when informational items are pieced together by the subject in question" (192). The key idea is "piecing together" a set of diverse informational items to form a coherent explanatory system. There are thematic parallels here with IDE's relational understanding. The latter also involves relations among diverse items (a kind of unification) and is associated with coherence. Those ideas are central to the IDE account and the associated view of unity via a network of models linked by bridging relations constrained to be coherent. Kvanvig sees the value of understanding as systematization and organization of an individual's knowledge, qualities that contribute to epistemic justification and are valuable in their own right. The connection to coherentist theories of justification is obvious and explicit (cf. BonJour 1985, 192–193). For Kvanvig and most other general epistemologists, however, the relations involved in understanding are all "internal"; i.e., held in the mind of a single cognitive agent. This is in keeping with epistemology's long-standing individualistic orientation.

Epistemic communities and their relations, including norms for collaborative activity, do not figure in general epistemological accounts. In contrast, on my view understanding is distributed across different scientific communities. The value of IDE specifically involves wider sharing of (some) varieties of understanding among members of different scientific communities. The latter idea has no counterpart in general epistemological theories of understanding. Of such theories, Linda Zagzebski's (2001) is perhaps the most akin to the view presented here. Zagzebski theorizes the 'unit' of understanding not as a single proposition, but as a whole field of knowledge—a domain of (practical) expertise (240–241). The value of understanding, she argues, derives from its contribution to expertise; i.e., skill or ability in some specific area. Within such an area, one who understands is able to explain the interrelationships of diverse elements within a field of knowledge, and to explain how some item "fits into the larger framework" (241). Zagzebski's view of understanding applies primarily to individual persons, but can also extend to groups—particular epistemic communities. Understanding so construed is an epistemic achievement, cashed out as reliable success in accomplishing the goals of the techne; "demonstrated [internally,] within a practice" (246). Although Zagzebski's philosophical project is motivated by very different concerns (in her case, history of theories of knowledge originating with Plato), there are a few points of contact with explanatory particularism.

Contra Zagzebski and de Regt, Stephen Grimm argues that understanding of natural phenomena is factive; a species of knowledge (2006, 2010, 2012). On his view, understanding is the goal of scientific explanation, both including a truth condition. This rebuts the classic philosophy of science distinction between subjective/psychological understanding and objective/factive explanation. Grimm's philosophical approach is broadly integrative; he seeks to bring epistemology and philosophy of science closer in the matter of understanding. In so doing, he offers an inclusive account, arguing that explanation is more psychological, and understanding more factive, than traditionally thought in philosophy of science.[34] Understanding, on his view, is an achievement that involves grasping an answer to a why-question. Explanation is associated with "explanatory inquiry," i.e., seeking an answer to a why-question. An acceptable answer—an explanation—must be true, but skill is also necessary for understanding. The latter's psychological achievement is "an act of grasping that is characterized by an ability to put the correct answer to use" in answering a range of contrastive ('why-this-rather-than-that?') questions (2010, 340). An individual scientist exercises this ability in accordance with epistemic values that fit their own experience. Truth requirements on the one hand, and cognitive achievement on the other, span the mind–world relation in the manner of general epistemology, while engaging important ideas from philosophy of science (e.g., the role of models in scientific practice, Woodward's interventionist theory of explanation). Grimm's account shows some of the same parallels with EP/EP_M as de Regt's. Other aspects of his view contrast more starkly with the account here. The factive truth condition brings in ontic considerations absent in my version of explanatory particularism. Grimm's view of explanatory inquiry also departs significantly from EP's tenets. Questions asked by an individual epistemic agent are often everyday, idiosyncratic, or scientifically marginal. EP-explanations, in contrast, are in the first instance major achievements of a scientific community. Most questions asked by individuals won't meet that bar, although some will connect to those achievements. So what Grimm means by "explanation" is a far cry from EP-explanation.

Understanding with idealized models

Catherine Elgin takes a different stance on the factive/nonfactive debate. She conceptualizes understanding as "primarily a cognitive relation to a fairly comprehensive, coherent body of information" (2007, 35). A cognitive agent who understands such a body of information grasps relations among those truths, and so is able to use the information to effectively reason further. So Elgin's characterization of understanding is typical of general epistemology, associated with coherence and relations among items of knowledge. But, like Zagzebski, she also emphasizes skill; understanding as a form of expertise, facility, and "adeptness in using" a body of information. What distinguishes Elgin's account from others in general epistemology is her concern with the problem of understanding via idealized models, and her solution, Goodman's analysis of "exemplification." The central question

[34] Gijsbers (2013) offers a similarly inclusive account, albeit from different philosophical motivations.

for Elgin is how idealized—false—models can provide understanding that is "true to the facts." (That question intrigues many philosophers of science as well.) Elgin argues that idealized models exemplify features shared with phenomena that are thereby understood (2017, 77). The exemplification relation has two parts: reference to a property (i.e., model–target relation) and instantiation of that property. Reference can be established by stipulation, but instantiation is "objective"; that is, the property has to really exist in both model and target. Other properties of the exemplifying model and its target are obscured or downplayed.

Elgin's account of understanding is focused on the model–world relation, rather than on relations between models as in IDE. For that reason, her account is somewhat orthogonal to the views presented here. But Elgin's view of understanding does parallel EP/IDE in its rejection of an ontic framework for theorizing understanding, in the idea that "systemic understanding of how things hang together is epistemically valuable for its own sake," and in blurring the line between pragmatic and epistemic aspects of inquiry in characterizing the role of models. The importance for understanding of seeing relations among items within a specialized subject matter is a shared theme. Furthermore, Elgin's account allows for considerable diversity across scientific fields—exemplification may play out differently in different contexts, responding to different scientific aims and values. Alongside these commonalities are significant contrasts. Elgin analyzes scientific understanding as a monolith, with no attention to diversity among specialized scientific fields (not all of which trade in idealized models, or use them in the same ways). As with other philosophical accounts of understanding, Elgin's has no counterpart to the social action framework, collaborative norms, or associated ideals of unity and shared understanding. That being said, Elgin's account is open to elaboration along these lines.

Angela Potochnik (2017) develops Elgin's ideas about understanding in another way, within her broader account of idealized models in scientific practice. On Potochnik's view, understanding is both epistemic and psychological. Models provide understanding relative to scientific aims, which may vary widely across contexts or projects. Because of the variety of (legitimate) scientific aims, different idealizations of the same phenomenon are acceptable under different circumstances. This in turn yields a variety of scientific explanations and forms of scientific understanding. However, Potochnik does theorize explanation and understanding in monistic terms, as grasping causal patterns. She defines 'understanding' as grasping a (simple) causal pattern, embodied in the phenomenon, that is focal to the cognizer(s). All scientific explanations, therefore, communicate information about causal dependence and its scope. A wide variety of causal patterns exist in the objective world, and myriad different idealizations can capture them. Science, therefore, doesn't generate unified understanding of a domain, but rather a fragmented disunified set of idealized representations that yield understanding. Potochnik's emphasis on diversity across the sciences is a clear parallel with EP.

However, IDE and the social organization of science don't figure in her view, nor do non-causal varieties of explanation. The crux of understanding, for both Elgin and Potochnik, are patterns and regularities picked out of the world's complex welter. Models contribute to understanding by simplifying—focusing inquirers' attention on certain patterns at the expense of others. The latter is a pragmatic necessity, given human cognitive limitations. A model's proper use is to constrain and channel human reasoning. Exemplification, highlighting certain features over others, is an effective means of doing so, leading to scientific success. Although Potochnik's account emphasizes diversity in scientific aims, and thereby in explanatory models across the sciences, her view of explanation and understanding partakes of traditional reductive (simplifying) ideas going back to Friedman.

Veritistic understanding

Michael Strevens (2008, 2013) defends a sophisticated causal theory of explanation, closely tying it to understanding. Regarding the latter, Strevens' account shares some important features with de Regt's, while contrasting sharply in others. Notably, Strevens' account of explanation and understanding of phenomena is "veritistic" (i.e., truth-oriented). His "simple view" is a powerful dialectical focus of recent debates: "An individual has scientific understanding of a phenomenon just in case they grasp a correct scientific explanation of that phenomenon" (2013, 510). Understanding-why and understanding-with are different epistemic achievements with different aims. The latter is skill-based and has only internal standards for correctness; i.e., all desiderata but correctness. In this framing, and the characterization of understanding with a theory as skillful expertise, Strevens concurs with de Regt. However, according to Strevens understanding why a phenomenon occurs or a state-of-affairs obtains has external (ontic, realist, factive) conditions as well. The skill involved in understanding with a theory—grasping a theory and its models—is inferential; there are no material aspects of understanding in Strevens' view. "Grasp" is the facility of making inferences about the facts grasped—an ability had by experts in some domain of facts/phenomena. Non-expert grasp is, it follows, feeble at best.

Strevens' substantive, monistic account of explanation informs his account of understanding. The key idea is that scientific theories and associated models afford understanding of their target phenomena only if they provide veridical "difference maker" (DM) models of those phenomena. A DM, in Strevens' theory, is a causal factor such that the phenomenon of concern would not occur if that factor were removed. A veridical DM model is one in which all the DMs that actually give rise to a phenomenon are adequately represented—complete representation of dependence relations for a target phenomenon. Idealized models present an obvious challenge for this view. Strevens' response is to interpret many cases of idealized assumptions in models as true claims about difference-making. Idealizations indicate what *doesn't* make a difference to the phenomena, revealing truths about it via contrast. In this way, idealizations improve an agent's grasp of an explanation of a phenomenon. So Strevens' account of understanding is factive and causal—in both respects, unlike EP/IDE. The only clear parallel is a relational aspect for both varieties of understanding:

> To understand why is to grasp among other things this state of interrelation [the structural relation that the other parts of the explanation must stand in with respect to one another and to the phenomenon to be explained], and thus to grasp a property of the whole explanation that goes beyond the parts. On the simple view, understanding why is indeed holistic (and understanding with still more so). (2013, 514)

Soazig Le Bihan (2016) offers an account of "modal understanding" that combines elements of de Regt's and Strevens' theories. She argues that an agent has modal understanding of a phenomenon if and only if that agent "knows how to navigate some of the possibility space associated with these phenomena, where by "possibility space" it is meant the set of possible dependency structures that give rise to all subsets of the phenomena and the relations between those structures" (127). The resemblance to Strevens' account is clear, However, Le Bihan, like de Regt, is neutral with respect to factivity for modal understanding. Boundaries of possibility space are another aspect of understanding, introduced by Le Bihan's perspective. As this short survey indicates, the dialectic around understanding in philosophy of science follows the familiar pattern of debate on one or two key issues: factivity of understanding, the scope of skills constitutive thereof, and relations between understanding and explanation. Much of this debate is driven and framed by concepts from general epistemology—unusually for recent philosophy of science. What tends to be missing, however, is engagement with scientific practice beyond a few familiar examples from fields that are well-studied by philosophers of science.

Kareem Khalifa (2017, 2020) takes a formal analytic approach to understanding, informed by general epistemology. He defines understanding in terms of explanation: an epistemic agent understands a proposition p just in case that agent grasps a correct explanation of p. So the two are very tightly associated. Khalifa's account is veritistic, in that (approximate) truth or correctness is the primary requirement for and sole epistemic value of scientific explanation and understanding (2020). He analyzes understanding in terms of minimal and comparative conditions, articulated as theses about individual epistemic agents:

(EKS1) S_1 understands why *p* better than S_2 if and only if:

(A) *ceteris paribus*, S_1 grasps *p's* explanatory nexus more completely than S_2; or
(B) *ceteris paribus*, S_1*'s* grasp of *p's* explanatory nexus bears greater resemblance to scientific knowledge than S_2*'s*.

(EKS2) *S* has minimal understanding of why *p* if and only if, for some *q*, *S* believes that *q* explains why *p*, and *q* explains why *p* is approximately true.

(2017, 14)

EKS1 is comprised of what Khalifa terms the Nexus Principle (A) and the Scientific Knowledge Principle (B). The former states that an agent's understanding is improved by grasping more explanatory information relevant to explanandum p; i.e., more factors that explain p, as well as how those factors hang together (5).[35] The explanatory nexus of p is the set of correct explanations of p, together with relations between those explanations. The Scientific Knowledge Principle states that an agent's understanding is improved the more their belief-formation process resembles "scientific explanatory evaluation" (10). "SEEing," Khalifa's acronym for this generic scientific practice, consists of three generic activities of scientists culminating in scientific knowledge about a phenomenon's explanation. EKS2 specifies the minimal threshold that epistemic agents must meet to have any degree of understanding. Beyond this threshold, understanding comes in degrees; a single general continuum from minimal, to everyday, to scientific, to ideal. A layperson's understanding is thus genuine but of lower degree than scientific understanding of the same p.

The SEEing process is Khalifa's analog of explanatory model-construction in my account. First, scientists consider a range of plausible alternative explanations for a phenomenon of interest. Plausibility is relative to a scientist's context, and these judgments involve explanatory values. Next, scientists compare alternative explanations in light of evidence and other methodological canons. These too vary with context. Finally, scientists form beliefs about the correctness of plausible explanations on the basis of the comparative evaluation. In "paradigmatic cases" there's a clear "winner," accepted as the correct explanation of a phenomenon. The more a layperson's practice of forming beliefs about how a phenomenon is explained resembles the scientific stages of SEEing, ceteris paribus, the better that person's understanding of that phenomenon. Although the analogy is loose, and our two accounts are developed using very different frameworks, Khalifa's account is liable to extensionally agree with EP/EP_M in most within-community cases. The three stages of SEEing (consideration, comparison, belief-formation) are very thin, allowing for very diverse explanatory practices and resulting models. Khalifa's own 2017 case study of Bjorken scaling, illustrating SEEing, also fits EP's

[35] Coherentist aspects of understanding are further emphasized in Khalifa's latest works; e.g., "understanding consists of a systematic, inferentially and explanatorily interconnected web of propositional and non-propositional representations that answer relevant questions" (2020, 952).

community-grain approach to explanatory achievement. The example concerns theoretical and experimental physics in the 1960s and 1970s. An important explanatory achievement, Bjorken scaling, was first innovated by one man, improved by another (Feynman) and validated by experiments involving many physicists' contributions at CERN.

However, in other ways, Khalifa's theory of understanding is an opposite extreme to the view proposed in this book. His approach is thoroughly monistic, analytically framed as necessary and sufficient conditions. He downplays subjective/psychological aspects; the 'grasping' relation is explicated in terms reminiscent of coherentist justification.[36] The cognitive abilities required for understanding are very commonplace; specialized expert abilities aren't required for understanding.[37] Khalifa's theory treats "scientific knowledge" monolithically, evaluates a "layperson" (non-expert) against a scientific standard, and does not engage social aspects of science. Dialectically, Khalifa's view has been honed by debates about factivity/nonfactivity of understanding and ideas from general epistemology. There are relatively few points of direct contact between our respective accounts. Although admitting a wide variety of explanatory styles, Khalifa's theory lacks the resources to engage science's dynamic community organization, and thus IDE. He endorses a contextualist semantics for EKS1 and EKS2, rather than any sociological units of analysis. Instead of social action norms, his account appeals to virtue epistemology and the primary epistemic value of truth. Khalifa's account requires that explanatory factors grasped so as to yield understanding be "correct"; i.e., approximately true and empirically adequate. My EP/IDE account eschews any such general requirement, allowing the standards of participating fields to determine this on a case-by-case basis. The closest substantive parallel is between Khalifa's notion of "grasping more of p's explanatory nexus" (which need not be causal) and my view of comprehensive explanatory models. The shared thought is that including more information about, or aspects of, some target of scientific interest, is better than less information or fewer aspects—a way to compare epistemic achievements with a common target. And yet! Khalifa's formal analytic approach, as in general epistemology, is very inclusive. He acknowledges the diversity of scientific explanations, leaving both explanatory values and ontological commitments open in his theory. His account is thin, deflationary, noncommittal with respect to most substantive issues (including scientific realism and antirealism). His general analytic approach, tending toward the formal, actually allows for more diverse forms

[36] But see Doyle et al. (2019)—Khalifa, with co-authors, admits nonfactive understanding under specific (rather stringent) conditions, while defending veritism and monism about epistemic value. There are some parallels with EP/EP_M in Doyle et al. (2019), who argue that understanding involves four plausible criteria; distinct "cognitive goods" that may be in tension: ease of calculation, highlighting what's irrelevant, grasping an explanation, or constructing new models (fruitful application). Idealizations strike a balance between these desiderata.

[37] In Khalifa's words, "Grasping isn't special" (2017, 79).

of understanding and explanation than most philosophy of science accounts today. Khalifa endorses explanatory pluralism and looks to scientific practice to settle substantive questions (e.g., 2017, 8, 21, 24, 108). In these respects, Khalifa and I are fellow-travelers in studying explanation, despite the considerable distance between our respective approaches.

Interdisciplinary studies in philosophy of science

Although philosophical studies of understanding do not (with very few exceptions) engage social aspects of science and interactions between distinct scientific communities, other recent strands of philosophy of science connect importantly with the IDE account. One such is distributed cognition accounts of science. While somewhat orthogonal to this project, it's interesting to see where and how they converge with the EP/IDE account. Extended cognition approaches (e.g., Toon 2015) characterize understanding as a state 'criss-crossing' mind, body, and world. A shared theme with EP/IDE is emphasis on wider environment, interaction, tools, and activities. However, the extended cognition approach highlights material aspects and embodiment of understanding in science. Distributed cognition accounts highlight concrete, material aspects of scientific communities, and the intermingled roles of material and cognitive aspects of the research process. Explanatory particularism is broadly compatible with these accounts. But rather than material aspects, the view put forward here focuses on social organization of science, interactions within and among scientific communities, and collaborative social norms. A few extended cognition accounts do emphasize scientific research teams and community interactions. For example, Hanne Andersen (2010) discusses "heterogeneous conceptual consensus . . . an overall conceptual structure [that] is jointly accepted by a group of scientists while detailed parts of it are distributed among group members with different areas of expertise" (2010, 248). Nancy Nersessian argues that bioengineering models are hubs for interlocking cognitive–cultural systems (2019, 2022). Moreover, Nersessian's "interlocking" relations are integrative, just like the bridging relations discussed here. She focuses on different epistemic communities (biology and engineering) and hybrid configurations of models integrating biological and engineering components. As with EP, Nersessian's conception of 'model' is inclusive, spanning in vitro, conceptual, and concrete models. Models in this inclusive sense are constructed gradually and iteratively, out of diverse resources—again, this is like EP. There are some significant differences, however. Rather than explanatory models specifically, Nersessian's account engages a broader array of scientific practices. Her analysis is centered on a few individual laboratories, rather than specialized epistemic communities. Norms for collaborative activity, again, are distinctive of the EP/IDE account.

The IDE account also complements philosophical studies of interdisciplinarity (see Chapter 4). It's a familiar point, of course, that participants in interdisciplinary research need to communicate effectively with one another, and that this requires mutual understanding of key terms and ideas. For example, an early influential work by Petrie (1976) recommends that interdisciplinary researchers learn key elements of one another's "cognitive maps" bearing on the project:

> The minimal constituents of the amount of learning needed of the others' disciplines seem to be the following: First one must learn the observational categories of the other discipline and, second, one must learn the meanings of the key terms in the other discipline ... learning at least a part of other disciplinary maps is a necessary condition for turning multidisciplinary work into interdisciplinary work. (11–12)

Recent work by Boon and van Baalen on interdisciplinary research (IDR) takes a similar tack, emphasizing the need for communication across disciplinary boundaries (2019). The requirements for and epistemic significance of IDE are rather different. To be sure, in some cases, mutual learning and extensive interaction across disciplinary boundaries are crucial for a project's success. I've characterized such cases as "transdisciplinary," involving early-stage high-degree connections between explanatory models constructed in different scientific communities. But other, "lower-effort," interdisciplinary projects don't require mutual learning of the other field's key terms and concepts. For those projects, the class of cases at issue in this chapter, vocabulary and translation exercises are both too much and not enough. They are too much because IDR of this sort doesn't require participants to learn very much about the other field at all. All that's required for a narrow sturdy bridge is a small point of contact; just one connection via a shared or contrastive concept (or method, or object, etc.). As the bridging requirements above show, even a single point of contact between models needn't accurately capture the other field's understanding of their connection. Yet in another way, learning the basic terms and concepts of another field also isn't enough. What matters is that the two models *connect* on both sides; that participants from both specialized fields understand how their own explanatory efforts make a path to others'. A translation manual across disciplinary boundaries, as such, doesn't achieve that. Participants' attitudes toward their explanatory activities are crucial, constructing an IDE places normative constraints on participants' attitudes toward their task and toward one another. The result is a connected overall model, which no single participant need understand in full specialized detail, yet which all participants can use to advance their relational understanding of how their own specialized model fits into a bigger explanatory picture.

6.4 Conclusion

This completes the IDE account, an application of EP's tenets and framework to IDR projects that involve explanation. Relations between different explanatory achievements are one important aspect of scientific practice, notably for interdisciplinary research and for engagements between science and other sectors of society. Most real-world problems for which scientific solutions are sought today—environmental sustainability, global health, anthropogenic climate change, and more—demand solutions that transcend boundaries of scientific specializations. EP_M's core ideas, the taxonomy of model–model relations, and norms for collaborative social activity are integrated to articulate guidelines for this kind of explanatory achievement. I focus on low-effort IDE, constructed from diverse specialized models that relate to one another via late-stage low-degree connections, usually indirect. This answers the challenge from Part I, showing how EP-explanations can provide understanding beyond the boundaries of their particular communities of origin. To be sure, scientific understanding advanced by an EP-explanation often does remain sequestered within a particular community. Disseminating it outward takes work, on both sides. A major obstacle to low-effort IDE is entrenched ideas about scientific hierarchy, prevalent in the sciences. In practice, explanations are often implicated in boundary-work distinguishing different scientific communities. The IDE account highlights those obstacles and indicates ways to overcome them. The latter are presented as specific guidelines for IDR projects, to be implemented by a "counselor" without a direct stake in the project.

The rest of the chapter examined the epistemic benefits and broader normative significance of IDE. The main epistemic benefits concern understanding. While EP entails particularism about understanding, the social action framework provides a general account of interrelated aspects: understanding of self, others, model, and target. For a group conducting or planning an IDR project, IDE's epistemic benefit is straightforward: the group gains an explanatory model of the target of interest. For individual members, however, the main advance is relational. The previous chapter gave minimal conditions for connecting different explanatory models via a narrow 'conceptual bridge' (i.e., low-effort IDE). From such connections, an individual researcher gains understanding of how their specialized explanation relates to those of other fields. They do not thereby understand the explanation as a researcher from another discipline would—but they can see "how to get there from here." An overall explanation comprised of interconnected specialized explanations gives individual inquirers *direction* for gaining a more comprehensive understanding of the object of inquiry. Connecting explanations in this way is a kind of collaborative activity; IDE are *collaborative explanations*. Norms for constructing a collaborative explanation are also good guides for research involving scientists and nonscientists, with emphasis on mutuality, absence of hierarchy, and openness. In this way, explanatory particularism offers guidance for collaborative projects across different communities.

Relational understanding also has broader significance for philosophy of science. Extrapolating IDE connections and their epistemic benefits to the wider dynamic network of interrelated epistemic communities, yields ideals of scientific unity and objectivity that are at once social and epistemic. Explanatory particularism, developed in this way, thus has some normative bite. On the positive side is EP's vision of as pro-social and collaborative. Unity of science is just the potential and actual connections between different explanatory models—narrow bridges that don't reduce the diversity of those models, nor devalue their various specialized achievements. Patchwork unity of science in turn suggests a particularist conception of scientific objectivity as shared understanding. While no single epistemic ideal exhaustively covers all the diverse scientific communities currently at work, this particularist ideal of objectivity is part of that complex landscape, worth attending to precisely because it's been left out of philosophical discussions so far. The ideal of shared understanding is one we can work toward, through interdisciplinarity building on minimal connections between explanatory models. This completes my application of explanatory particularism to the case of interdisciplinary research. There is more to say about the abstract norms of unity and objectivity, their pro-social grounding, and implications for science–society relations. I will leave those explorations for future work.

PART III
CASE STUDIES

7
Integrated Understanding for Immunology and Epidemiology

This last part of this book uses case studies from scientific practice to illustrate EP's approach to studying scientific explanation, and the IDE norms' application to particular interdisciplinary contexts. Because EP's approach requires empirical groundwork, each case study is rather lengthy. Even so, a full, substantive EP account is beyond the abilities of any single scholar. Instead, these two cases fill in small pieces of a much larger picture. To illustrate bridging connections and requirements, I'll examine one case of successful IDE and one unsuccessful case (this chapter and the next, respectively). The concluding chapter indicates a number of other such cases, building on the work of other historians and philosophers of science, as well as scientists' own insights.

The first case is a success story. Immunology and epidemiology are distinct epistemic communities with very different explanatory modeling practices. Both are life sciences with strong ties to medicine. They share the downstream practical goal of improving human health. But their explanatory targets differ. Very roughly, immunology aims to understand organisms' immune systems and responses so as to improve human health, while epidemiology aims to understand causal patterns of health and disease in populations so as to improve human health.[1] Their purposes overlap but are not the same. Their respective explanatory models differ too—not only in virtue of having different targets, but also because of technical specialization. Immunology has its own vocabulary and concepts, largely inscrutable even to other life scientists, grounded in distinctive experimental systems.[2] Epidemiology's technical concepts and methods are tied to advances in mathematics and (more recently) computer modeling and simulation. Yet for all this, immunologists and epidemiologists collaborate effectively in many contexts. The IDE account shows how and why this is so. I'll first sketch each community separately, using the EP/EP_M framework (see Part I). I'll then present an explanatory model from each, and posit a minimal bridging relation between those models. This illustrates the IDE account from Part II. For simplicity, I'll focus on just one bridging connection. In reality, these two fields exhibit richer and more varied modeling ties, as well as

[1] This is not to say that either field lacks applications to non-human organisms. But the main goal of each is medical and thus human-oriented.

[2] See, e.g., Yong (2020).

Explanatory Particularism in Scientific Practice. Melinda Bonnie Fagan, Oxford University Press.
 DOI: 10.1093/9780191998874.003.0008

links with other epistemic communities. Successful IDE in this case affords integrated understanding, in the manner discussed in Chapter 6. This simple example also yields some insights, for philosophy and for science. I survey these near the end of the chapter.

7.1 Social epistemic organization

Filling in the EP_M framework sketches a field's social epistemic organization in terms of epistemic community S, model M, target X, and purpose P. In this schema, the field is conceptualized as an epistemic community S, specialized and distinct from others in explanatory aims and style. At this relatively coarse grain, S, X, and P are closely intertwined. Filling in EP_M's schema yields a broad-strokes sketch of each field. For the two at issue, several interesting contrasts emerge.

Immunology

Sociological indicators pick out immunology as a scientific community: university department designations, specialized training programs and accreditation, regular scholarly meetings, dedicated journals, international professional societies, and more. All these are hallmarks of established scientific disciplines.[3] Like many such, immunology dates from the nineteenth century.[4] Its origins straddle the science–medicine divide, and the field today has a foot in both sectors. Immunology departments (at least in the United States) are often housed in medical schools, with the National Institutes of Health a major funding source, and most immunologists view their field as medically relevant. This disciplinary location, spanning basic science and medicine, gives immunology's purpose P a dual character. On the basic science side, the goal is to understand biochemical and physiological mechanisms underlying immune phenomena. The clinical side aims to leverage that understanding to improve human health and well-being. Overall purpose P combines both aspects: understand immune phenomena in ways that can be applied to improve human health. The scientific community organized around this complex goal is large and heterogeneous.

[3] This is a coarser sociological grain than most studies of EP-explanation (see Chapter 1), but the EP_M framework applies nonetheless.

[4] On history of immunology, see Fleck ([1935] 1979), Paul (1983), Mazumdar (1989), Silverstein (1989), Löwy (1992), Swiatczak and Tauber (2020). Of these, Löwy's approach is the most similar to that taken here. She focuses on the idea of "the 'immune system' as a fundamental regulatory mechanism maintaining integrity of the 'self'" (1992, 387). This loose boundary concept, she argues, effectively linked clinicians and bench biologists under a single umbrella with shared explanatory aims.

Target X is that which members of a scientific community aim to understand. In any active scientific community, X is somewhat inchoate. For immunology as a whole, just a few features are settled. First, X is a real thing or process in the world. Second, this worldly X is localized within bodies of living organisms (arguably, all and only living organisms). Lastly, X is implicated in organisms' experiences of health and disease. X so conceived is called *the immune system*. At this coarse grain, that's about all we can say: immunology is a scientific community aiming to understand the immune system, specifically its role in health and disease.[5] This suggests some overall explanatory model M of the immune system (of some generic organism). But in fact we don't have any such model. That's not to say there are no explanations in immunology. Just, there is no single model M of the immune system as a whole, for any organism of any species. Instead, we have many different explanatory models that detail some aspects of the immune system and its workings, for some kinds of organism—fragments of a more complex whole that exceeds our current grasp. This is why EP-explanations (at least in this case) are better studied at a finer sociological grain than the entire discipline. The immunology community subdivides into numerous smaller epistemic communities, each constructing explanatory models of some fragment or aspect of the immune system for some organisms, strains, and species. It's at this finer grain that EP-explanations—major explanatory achievements for immunology—show their features, advantages, and associated values.

The concept of the immune system, rather than mapping onto a single explanatory model M, instead plays a vital social epistemic role: unifying various explanatory achievements in immunology. Discipline-wide community S has overall purpose P of understanding the immune system (target X) in clinically useful ways. But that understanding is not achieved by a single model M. Explanatory achievements in immunology take the form of more limited explanatory models constructed by subgroups within that overall community. The field as a whole is subdivided into numerous specialized communities aiming to understand some fragment or aspect of the immune system in some class of organisms. There is, to my knowledge, no canonical breakdown or classification of immune phenomena, and so no standardized list of specialized subcommunities. Different practitioners subdivide and prioritize the main areas differently. Breakdowns are often informed by P and X: subdivided by organism type, organ, disease entity, or some combination thereof. I think most would agree that tumor immunology, transplantation immunology, immunogenetics/genomics, autoimmunity, and studies of antibodies and T cells are distinct, specialized foci of immunology research. Sections of the 2022 American Association of Immunologists annual meeting

[5] The basic science aspect of purpose P (biochemical and physiological mechanisms underlying immune phenomena) suggests further features of X. In my view, though, these are better conceptualized as features of the explanatory style characterizing immunology models M.

include tumor immunology, studies of neonatal exposure, hematopoiesis (blood cell development, with further subdivisions by developmental pathway; B, T, and myeloid lineages are major foci), inflammation, autoimmunity (with further specialization as to organs; pancreas and brain are major foci; lupus amounts to its own subspecialty), and various host–pathogen interactions individuated by pathogen (HIV, coronaviruses, etc.).[6] A member of the overall immunology community at a given time works in, at most, a few of these areas.

Accordingly, the immune system as we understand it resists tidy summary. Comprehensive depictions of an organism's immune system are rare in immunology publications. When they do appear, such overviews are typically introductory treatments for students and other non-experts. Even these introductions are dauntingly complex, moving from one to the next of main parts or phenomena of immunity (e.g., antibody–antigen interactions, molecular genetic generation of antibody diversity, developmental pathways of blood cell lineages with focus on B and T cells, antigen processing and presentation, the major histocompatibility complex, T cell receptor diversity). An introductory overview can quickly become an overwhelming flood of jargon-ridden detail.[7] Explanatory advances in the specialized research community are far more detailed and technical. One source of complexity is that the immune system spans multiple scales of biological organization, from molecules to whole organisms.[8] At each scale, the immune system produces or constitutes a variety of biologically significant phenomena: allergy, autoimmunity, tolerance, chronic inflammation, responses to infection (bacterial, viral, parasitic, etc.), prospects for cancer immunotherapy, and more. Diagrammatic representations of the immune system use insets and perspective shifts to capture these multiple scales. Within each scale (especially cellular and molecular) immunological models are densely and intricately detailed. Another source of complexity is the nature of organisms. The immune system is not isolated—it is an integrated bodily system that, alongside nervous, circulatory, and endocrine systems, partly constitutes a living organism as such. Its activities consist of multiple effector pathways operating at various timescales via interactions of diverse cells and bioactive molecules in various tissues and organs.

[6] A recent *Wikipedia* summary lists the following areas of immunology research: classical, clinical, computational, cancer, circadian, diagnostic, evolutionary, intravascular (blood), mucosal, nutritional, reproductive, systems, testicular, tissue-based, and tumor immunology; immunomics, immunoproteomics, immunophysics, immunochemistry, ecoimmunology, immunopathology, psychoneuroimmunology, ocularimmunology, immunotoxicology, palaeoimmunology, immunodermatology, osteoimmunology, neuroimmunology, and study of respiratory tract antimicrobial defense systems (*Wikipedia*, accessed 11/2022).

[7] There are many excellent introductory treatments of the immune system aimed at a scientific audience, but none is both comprehensive and simple. These include Abbas et al. (2021), Punt et al. (2018), and (although somewhat dated), essays in *Scientific American*'s Special Issue (September 1993). The main anglophone professional-grade textbook is William Paul's *Fundamental Immunology* (2012).

[8] The term "scale" is used here in place of "level," the latter having been cogently criticized as vague and laden with unwelcome metaphysical baggage.

It's highly decentralized, involving both specific and non-specific pathways, and multiple layers of feedback and regulation. As we understand it, an organism's immune system is multifariously complex.

Yet another source of complexity is that details of an organism's immune response vary according to pathogen—and those details matter greatly for immunology's explanations.[9] The exact shape and charge distribution of particular molecules—antigens, cell receptors, bioactive secreted small molecules, and more—interrelated with the dynamics of cell populations mediated by cell reproduction and development, and with intracellular processes of genetic rearrangement, signal transduction, lysis, etc.—all these details are central to models of how the immune system works in response to a particular pathogen or type of pathogen. Therefore, much immunology research is pathogen-specific; i.e., research teams (individual laboratories or subgroups within a laboratory) often focus on a single pathogen, human or mouse response to that pathogen, and ways to modulate that response to improve health outcomes. Some pathogens are sufficiently complex or recalcitrant to engender sub-fields of their own (e.g., HIV, forms of cancer). The immune system "at rest" is also dynamic and constantly active. So all the detailed processes of immune response have counterparts in "no response" states. And of course, the immune system develops as part of an organism's body; developmental processes are thought to be central in establishing immune "tolerance"—i.e., lack of targeted immune response to an organism's own molecules, cells, tissues, and organs, as well as microbial symbionts and other "external" features. Autoimmune disorders are conceptualized as breakdowns of tolerance—and potentially as complex and varied as those mechanisms. For most research projects in immunology, there is no abstracting away from molecular and cellular details.

All this complicates immunology's social epistemic organization, such that our understanding of the immune system is patchy, distributed, and partitioned among many distinct, byzantinely detailed explanatory models. In this sense, immunology research has a particularist character, reflected in the field's social epistemic organization. Overall, immunological understanding is distributed, decentralized, and dynamic. This mirrors the target of immunological explanation, as conceptualized in those explanations.[10] The image of the immune system that emerges from these patchy, distributed, models of particular immune phenomena and/or pathogen responses is exceedingly complex. It includes a menagerie of different immune cell types, entangled molecular signaling pathways, and elaborately orchestrated immune responses and host–disease interactions. New explanatory models are built on this hard-won, if patchy, understanding of the immune system's workings.

[9] Variation in response to a pathogen across humans is much less emphasized, emerging only in clinical trials that are outside immunology proper.

[10] This mirroring suggests "co-construction" of science and its object of inquiry.

Epidemiology

Epidemiology, as the name suggests, originated as study of epidemics. But its scope has since broadened to, as one influential practitioner puts it, "distribution and determinants of health-related states or events in specified populations, and the application of this study to the control of health problems" (Last 2001, 61).[11] This statement suggests a three-fold purpose: describe health-related phenomena in populations, identify their causes, and use these results to benefit population health. Again, EP_M's framework of community S, model M, target X, and purpose P affords a sketch of the field, revealing its social epistemic organization. Epidemiology has the familiar sociological markers of a scientific discipline: research departments in university and clinical settings, standardized training and accreditation, scholarly societies, dedicated journals, and regular professional meetings. Like immunology, then, the field as a whole can be identified as a scientific community S. Also like immunology, this community emerged at the interface of science and medicine (albeit somewhat later).[12] And so too epidemiology's overall purpose P has epistemic and practical aspects. The epistemic goal is to understand disease etiology, while the practical goal is to intervene on specific populations so as to improve health. But within these broad similarities to immunology, some contrasts can be seen. Epidemiology's practical goals are more immediate and externally-directed than immunology's. The former has close ties to health policy and thus, at least the potential for direct social impact. Links between immunology and broader social outcomes are more attenuated. Other fields and sectors of society have always been significant for epidemiology.[13] From the fields' origins at the interface of science and medicine, its relations to neighboring fields have continuously shifted, with boundaries often blurring. Epidemiology's outreaching, interstitial character contrasts (at least in degree) with immunology.

Epidemiology's overall purpose P of course bears on target X: causal etiology and distribution of health and disease in specific populations. This is a real-world target, localized to populations of living organisms. More specifically, X is, collectively, phenomena of health and disease in these populations. This is a very broad, diffuse target, as there are infinitely many ways to delineate populations of living organisms. Phenomena of health and disease are extremely diverse as well, affording many interpretations and modes of study. It thus seems that epidemiology's target X is unspecialized, accessible to commonsense, and shared

[11] See also Frost (1928), Garfield (1988), Langmuir (1976), Lilienfeld and Lilienfeld (1980), Morabia (2004), Kaplan and Valles (2021).

[12] While "epidemiology" has denoted a medical specialization since 1850, early practitioners were relatively isolated thinkers within medicine. Sociological markers of disciplinarity accrued gradually, with full disciplinary status acquired in the post–WWII era. See, e.g., Amsterdamska (2005, 29); Magnello and Hardy (2004, 221), Morabia (2004, 121).

[13] In the words of bibliometrics pioneer Eugene Garfield, "epidemiologists are being called into other areas to work with laboratory investigators, clinicians, and public-health administrators" (1988, 288).

with other epistemic communities (including many outside science). There is no counterpart to 'the immune system'– no concept delineating the field's distinctive explanatory object, no overall coordinating concept. No specialized subject matter demarcates epidemiology's 'turf' from other sciences or sectors. Many epistemic communities are—reasonably!—interested in understanding causes and distribution of population health and disease. Unsurprisingly then, epidemiologists often make demarcation efforts against neighboring fields or sectors. Commentators have noted "long-standing concern with the status of epidemiology as a science" and its relation to other specialized expert communities (Amsterdamska 2005, 18). Concern arises because epidemiology's boundaries are fuzzy and porous. This is the flipside of its direct impact and engagement with health policy and social effects. Because epidemiology is closely tied to many other fields, with purposes overlapping those of policy-making communities, its explanatory advances often make an impact beyond community borders. Accordingly, those borders are perennially fuzzy, fraught, and contested.[14]

In this sense, epidemiology's social epistemic organization is more decentralized and particularist than immunology's. Instead of an internally-defined target system, epidemiology's methods actively connect with other scientific communities and other sectors of society. The field answers questions posed by other fields or sectors of society, often using data from other fields in constructing those answers. Practical outcomes also rely on other sectors, chiefly in policy areas. In this sense, epidemiology is a dependent epistemic community, inherently interdisciplinary.[15] Projects can be undertaken by lone researchers or, more commonly, small teams. Epidemiology research projects usually have circumscribed, well-defined questions, targets, and problems. The field does exhibit some social epistemic structure beyond particular research questions, of course. Like most scientific disciplines today, it subdivides into a number of distinct specializations, each focusing on a different class of health-related phenomena. Bibliometric analysis in 1988 identified major "research fronts" concerned with specific diseases (e.g., hepatitis, measles, AIDS) as well as multiple clusters focused on cancer. Other subject area groups have since coalesced into distinct subspecialties: genetic, psychiatric, reproductive, pediatric, environmental epidemiology. Broader clusters emphasize social concerns: healthcare inequalities, epidemiology of workers and the unemployed, community healthcare.[16] Notably, most of these sub-areas of epidemiology

[14] Some epidemiologists respond by positioning their field as distinctively integrative, connecting different levels of causality and of reality (see Section 7.4).

[15] All epistemic communities rely on others; this is a difference of degree. In epidemiology this is more pronounced than in, say, immunology, which tends to generate its own data and is less constitutively concerned with practical outcomes.

[16] Currently, the Department of Epidemiology at Harvard's T.H. Chan School of Public Health, a prominent US center for the field, distinguishes twelve areas of specialization: cancer, cardiovascular, clinical, environmental, genetic, infectious disease, neuro-psychiatric, pharmaco-epidemiology, reproductive/parental/pediatric epidemiology, epidemiology of aging, and epidemiological methods (accessed 5/20/23 <https://www.hsph.harvard.edu/epidemiology/areas-of-interest/>).

are linked directly to other 'free-standing' scientific fields. Like epidemiology as a whole, its sub-fields are closely tied to one (or more) other specialized epistemic communities.

Epidemiology's specialization is concentrated not in its target but its methods, which shape its explanatory achievements. Epidemiology's epistemic core is a toolbox of methods for constructing models that mediate between observations about populations and health outcomes for those populations. This brings us to model M in the schema. Again, there is no overall model M for epidemiology. The field doesn't aim at theories in the traditional philosophical sense, but seeks causal interpretations of population-level measures of association between variables.[17] Its explanatory achievements are for the most part local and tied to particular research questions. Causes sought are those amenable to intervention by health policy and/or individual decision-making. This is explicit in purpose P. The goal is to understand what is happening (causal dependency relations) in a specific population or class of populations, and from this offer guidance for how to intervene on that population (or class) to improve health. So major epistemic preoccupations of the field are the nature of and evidence for causality. Identifying risk factors (in a population, for some condition) is a related theme. Core methods rule out or control for biases and confounding causal factors (insofar as possible). The main mode of research design is *comparison*—among individuals and across groups. Causality at both scales is relevant. However, epidemiological results tend to be about populations; e.g., comparing differences between groups to find "average risk at the level of the individual" (Paneth et al. 2004, 292). Even when the goal is to identify "factors predisposing to disease operating at the level of the individual" (Paneth et al. 2004, 292), such factors are revealed through studies of populations. Individuals are grouped into case-categories for comparison, and statistical methods applied to data collected from different groups. Methods of inference use group-level concepts and methods; e.g., frequencies, statistics, average risk, odds ratios, rates of disease incidence.

Lacking a distinctive specialized subject matter, epidemiology is instead unified by its combination of methods, purposes, relations to other fields, and explanatory values.[18] This specialized mixture yields explanatory models M; again, many diverse epistemic products rather than a single overall explanatory model. But, because there is no unifying conception of populations or health/disease etiology unique to epidemiology, the diverse causal models constructed in the course of epidemiology research do not generally coalesce or combine. (This happens only

[17] See, e.g., Broadbent (2013), Tulodziecki (2021). See DiMarco (2022) on concerns about "stability" of causal relations in epidemiology.

[18] Alfredo Morabia (2004) argues that principles of population thinking and group comparison are central for epidemiology, comprising its core theory. What distinguishes epidemiology as an epistemic community positioned 'between' medicine, statistics, and public health policy, he proposes, is the specific integration of population thinking and group comparison (6).

in the context of specific research projects that build on earlier models—there is no community-wide model construction process, even behind-the-scenes.) Research groups (or lone researchers) construct explanatory models of particular, specified populations, using data drawn from many sources. Results are often disseminated beyond the borders of epidemiology's own specialized community, rather than forming (even loosely) an interconnected array of models like that of immunology. The main unifying, specialized element is a shared toolkit of central concepts and methods; "a coherent set of methods and concepts spanning the different circumstances a researcher faces when trying to establish the causal nature of an association" (Morabia 2004, 123). This basic toolbox of methods and concepts is taught in accredited epidemiology programs, then adapted by individuals or small teams of researchers for particular populations. Unity of method shouldn't be overstated; epidemiologists use diverse methods of causal inference in a range of study designs.[19] But, broadly speaking, understanding in epidemiology advances by adapting a standard (albeit heterogeneous) toolkit of methods and concepts to a particular population and health issue.

Summary

Immunology and epidemiology are distinct specializations. They share medical origins and purposes but aim to understand different explanatory targets via different model-construction practices. They're also organized differently, as epistemic communities. Immunology is loosely unified by an overall target-concept (the immune system), while epidemiology is more diffuse, sharing a toolkit of primarily statistical methods. Explanatory models in immunology describe and manipulate various aspects of immune phenomena in living organisms. The field is divided into multiple sub-areas, each targeting a major category of immune phenomena; e.g., tumor immunology, studies of blood cell development, autoimmunity, and tolerance. Each specialized sub-community constructs richly detailed mechanistic models, with diverse ties to one another. So far, what's been achieved is a collection of explanatory models of parts of the immune system, some much more detailed than others. The upshot is an incomplete, yet impressively detailed in patches, picture of the immune system.

Epidemiology, in contrast, is more problem-oriented and applied. A core set of methods and concepts is modified on a case-by-case basis to answer health-relevant questions about specific populations. Compared to immunology, epidemiology is more outward-looking and epistemically dependent; interstitial and inherently IDR-oriented. All epistemic communities rely on others, of course. In

[19] Pluralism of this sort is an enduring feature of epidemiology research (Broadbent 2013, Valles 2018).

epidemiology this is more pronounced than in, say, immunology, which tends to generate its own data and is less constitutively concerned with practical outcomes. Epidemiology's mediating position, suspended between biology, medicine, and public health policy, influences its specialized approach to explanation and understanding. Immunology has ties to medicine and clinical outcomes, but is overall more internally-oriented. In different ways, both fields have features that lower barriers to IDR and encourage attitudes conducive to bridging models.

7.2 Explanatory models

So far, I've introduced the two scientific communities in this case, sketching their overall epistemic organization of each in terms of explanatory purposes, targets, and models. The last embody key features of these communities' respective specialized explanatory styles. This section examines an explanatory model from each field in-depth. Both are explanatory works-in-progress poised to figure in IDE projects. While neither is as such a major explanatory achievement, each builds on EP-explanations in their respective community. Both models were constructed during the recent COVID-19 pandemic. Examining them offers a window into two sciences concerned with addressing that recent health emergency.

Explanatory model 1: immune response to SARS-CoV-2

As discussed above, immunology explanations comprise a family of models, each targeting a different aspect of some organism's immune system. As currently understood, that system's paradigmatic activity is response to an external antigen—e.g., a bacterium, virus, or allergen. ("Antigen" means anything to which an organism's immune system responds. That response is understood as emanating from a molecular interaction; specifically, a binding interaction between some part of an immune molecule and a corresponding, complementary, part of the antigen. More on this shortly.) Small differences of detail in organisms' immune responses can make for major differences in health outcomes. So for a new pathogen, like SARS-CoV-2, the human immune response needs to be worked out in detail, rather than extrapolated from responses to similar viruses (like SARS and MERS). In early 2020, as SARS-CoV-2 began to infect the human population worldwide, this motivated construction of new explanatory models. Complicating this task is the diversity of health outcomes for humans infected with SARS-CoV-2. The same virus (roughly speaking) can cause dramatically different health outcomes for different persons. The combined epistemic and practical goal is obvious: understand the human immune response to SARS-CoV-2, to identify targets for medical intervention. In 2020, Alessandro Sette and Shane Crotty, researchers at La

Jolla Institute for Immunology, reviewed available data and proposed a model explaining different health outcomes after SARS-CoV-2 infection.[20] Although not a full-fledged EP-explanation, their model is a simple and accessible treatment of immune response that builds on a rich background of more technically detailed explanatory models from the immunology research community.

A key part of this background is understanding viral life cycles and impacts on hosts. The molecular structure and genome sequence[21] of SARS-CoV-2 were quickly elucidated and made available for other researchers in January 2020. At the stage capable of infecting a new host (fully assembled "virion"), SARS-CoV-2 consists of various nonstructural proteins and four major structural proteins. The latter enclose an mRNA genome that encodes all the virus' proteins. Although a new virus for humans, SARS-CoV-2 resembles other previously-characterized viruses, and is grouped (along with SARS, MERS, and influenza viruses) in family *Coronaviridae*, named for the crown-shaped spike used to enter human cells: spike protein. Many copies of the spike protein protrude from a virus' surface. On each spike, a small region termed the "receptor-binding domain" does exactly what the name says: it specifically attaches to a complementary region of a protein on the surface of some human cells (angiotensin-converting enzyme-2 receptor; ACE-2 receptor for short). Binding depends on three-dimensional complementarity of shape and charge in small regions of two (or more) different bioactive molecules.

This biochemical specificity is the cornerstone of molecular interactions in living things, as we currently understand them. It's what holds together DNA's double helices, allows viruses to infect cells, and enables the immune system to "recognize" infection. Antigen-receptor binding is fundamental to any immune response. An *antigen* is that to which an immune system responds, conceptualized in molecular terms (e.g., a short, folded protein region; a structural motif on bacterial cell surfaces). Response to an antigen begins with binding between the antigen and a specific molecule, either free-floating or attached to a cell's surface. These binding interactions are based on complementarity of molecular shape and charge, visualized as "lock and key fit." Such interactions are well-understood biochemically; indeed, they are the core of structure/function explanation in molecular biology. Specific binding between SARS-CoV-2's RBD and human cells' ACE-2 receptor allows the virus to 'dock' onto human cells and offload its viral genome into them. Once inside, the viral genome coopts the infected host cell's 'machinery' for synthesizing proteins, copying itself and making the protein components for more infectious virus. Viral genomes and components then assemble into the next generation of infectious virus, killing the host cell when they burst out. If the process

[20] This section draws on Sette and Crotty (2021), although simplified considerably.

[21] Of course, viruses can evolve rapidly within an infected organism, and the SARS-CoV-2 virus had enormous scope for evolution due to the scale of infections worldwide. Genome sequences diverged rapidly into many different strains, a process that continues to produce new COVID-19 variants. For simplicity, I'll bracket this aspect of SARS-CoV-2, which was less of an issue in early 2020 in any case.

continued unchecked, an infected organism would eventually lose most of its ACE-2-expressing cells, with disastrous effects for the whole organism. But viral infection doesn't proceed unchecked, because organisms have immune systems that respond. Although we don't fully understand that response, we do understand some aspects of it, via explanatory models constructed over a century of immunological work. These models continue to advance our understanding of the immune system. Sette and Crotty's model is an extension of that explanatory edifice. What follows is a very simplified summary of that understanding, eliding many specialized technical features.[22]

The immune system[23] is conceptualized as having three main parts: (1) anatomical barriers separating an organism's body from its environment (e.g., skin); (2) multiple cell types and their inflammatory products that mediate an *innate* immune response; and (3) multiple cell types and associated biochemical, genetic, and developmental processes that mediate an *adaptive* (aka acquired) immune response. In practice, immunological models focus on (2) and (3); rapid-onset and longer-term responses, respectively, to breaching of (1), a standing anatomical condition.[24] Immune responses are the immune system's paradigmatic activity. Innate and adaptive responses are conceptualized as two branches or arms of one immune system, the anatomical component relegated to background. The two branches are interlaced and mutually-supporting. Innate immunity involves a rapid, generic response to some pathogens (e.g., molecular motifs found on the cell surfaces of many disease-causing bacteria), typified by inflammation. Adaptive immune responses take more time and are directed against specific antigens. Adaptive immunity is conceptualized as having two main arms: humoral (typified by soluble antibodies) and cellular (typified by T cells). So the immune system is understood, in explanatory models, as having a branching structure of diverse components and processes, which work together to maintain or restore an organism's overall health.

Human organisms manage viral infections by a combination of innate and adaptive immunity. Innate, rapid responses begin within a few hours of infection. If a pathogen, such as a virus, is detected in the body, then molecular "alarm signals" (e.g., interferons) are synthesized by cells of this branch. This rapid innate response combats virus in several ways: inhibiting viral replication inside infected cells, recruiting a variety of effector cells to the area of infection to create

[22] One important caveat: although single-cell organisms have their own intricate molecular defenses, immunology concentrates on the immune responses of multicellular organisms—specifically vertebrates, more specifically mammals, and most of all *us*. Mammalian immune responses have broadly similar patterns, although the specifics vary widely by species, pathogen, and even across individuals of the same species.

[23] This overview applies to metazoans, primarily mammals. Our understanding of the immune system is not evenly distributed phylogenetically; most of our explanatory models target immune phenomena relevant to humans.

[24] Immune system activities are not limited to responses to "external threats." Phenomena of tolerance, autoimmunity, and (more recently) cancer surveillance and eradication are also central to the field.

an "antiviral environment" in nearby tissues, and triggering an adaptive immune response. The adaptive response is more specific, hinging on recognition of viral antigen by antibody and T cell receptors. Three main cell types are involved in this response: antibody-producing B cells, CD4+ T cells, and CD8+ T cells.[25] T and B are distinct lineages of white blood cells that continuously develop (alongside red blood cells and other immune effector cells) from stem cells in bone marrow. B and T cell development involves stages of programmed gene rearrangement and mutation, designed to produce an extremely wide range of specific receptor proteins that appear on those cells' surfaces. An immense diversity of antibody and T cell receptors is produced (and continuously regenerated) within each organism. Receptors whose specific complementarity matches molecular segments of an organism's own cells and tissues are eliminated, deactivated, or suppressed ("tolerance"). The rest ("naïve" B and T cells) circulate throughout the body, each displaying receptors of a particular molecular specificity. An adaptive immune response is initiated if specific B or T receptors match molecular features of a new, "non-tolerated" antigen. For most first-contact viral infections, the adaptive response takes 6-10 days to prepare, as cells divide and develop to form specific effector populations. During that interval, innate immunity keeps the viral infection in check while the adaptive immune response develops.

For an effective adaptive immune response, the most important matching is between an antigen and proteins on the surface of B and T cells: antibody (aka immunoglobulin) proteins, and T cell receptors, respectively. Specific binding of antigen to B and T cell-surface receptors triggers a cascade of further molecular and cellular activities, including cell reproduction and population growth enabling an effective immune response. In brief, receptor-antigen binding initiates a process of clonal expansion of *virus-specific* B and T cells: these cells rapidly divide, producing a large population of offspring B or T cells with the same receptor specificity. Alongside this clonal cell reproduction, B and T lineages differentiate into other adaptive response cells, enabling effector mechanisms that engulf, surround, or break down virus-infected cells. In this way, the adaptive immune system is directed toward eliminating a specific virus antigen from an organism's body.[26] The main effector T cells are "helper" CD4+ T cells (which prime specific B cell responses, along with other roles) and cytotoxic CD8+ T T cells (which recognize and kill virus-infected cells).[27] Effector B cells secrete large amounts of

[25] These two categories of T cell refer to specific receptors on those cells' surfaces, which participate in binding interactions that have various 'downstream' effects

[26] This description is very simplified. For one, many different cell types are implicated—to varying extents—in immune responses. These include dendritic cells, monocytes, neutrophils, macrophages, and various B and T cell "subsets". The full immunological explanation of the immune system is much more complex and detailed—a more complete description of interactions among cells, molecules, tissues, and organs over the course of immune response.

[27] Helper T cells typically exhibit the CD4 surface receptor; cytotoxic T cells the CD8 receptor. Hence the CD4+ and CD8+ designations above.

virus-specific antibody, which help to clear circulating virus particles outside cells. The various antigen-specific effector cells work together to efficiently control the virus. Antigen-specific T-cell responses are often crucial for controlling disease and clearing virus, both via neutralizing antibodies (which act outside cells) and cytotoxic T cells (which kill infected cells, indicated by specific viral peptides displayed on their surface). One pathway of cell development initiated as part of an adaptive immune response leads to "memory" B and T cells. These are, as the name suggests, long-lived B or T cells with receptors specific for the antigen to which an adaptive response has been mounted. Unlike "naïve" B and T cells, memory cells of that specificity are more numerous and require less developmental time to initiate a second (or third, etc.) adaptive immune response.

Sette and Crotty's model is a hypothesis about how SARS-CoV-2 interacts with the human immune system. Briefly, the virus evades or delays the initial innate immune response, allowing a virus a longer lead-time to reproduce before adaptive immunity is primed and ready. The longer the delay, the more severe the symptoms, as the virus rapidly infects and destroys human cells. Severe cases can result from longer delays in starting an adaptive response. In some such cases, the innate immune system attempts to fill the gap, resulting in "innate cell immunopathology and a plasma cytokine signature of elevated CXCL10, IL-6, and IL-8" (i.e., a "cytokine storm," 863). This longer, exaggerated innate response leads to cell and tissue damage, notably in the lungs. Their model is presented schematically as curves showing relative amounts of specific immune response components over time (Figure 7.1). Three kinds of immune response are distinguished: generic, average, and severe. For each, the magnitude of key components within an organism's body is graphed vs. time, with intervals of the latter marked as innate and adaptive. Amounts of viral load, innate effectors, T cells, and antibodies are represented as curves over time. The curve for generic infection shows an early peak for innate immunity, followed by sigmoid curves of T cells and antibodies, which plateau at high amounts. An average response to SARS-CoV-2 shows a lower and delayed innate peak, followed by slightly delayed but qualitatively similar curves for adaptive components (the amount of antibody being somewhat depressed). Severe SARS-CoV-2 infection shows a protracted and larger innate response, a similarly delayed and enlarged antibody peak, and very low amounts of T cells. Viral load increases to a high plateau associated with severe disease.

These simplified representations flatten an organism's complex, multi-scale immune response into patterns of population demographics involving a pathogen (virus) and an organism's cells and proteins that work to eliminate that pathogen (innate effectors, specific antibody protein/B cells, specific T cells). The curves are abbreviated summaries of more complex molecular mechanisms discovered and represented in detail by other research groups in the wider immunology community. An actively pathogenic virus increases its population within an infected organism by coopting the body's own cells to reproduce. The two immune branches

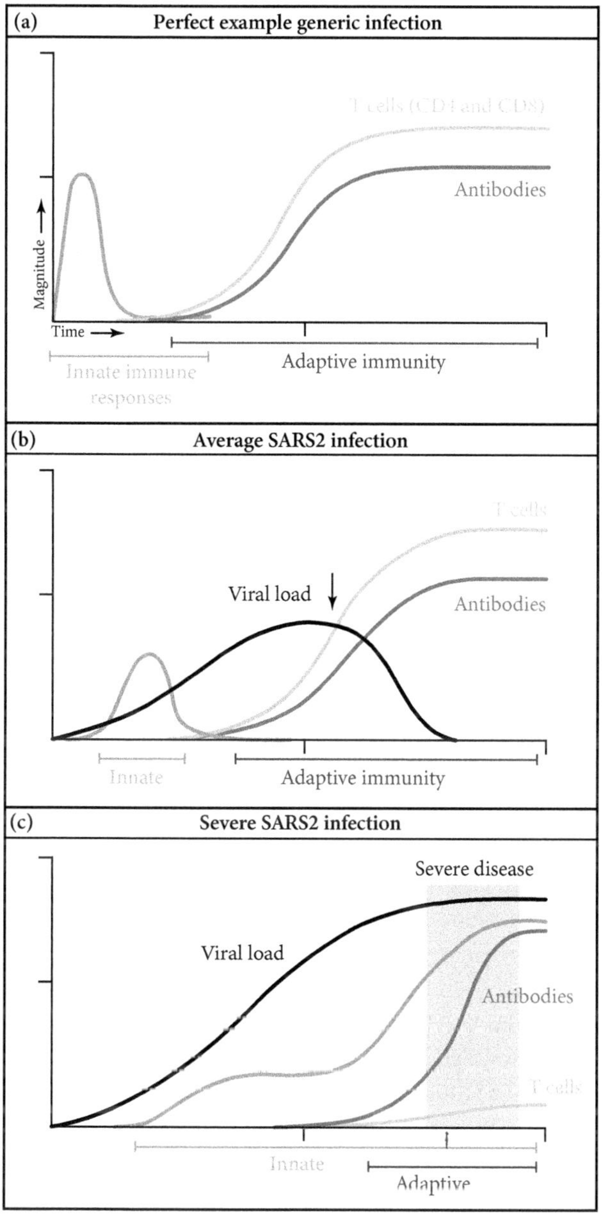

Figure 7.1 Sette and Crotty's integrated model of SARS-CoV-2 infection and immune response

Source: Reprinted from Cell 184, Sette, A, and Crotty, S, Adaptive immunity to SARS-CoV-2 and COVID-19. Pages 861–880, 2021, with permission from Elsevier.

work against the virus' reproduction by increasing the population size of molecular and cellular "species" that can act against the virus: molecular signals such as cytokines, innate immune cells, antigen-specific B and T cells—the latter being crucial for eliminating viruses such as SARS-CoV-2. An immune response so understood is sub-organismal reproductive competition; it works by way of targeted population increases of specific kinds of cells and bio-active molecules. The 'demographics' of these sub-organismal populations over time are crucial in determining the outcome of a given immune response; e.g., whether an infection is quickly eliminated or grows into disease symptoms for an organism. This model of SARS-CoV-2 interaction with the immune system is, interestingly, the converse of how vaccines are understood to work. Immune memory is the basis of vaccine effectiveness. The second, third, and later adaptive responses to a pathogen are faster, of greater magnitude, and more efficient than the first. A vaccine is an antigen designed to elicit a primary immune response without any serious disease symptoms for the organism. That organism's immune system then has a quicker, expanded, specific adaptive response to that antigen. COVID-19 vaccines are designed to produce exactly the opposite of this model's hypothesis about severe disease after SARS-CoV-2 infection.

Explanatory model 2: evaluating plans for COVID-19 vaccine distribution

Epidemiology models are various, targeting a wide range of health-relevant situations, a potentially infinite variety of populations, and ranging from highly abstract to richly particular.[28] One popular approach uses statistical tools to infer causal relations from datasets, yielding a causal model of 'what happened on the ground' in, e.g., a particular outbreak. Other models have broader scope, seeking causal associations between features of a disease and contextual variables such as population density, farming practices, etc.[29] Still others mathematically represent health-related situations and use computer simulation to trace their dynamics and outcomes (e.g., infectious disease spread in a population). The latter are increasingly prominent in health policy planning/guidance. One such model was constructed in late 2020 by an interdisciplinary team of epidemiologists, computer scientists, and applied mathematicians. The team was comprised mostly of

[28] See Leach and Schoones (2013) for a general breakdown of epidemiology modeling types; Araz et al. (2020) for a recent discussion of modeling methods.

[29] A number of epidemiologists, from the 1970s to present-day, have advocated change in epidemiological modeling practices and associated values (e.g., Susser 1973, Leach and Schoones 2013, Araz et al. 2020). If taken up, this strand of "participatory modeling" would border on anthropology and ethnography, adding rich social accounts and attending to health- and disease-relevant demographic factors.

members of academic and research units at the University of Colorado, Boulder, with other participants from Harvard's School of Public Health and the University of Chicago's Department of Ecology and Evolution (Bubar et al. 2021). This research team composition reflects features of epidemiology discussed above: extensive ties to other fields and highly technical methodology. Kate Bubar is first author; for simplicity I'll refer to the team as 'Bubar and colleagues' or 'Bubar et al.' Their goal was to answer a then-urgent question of COVID-19 health policy: how to prioritize vaccine doses when supply is limited? At the time, the first COVID-19 vaccines were soon to be available to wider swathes of the population of (some, privileged) nations. But supplies of those vaccines were still limited. A key policy question was how best to prioritize vaccination within a population, so as to minimize death and/or severe disease. Is it better to directly protect the most vulnerable members of the population with vaccination (e.g., the elderly), or to indirectly protect them by vaccinating less vulnerable groups (e.g., children, as for influenza)?

To answer this question, Bubar and colleagues used a mathematical model, comparing simulated outcomes of different COVID-19 vaccine prioritization strategies in terms of cumulative disease incidence, mortality, and years of life lost (Figure 7.2). Their model built on existing explanatory achievements in the epidemiology community. One such is the compartment model approach: abstractly modeling a population as subdivided into "compartments" representing different states of health or disease: susceptible, exposed, infectious, and removed (SEIR).[30] Ordinary differential equations represent rates at which members of the population transition between compartments. The form of these equations reflects understanding of causal relations pertaining to COVID-19, gleaned from earlier

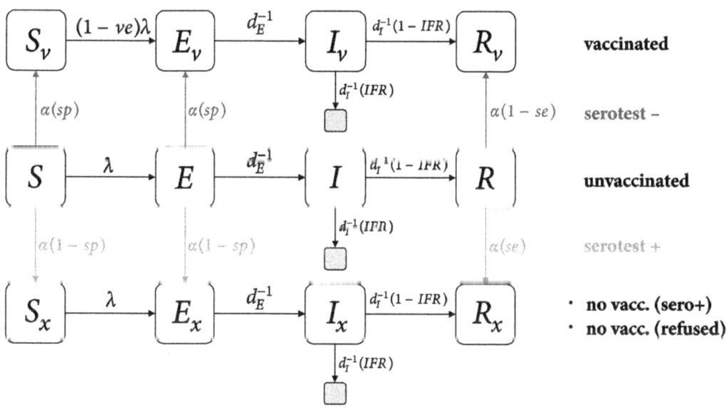

Figure 7.2 Bubar and colleagues' model of COVID-19 infection and vaccine distribution

Source: reprinted from Bubar et al. (2021).

[30] See Anderson and May (1985).

epidemiological studies and public health data-collection during the pandemic.[31] One important parameter is contact rate, specifying how frequently individuals in the population interact. Values for this were based on demographic evidence—age-stratified surveys of social contacts in nine national contexts. Age demographics were taken from the 2019 UN World Population Prospects; age-stratified contact matrices from 2020 *medRxiv* projections. Basic reproductive number (R_0) is also a key quantity, defined as the average number of secondary cases in a susceptible population that stem from one infected person (a primary case) over the entire infectious period. This quantity is relative to a population at a particular place and time, its magnitude determined by many different interacting factors, ranging from disease course within a patient to population birthrate to patterns of socialization. In practice, R_0 is estimated from a few parameters, as it's rare that all relevant factors are known and measured. In Bubar et al.'s model, R_0 was calculated as a matrix of transmission probability, country-specific contact rates, and infectious period, stratified by age.

In the model, a susceptible individual who contacts an infected individual transitions to the exposed compartment, where they remain for a latency period. Then they transition to an infectious state for a defined time-interval, followed by recovery or death. The population is unspecified; a generic nation grappling with the COVID-19 pandemic. Parameters include: latent period, infectious period, number of individuals in each age group, relative susceptibility to infection for age-i individuals, infection fatality rate, and age-stratified country-specific contact matrix (including home, work, school, and a catchall "other" category). Other relevant factors describe details of association between disease agent and host, at both individual and population levels. Omitted are descriptions of the course of infection within an individual, host population demography, duration of protection, age distribution of contact intimacy, and heterogeneity in genes, location, or behavior. Public health data from many nations showed that age is the main predictive determinant of health outcomes following SARS-CoV-2 infection—that is, disease effects vary more with age than with other variables considered. Accordingly, the model's SEIR compartments are further stratified by age. The "force of infection" for susceptible individuals at age i is calculated in terms of the probability of transmission given contact, number of age j individuals contacted per day (summed across ages), and the probability that a random age j individual is in the infectious compartment.

Once the model was constructed and values of key parameters filled in, disease dynamics over one year were simulated. The continuous time-course simulation tracked cumulative infections and mortality in the whole population,

[31] So the model includes an implicit theory of disease for COVID-19. Another strand of the project, here glossed over for simplicity, advocates serological testing alongside vaccination, as a way to mitigate healthcare inequities.

as well as calculating years of life lost to deaths at specific ages. Then the basic model was adapted to represent five different possible vaccine distribution policies, prioritizing different age groups. Simulation results for these scenarios were compared to one another and to the basic case, as percent of total population infected and cumulative mortality rate. A range of different transmission rates of COVID-19, represented as R_0 values from 1.1 to 2, were considered. Variation in vaccine performance was also studied; e.g., how effectively it blocks illness or transmission. A vaccine's efficacy for an organism is determined by internal, physiological factors, many immunological. But the latter don't appear in the model. Instead, the magnitude of efficacy is what matters, not the mechanism that produces it. Estimating this (along with safety) is the main task of clinical vaccine trials. In Bubar et al.'s model, efficacy was treated in two ways: perfectly protecting a fraction of individuals who receive it, or reducing the probability of infection for all vaccinated individuals by the same amount. A number of different vaccine features and scenarios were simulated.[32] Vaccine hesitancy was represented by allowing at most 70 percent of any age group to be vaccinated. Vaccine supply was also varied (inoculating 1 percent to 50 percent of total population). So researchers considered a wide range of different social demographic conditions, as well as a wide range of vaccine features, to aid vaccine prioritization decision-making in different national contexts.

The main result was that prioritizing adults 60+ is the best way to reduce mortality when transmission is high; at lower rates prioritizing the 20–49 age group is best. Prioritizing the latter group also minimized infections for all values of R_0 tested. Details about the population and disease dynamics (e.g., vaccine rollout speed, contact patterns, age demographics) made a difference to outcomes, but the main conclusion was fairly robust.[33] Importantly, this result is not a prediction, forecast, or precise estimate. Rather, the goal was to evaluate possible vaccine prioritization strategies within a framework of simplifying assumptions, and identify robust results by varying a few elements. The model is simple and idealized, aiming to "provide a useful guide" in managing population health (Anderson and May 1985, 324). It offers a flexible framework that could be adapted to more specific questions in different contexts. Some epidemiological models do afford specific forecasts or estimates—but these are constructed to represent the time-course of disease dynamics in a specific unfolding epidemic. In such cases, a SEIR model

[32] These include protection from symptoms/mortality together with or separate from transmission block; variable degrees of transmission blocking; variable degrees of (age-dependent) vaccine efficacy; pre-transmission or continuous rollout vaccination. Most conditions assumed 90 percent efficacy, but the team also simulated the effects of a vaccine with lower efficacy in aged populations (60+, 70+).

[33] "[W]e conclude that for mortality reduction, prioritization of older adults is a robust strategy that will be optimal or close to optimal to minimize mortality for virtually all plausible vaccine characteristics" (918). Robust results: "maximum impact prioritization strategies were broadly consistent across countries, transmission rates, vaccination rollout speeds, and estimates of naturally acquired immunity, this framework can be used to compare impacts of prioritization strategies across contexts" (916).

is filled in with real time-course data about an epidemic: cases, deaths by day, week, or month by demographic group. Model-construction is iterative: first it's calibrated with data from a specific epidemic, then simulation outcomes matched to real-world data, and results used to refine the model's parameters. Bubar and colleagues' model doesn't include fine-grained data of this sort.

Indeed, their model omits many relevant factors: co-morbidities, living conditions, occupations, and all sources of heterogeneity within age-stratified compartments. Some of the model's assumptions are now known to be false (e.g., recovery doesn't block reinfection). Uncertainty is pervasive throughout. Yet the model may be an explanatory achievement regardless. It's not intended as a detailed, accurate representation of any population facing COVID-19. Rather, the model is used to highlight one variable (age); policy options are considered through this lens. Simulations implementing the model effectively try out sets of interrelated assumptions and data, as an aid to everyday reasoning about what to do. These simulations go beyond our ordinary reasoning capacities about interrelated causes and effects, offering guidance about outcomes that can help minimize harm (disease and death) in the context of the COVID-19 pandemic. It doesn't replace other forms of deliberation, but supplements them. Within the framework of assumptions made, the model is systematic, rigorous, and quantitative, showing how multiple factors and tradeoffs yield outcomes from a range of different starting conditions. The (potential) explanatory achievement of this model is part of a wider community context. Different vaccine epidemiology models explore different alternatives—variables such as timing of vaccination; different sets of alternative possibilities demand different models. At the grain of the whole community, these diverse models dissect disease dynamics, predict/project next stages, identify causal factors or roles, and evaluate different strategies (as in this example). The different models (model frameworks) are constrained to be compatible with one another. New studies pinpoint an unstudied area, fill in a gap in the mosaic of exploratory/descriptive models of infectious disease or other health-related phenomena. There is community-wide exploration of possibilities and causal links; various models casting a net of simulation to advance understanding of health-related situations under considerable uncertainty. In the vaccine epidemiology community, this model fits into a broader model network, just as cell and tissue engineering models do in their epistemic community (see Chapter 2).

Summary: contrasts

Both models were spurred by the COVID-19 pandemic, aiming to advance understanding so as to benefit human health with respect to that disease. They show some similarity in form and content. Both are abstract, with key results presented as curves showing aspects of disease progression over time. But there are significant, obvious contrasts. The two models target different phenomena: immune system responses to SARS-CoV-2 infection for the one; outcomes of different COVID-19 vaccine prioritization policies for the other. Model-construction is also starkly different. Sette and Crotty's model is based on decades of experiments on mice and human tissue/cell cultures. They synthesized key aspects of this background explanatory edifice and new data from COVID-19 patients as well as controls. Bubar and colleagues' model uses population demographic and health data gleaned from a variety of sources. The models themselves take different forms. The immunology model distills and summarizes more complex molecular mechanisms, to highlight the timecourse of immune response within a human organism. This advance in understanding rests on a background of (partially understood) immune system workings from decades of experiment. The epidemiology model consists of equations implemented in simulation, based on demographic and public health data, as well as assumptions. It exhibits a number of qualities associated with explanatory power in philosophy, being quantitative, abstract/minimal, (arguably) robust, inviting broader application and fruitfulness. It's not a mechanistic, push–pull, step-by-step sequence yielding an outcome, like the molecular and cellular mechanisms of the immune system. The SEIR model is a form of scaffolding for decision-making involving complex, interrelated assumptions amid considerable uncertainty.

7.3 Bridging to IDE

The two explanatory models are very different, yet clearly relevant to one another. Both are responses to the COVID-19 pandemic, alongside products of many other epistemic communities (scientific and otherwise).[34] From 2020 onward, immunologists responded effectively to epidemiological data and vice versa, contributing to rapid production of safe and effective vaccines. The two models presented here are part of that broader effort. The immunology model aims to explain different observed outcomes of SARS-CoV-2 infection in humans; some people experience severe symptoms, while others don't.[35] The epidemiology model aims to advance understanding of alternative COVID-19 vaccine prioritization strategies. Each model targets a phenomenon within their respective field's domain of expertise: immune response mechanisms and social interactions, respectively, that impact population health during the COVID-19 pandemic. Yet there is relatively little knowledge-exchange or mutual specialized understanding between

[34] See Fagan (2023) for some other examples of collaborative projects involving immunology and epidemiology during this period.

[35] It's a "proposed explanation" because it's not (yet) proceeded through criticism within the relevant epistemic community, so as to be recognized as a major achievement advancing understanding—a full-fledged EP-explanation.

the two communities. Model-building methods and technologies overlap very little; one based on experiments and intricate molecular/cellular mechanisms, the other on sophisticated equations representing the dynamics of population–disease demographics. The specialized concepts, norms, and standards for constructing an explanatory model are very different in the two fields, as are resultant models. Collaborative interactions between immunology and epidemiology fit the profile of low-effort interdisciplinarity.

The IDE account frames these collaborations as a network of models variously linked by narrow conceptual bridges (see Part II). Connecting two disparate explanatory models requires effort—relatively low effort for IDR, but effort nonetheless. Some adjustment is needed on each side. Of course, details vary widely across cases. A minimal condition for bridging models in this way is coherence at three points: near-side, far-side, and middle (Chapter 5). On the near-side, adjustment to the model must cohere with its field's explanatory style. This rules out adjustments that present a mystery, non-sequitur, or direct explanatory challenge to community in which that model was constructed. Second, the adjustment should cohere with the other specialized explanation; i.e., the far-side. Here the coherence requirement is even more relaxed, because members of one specialization are largely ignorant of the standards of others. All that's ruled out is any obvious incompatibility between the bridging idea and the other's model. Finally, the two reciprocal bridging connections mustn't directly conflict with one another: middle coherence. I'll work through these conditions for the two models presented above.

Point of contact

What element(s) of each model to adjust? There are many options here. One straightforward way to connect two distinct models is to identify a point of contact between them; of either similarity or difference (see Chapter 4). In this case, the models share a point of similarity: the concept of an individual human organism. That concept figures in both models, but not centrally. In the immunology model, individual human organisms contain immune system responses to COVID-19. An immune system is of an organism, comprised of many interacting parts at a range of biological scales. In the model, SARS-CoV-2 inhibits an organism's innate immune response, with knock-on effects for adaptive immunity. Individual organisms aren't directly represented in the graph depicting three grades of SARS-CoV-2 infection/response (Figure 7.1). That graph tracks key components of immune response over time. These responses are within an organism's body; the organism experiences disease effects, mild or serious. Data-collection underpinning the model consists of samples from individual organisms—human COVID-19 patients. But organisms as such are peripheral in the model of SARS-CoV-2 infection and immune response. The immune system internal to the organism, responding to infection, is central. In the epidemiology model, individual human organisms are units contained within compartments, transitioning between these health-relevant states. They aren't directly represented in the model's equations. SEIR models are first and foremost of a population, subdivided into compartments representing different states of health or disease—in this case, age-stratified as well (Figure 7.2). Age-stratified interaction rates are based on data-collection from individual humans in nine different national contexts. But those individuals don't figure directly in the model. So in both models, individual human organisms are in the background, container and contained, respectively. The complementarity of these concepts is well-suited to bridging the two models. So is their peripherality—adjusting this element of each model is relatively low-stakes.

Near-side coherence

Next, we need some extension to each model, that makes its limits explicit and gestures beyond them while still comporting with the community's explanatory style. The near-side coherence requirement rules out only anomalies or direct explananda. In the immunology model, the individual human organism is the container for infection and immune response, experiencing overall health effects of these processes. A simple adjustment is to highlight the boundary between an individual human organism and its environment, acknowledging that phenomena outside it can be relevant to an organism's immune response. Visually, draw a boundary around the immune system and arrows showing causal impacts from outside-in. This doesn't introduce anomalies for immunology. An anomaly, recall, is completely outside a field's explanatory framework; a non sequitur that doesn't make sense from within it. The organism–environment boundary isn't anomalous for immunology. On the contrary: the boundary provided by skin is recognized as one of three major parts of the immune system. In some immunology subcommunities, it's a major focus. Sette and Crotty's SARS-CoV-2 response model focuses on the two-pronged innate/adaptive immune response, so the organism–environment boundary in the background. But making that boundary salient is no non sequitur. Recognizing that phenomena outside that boundary can affect an organism's immune response is similarly unproblematic. Infections come from somewhere, and any individual organism's immune response is influenced by many factors, internal and external. These points are familiar for immunologists

Yet recognizing them doesn't present a new explanatory task for this model; nothing that cries out for explanation in terms of an organism's adaptive immune response to SARS-CoV-2 infection. Sette and Crotty's model explicitly targets how SARS-CoV-2 infection proceeds within an individual human organism. Acknowledging an external source of infection and perhaps other factors contributing to different immune responses and downstream effects does not alter that target. There's no call to explain external factors in terms of antibody, T cell, and cytokine levels within an organism. And similarly for the organism–environment boundary. An individual human organism contains and is effected by operations of that individual's immune system. That system is modeled as a collection of effector pathways operating at various timescales via interactions of diverse cells and bioactive molecules in various bodily tissues and organs, subject to multiple layers of feedback and regulation. Some sub-fields focus on the organism–environment boundary, but studies of the workings of innate and adaptive immunity within an organism typically don't.

This adjustment does require some humility from model-builders, recognizing limits of their model. It also demands openness to explanatory contributions from others—there are immune phenomena not covered by the model, that others may understand better. Such attitudes regarding immune explanations are not rare in the immunology community. They are encouraged by immunology's dispersed social epistemic organization; explanatory efforts distributed among different subfields. No single immunology model covers all relevant immune phenomena. No matter how advanced one's understanding and control of some aspect of immunity, there is always more going on in the immune system than is covered by one's own expertise. In this way, immunology's social epistemic organization fosters a sense of individual scientists' work as part of a broader effort, beyond the reach of their or any single model. Immunologists tend to be aware of the limitations, incompleteness, and partiality of explanatory models they construct. Attitudes of openness and humility about explanatory models are commonplace in immunology. This lowers barriers to IDR, since immunologists are accustomed to the idea that understanding the immune system demands contributions from outside their own specialized area, from researchers with different areas of expertise.

A corresponding adjustment to the epidemiology model is to highlight the organism–environment boundary, and acknowledge that what falls inside it is relevant to the population's movement among compartments. In Bubar et al.'s SEIR model, individual human organisms are interchangeable units contained within different health-relevant compartments of an overall population.[36] These units are modeled in terms of an age-stratified population, transitioning between compartments according to rate equations. Parameter values for these equations are gleaned from demographic data and information about COVID-19 disease, features of hypothesized vaccines, and patterns of human behavior during the pandemic. Individual human organisms don't explicitly appear in the equations. Yet they are presupposed in the SEIR model, as 'atoms' comprising the population. So there's a corresponding, complementary adjustment: conceptualize individual human organisms as having internal parts and processes relevant

[36] Their full model does include serological data from individual patients—measurements of immune components in blood. I've omitted this aspect of Bubar et al.'s model, for brevity and ease of discussion. Including it actually makes the task of bridging models easier, as there's more overlap.

to population health and vaccine policy outcomes. That is, draw organism–environment boundaries for each individual in the overall population, and causal arrows from inside-out. There's no change to the model's equations or simulation results, though. The adjusted model is of a population of individuals impacted by COVID-19, affecting different groups in the population differently as they transition between health-relevant stages: susceptible, exposed, infected, removed. All that's new is the idea that health- and vaccine-relevant interactions occur within individual human organisms, as well as between them—some physiological 'inner life' for individual human organisms.

This conceptual shift doesn't introduce a mystery, non-sequitur, or direct explanatory challenge for epidemiology models. It's not a direct explanatory challenge, because the model's target—different possible COVID-19 vaccine prioritization strategies and their outcomes for population health—is unaltered, and organisms' internal physiology is outside the scope of that population-level study. Yet neither is the adjustment anomalous. Epidemiologists know perfectly well that individual human organisms have an internal physiology. Features of organismal infections and immune responses do figure prominently in some epidemiology models. SEIR models, however, are abstract and simple, omitting many details—in this case, everything about individual differences not related to age or nationality. One aspect of understanding the model is understanding what it leaves out. There's nothing anomalous about recognizing that individual human organisms have an inner biological life relevant to health; e.g., dynamic interactions with a pathogen in their body. Indeed, an internal physiological story can be told about a representative individual in each compartment—what makes them susceptible? What changes with exposure, and what is involved in the further transition to "infected"? The final stage, "removed" (dead, recovered, or otherwise not susceptible) involves other (disjunctive) physiological states. There's a story to be told about all of these, although in the epidemiology model itself the details of that 'inner life' remain a black box. The adjustment highlights the organism–environment boundary and identifies organisms' internal biological life as one of the many aspects of human health omitted in the model's equations. Recognizing that relevant factors in the real-world situation are omitted from the model is an aspect of understanding the model itself. Bubar and colleagues' model remains abstract and simple, like other models using the SEIR approach. All the adjustment involves is highlighting one area of omission, and noting its relevance to the target to be understood.

Attitudes supporting this adjustment are widespread in epidemiology. As a deeply applied field, research is often dependent on other epistemic communities for data, techniques, and achievement of practical goals. So the insight that an epidemiology model is not a solo effort at understanding should not be jarring. Other fields—and other epidemiological modelers—contribute to understanding the same target: this is a given of epidemiology in practice. The specialized features of SEIR models also imply their incompleteness and partiality—to be an expert regarding these models includes recognizing their limits. In this way, part of the specialized expertise involved in constructing and using this explanatory model conforms to norms of humility and openness: one must acknowledge that this single model is not absolute, complete, or final on its own. There is more explanatory work to be done than it can provide.[37] Acknowledging a model's limits is an aspect of expertise in this modeling approach. So the near-side adjustment is not a heavy lift. Attitudes supporting near-side coherence in bridging models parallel the first-layer relation in collective intention. Each agent intends to contribute to the shared goal, recognizing that the activity isn't their solo project. Both sides of this IDR connection acknowledge the limits of their field's model, indicating what is just outside its boundaries.

Far-side coherence

The bridging adjustment to each specialized model must also cohere with the other, on the far-side. This corresponds to the second-layer relation of collective intention; a member of one field attempting to see things from the other side. (They usually can't, but the attempt is important.) Far-side coherence requirements are exceedingly relaxed. This is because members of one specialization are largely ignorant of others' explanatory standards and values. Low-effort IDR doesn't require anyone to overcome this mutual ignorance. However, rather than projecting one's own standards on the other, far-side engagement should be openly naïve and non-expert: some commonsense statement of the relevance of another field's model to one's own. So the adjustment on this side must make an intelligible connection between the bridging extension and the other field's contribution, and no more. Some acknowledgement that the other community's explanatory efforts contribute to one's own explanatory goal; there's something relevant in the other field's domain, for one's own modeling project.

For the SARS-CoV-2 immune response model, this condition is satisfied if an individual human organism's *social* environment includes immune-relevant interactions with other such organisms. On the near-side, immunologists draw an enclosing boundary around an organism's immune system, acknowledging that the system is affected by phenomena outside. On the far-side is some tie to epidemiology—the relevant external phenomena include individual human organisms' social environment. A commonsense statement of this idea is "Social contacts between individuals affect SARS-CoV-2 infection and immune response." Visually, this amounts to cause-effect lines crossing the organism–environment boundary in both directions. So the adjusted immunology model gestures at causes and effects relevant to infection and immune response, emanating inward and

[37] Some commentators note, however, that epidemiology modelers using this approach don't exhibit humility; rather the opposite (see Section 7.4).

outward from an individual human organism. Details of these causes and effects aren't included—that's the epistemic territory of epidemiology. Far-side coherence demands only that immunologists recognize these epidemiological phenomena as relevant to SARS-CoV-2 infection and immune response.

This is a very modest requirement. It's obvious that patients immunologically responding to SARS-CoV-2 infection are infected via some causal pathway, and that they may infect others. Highlighting these cause-effect relations reaffirms immunology's longstanding ties with clinical medicine. That commitment in turn encourages accepting population health sciences as relevant to the SARS-CoV-2 immunology model.[38] With the onset of the COVID-19 pandemic, it's hard to deny that transmission of SARS-CoV-2 between individual human organisms is an important causal relation we should try to understand. It's common knowledge among life scientists that epidemiology is the field that studies diseases in populations—including transmission between individual organisms. (That characterization became almost inescapable during 2020, as the pandemic intensified.) So there's little reason for immunologists to balk at recognizing epidemiology models as relevant to phenomena they study—particularly immunologists studying SARS-CoV-2 infection. Members of many epistemic communities today recognize models of disease transmission and effects on population health as important contributions to understanding the COVID-19 pandemic. That doesn't mean they uncritically accept any proposed epidemiology model as an explanation. It does mean they recognize an explanatory role for some epidemiology models—that epidemiology models have some explanatory contribution to make, which is relevant to their own explanatory efforts.

Similarly for the epidemiology model's far-side connection. Again, far-side coherence demands only some intelligible commonsense statement of how immunology models are relevant to the epidemiology model. On the near-side, epidemiologists draw an organism–environment boundary and acknowledge that phenomena inside it are relevant for population health compartments and transitions among them. On the far side is a tie to immunology: the relevant internal phenomena include an individual human organism's immune system and its response to COVID-19 and vaccines. This adds causal relations crossing the organism–environment boundary, such that immune phenomena internal to the organism are relevant to the model. Both a disease and a vaccine do something inside human organisms, which causally affects their assignment to or transition between population compartments. In this way, internal physiological interactions have causal effects on the model's equations and parameters. This amounts to drawing cause-effect arrows from within individual human organisms to the model's compartments and the terms of rate equations specifying rates of transition

38 See Fagan (2023) for more on this connection.

between them. This far-side connection to immunology doesn't go into any detail about immunology aspects; all that's needed is a vague idea of the immune system having causal effects on elements of Bubar et al.'s model. This commonsense far-side connection can be stated as: "Individual immune system responses affect compartment occupancy and transition rates in a population."

The adjusted epidemiology model gestures at causes and effects relevant to population compartments, emanating from within individual human organisms. Details of these causes and effects aren't included; that's immunology's epistemic territory. Far-side coherence demands only that epidemiologists recognize these immune phenomena as relevant to COVID-19 vaccine strategies in an age-stratified population. This is a modest requirement, easily satisfied in practice. Vaccines are administered to individual members of a population, and affect the immune response of those members relative to a specific disease—in this case, COVID-19. What makes an individual human organism a member of one or another compartment in this SEIR model depends not only on their social interactions with others in the population, but also on their individual immune system. The issue of age-stratification with respect to COVID-19 and vaccine prioritization outcomes is often interpreted in terms of immune system performance; older adults, all else equal, tend to have less effective immune responses to infectious disease. Epidemiologists need not be familiar with models of the immune system and its workings, nor of a vaccine's effects on an individual's immune response, to acknowledge that these models are relevant to SEIR categories and rates of transition among them in a population afflicted with COVID-19. There's little reason to think epidemiologists would balk at recognizing immunology models as relevant to their own, especially those modeling population effects of vaccines.

So the far-side situation with each model is symmetric in some respects. Both epistemic communities in this case have reasons to adopt an 'other-oriented' attitude toward the bridge—to reach out to the other. Both can acknowledge causal relations emanating from the other side of the organism–environment boundary, as potential inputs to their field's model. The nature of those inputs—causal links to the other community's model—is left vague. This is the non-expert aspect of the far-side connection. As detailed above, causality is represented very differently in the two models. Causal relations in the immune system are modeled in terms of specific binding relations between molecules, alongside differentially-activated pathways of cell reproduction and development. Causal relations in the SEIR model are represented as rate equations within and between population compartments. Neither far-side connection is specialized in these ways. The coherence requirement is just commonsense intelligibility: a vaccine does something inside organisms that receive it; an organism's social environment includes social interactions that affect its immune response. The non-expert aspect helps mediate between the two specialized models with their different views of causality. Each field can see something relevant in the other's domain, although what it is remains something of a black box. The far-side of a model's connecting bridge is non-specialized, vague—naïve, from the other's perspective. That's fine. Indeed, imposing one community's specialized ideas on another can hinder bridging efforts.

Middle coherence

The third and last coherence requirement captures the previous point. Middle coherence demands that reciprocal bridging adjustments to-and-from each model don't directly conflict. Again, this is a very minimal constraint. Reciprocal extensions between models from different communities will rarely if ever coincide. But that's fine. What's required is that an extension from one side not undermine the reciprocal connection in the other direction. The bridge allows traffic in both directions; no contradiction arises in either crossing. In this case, models' extensions 'cross' by way of simple causal connections. From immunology, this is expressed as something like "Social contacts between individuals affect SARS-CoV-2 infection and immune response." From epidemiology, "Individual immune system responses affect compartment occupancy and transition rates in a population." There's no obvious inconsistency in these two statements. Each posits a causal connection as an 'input' into that field's model. For immunology models, social interactions with others affect an individual human organism's internal immune processes. On the epidemiology side, within-organism processes of COVID-19 and its vaccine have effects on the terms of the compartment model. So causal processes emanating from the other side give inputs to each model, in ways that fit with each. This securely connects the two models, with no obvious contradiction.

Are there less-obvious contradictions, though? There are ways of construing causality that make the two statements incoherent—ideas about simultaneity or causal priority, perhaps. If, say, epidemiologists were committed to instantaneous population effects of infection and vaccination—no time for internal physiological processes involving the immune system—then this bridge wouldn't satisfy the middle coherence requirement. Here philosophical theories of causality can play a clarifying role. Per explanatory particularism, talking with scientists about how they conceptualize these connections is informative too. In this case, immunologists I've spoken with use temporal order to distinguish the roles of each field. First, organisms encounter a pathogen or get a vaccine, and later internal physiological effects come about. Those immune responses in turn affect transmission in a population, and accordingly compartment membership and transition rates. The time-separation of these processes is reflected in temporal ordering of each field's role in, say, vaccine development.[39] Developing a vaccine comes first,

[39] Robert Fujinami, personal communication (3/4/22); see also Anderson and May (1985, 328).

then issues of community management take priority. Each field takes priority for part of this process, receding into the background for others. Both roles are acknowledged as crucial.

Attitudes toward the bridge in each community comport with and support IDE coherence requirements. Members of each have standing reasons to recognize the other's contributions to understanding aspects of the COVID-19 pandemic, and to see efforts at bridge-building on par with their own. That doesn't mean that immunologists should consider epidemiologists as their epistemic superiors, or vice versa. Nor that epidemiologists consider immunologists to have the same areas of expertise as themselves, or vice versa. Nor yet that immunologists think epidemiologists need to gain expertise in their field to be valuable collaborators, or vice versa. Minimal bridging conditions characteristic of low-effort IDR ask for little on either side of the connection. All that's needed is that explanatory modelers from each field see how the other's model connects with their own, and treat the connection as mutual. Norms of openness and non-exclusivity, equality and humility, and respect for contributions emanating from the other field—all tend to apply to the bridging connection in this case. Motivation for these collaborative attitudes is grounded in each field's community organization and explanatory style. Immunology's distributed epistemic organization accustoms community members to the idea that some important immune phenomena are outside their own models' scope: openness and non-exclusivity. Similar attitudes are encouraged by epidemiology's mediating, interdisciplinary position with respect to other fields and social sectors. Building on this, members of each field are well-positioned to recognize members of the other as contributing to a shared explanatory project. Understanding the COVID-19 pandemic is an overall goal animating many projects within a wide range of epistemic communities. Both communities considered in this case have reason to recognize the other's explanatory involvement. Immunology is concerned with viral and immune response processes within an individual human organism; epidemiology with virus transmission and vaccine effects in a population comprised of such individuals. Bridging connections between the two models are via two simple, complementary causal relations, which effectively stitch together two distinct epistemic products. This causal back-and-forth involves no inconsistencies, allowing members of each community to see the others as equal partners in the collaborative project—each model offers inputs to the other and generates outputs that traverse to the other in turn.

7.4 Collaborative insights

This lone bridge is not the only interdisciplinary connection between immunology and epidemiology. A number of insights can be gleaned, however, even from this simplified example. These results feed back into scientific, philosophical, and lay communities, exemplifying how particularist case studies reflexively enhance ties between distinct epistemic communities.

One such insight builds on a result from Chapter 4: there can be IDR without extensive communication, knowledge exchange, or (much) mutual understanding. To establish a bridging connection, neither side needs specialized, expert knowledge about the other. Members of different epistemic communities need not interact regularly, agree on key concepts, or learn anything about the other field. Distinct, specialized explanations can remain esoteric and impenetrable beyond community borders. Only the connection itself, extending from one model to the other in both directions, need be understood in accordance with coherence requirements. Mutual understanding applies only to the connecting bridge, and even that constraint is minimal. Far-side connections are non-expert, demanding commonsense intelligibility only. All this fleshes out the "low-effort" aspect of collaborative explanation. Because the bridging connection between models is so modest, demanding no more than commonsense understanding of another field, bridging connections are not difficult to conceptualize. Because the connection is narrowly focused, on a peripheral element of each model, there is less opportunity for interfield disputes or misunderstanding to block collaborative efforts. The non-expert aspect of bridging models may even encourage collaboration. The idea that non-expert, naïve attitudes towards another field can be advantageous for IDR is worth pursuing—a task for future work.

Second, this case shows a key role for philosophical theories of causality and causal explanation in IDE projects.[40] Although these theories predominate in philosophy of science today, they are rarely if ever applied to IDR. Explanatory particularism repurposes philosophical theories as heuristics for studying explanation in scientific practice. In IDE cases, these theories can serve as 'probes' to try out possible connections between distinct communities, to facilitate IDR works-in-progress. The success story of immunology and epidemiology's collaborative efforts, illustrated above, also illustrates the value of causal theories in this heuristic role. Middle coherence between these two models constrains ideas about causality involved in the bridging connection. Causality as represented in the two specialized models is quite different. The disparity is finessed by commonsense far-side connections involving vague and unspecialized causal concepts. This is hardly the last word, however. Philosophical theories of causality are key resources for clarifying this situation, and the causal concepts at work in distinct specialized models and connecting bridges more broadly. On the immunology side, some explanatory models are a good fit for philosophical accounts of mechanistic explanation. However, Sette and Crotty's model might be closer to a "mechanism

[40] See Fagan (in press) for more discussion of this issue.

sketch." What that means for the causal concepts involved is a further question. On the epidemiology side, it's unclear (to me, anyway) what notion of causality is presumed by sets of equations comprising a SEIR model like Bubar and colleagues'. Both models and their connecting bridge are ripe for analysis using philosophical theories of causality and causal explanation. Rather than imposing philosophical theories on science 'from the outside,' such theories serve as heuristics for studying explanation in scientific practice. In this way, philosophical theories can be resources for enhancing collaboration and encouraging (low-effort) IDR.

Although immunologists and epidemiologists collaborate effectively in many projects, not all IDE cases proceed smoothly. Biomedicine includes many such hard cases. For these, the IDE account, with causal theories playing a central heuristic role, offers some guidance. Different views about causes of disease underlie some longstanding controversies—for example, over explanations of AIDS in different geographic and cultural contexts. Katherine Furman (2020) critically reviews the fraught debate over causes of AIDS (acquired immunodeficiency syndrome) in regions of Africa. It has long been known, via explanatory models from virology and immunology, that this pathological state of immunodeficiency labeled is caused by a virus, HIV (human immunodeficiency virus).[41] But, Furman argues, our detailed models of HIV do not explain every aspect of AIDS, the disease. A strictly mono-causal HIV explanation of AIDS doesn't account for its geographic incidence and prevalence in, e.g., South Africa. One well-confirmed model offers an explanation in terms of socio-economic factors. Briefly, people who are poor and malnourished have less effective immune function. Immune function is complex, but there is evidence that people living in poverty (or malnourished for other reasons) spend more time in the "high viral load" phase of HIV infection than people who enjoy better conditions. The higher the viral load, the more likely a given sexual encounter will lead to a new HIV infection, and eventually to AIDS. This socio-economic explanation fits into a multi-causal model of AIDS. Multi-causal etiology of this sort is not unique to AIDS but extends "to all microbially-based diseases that also have a social component" (Furman 2020, 109)—including COVID-19. Arguably, all infectious diseases of humans have multi-causal origins.

However, in practice, many scientists associate any appeal to socio-economic factors as denial of HIV's role; viral and social explanations are cast as competitors. HIV denialism shares this adversarial framing. Debate about AIDS incidence in African nations is further complicated by histories of racism and colonialism. The IDE account cannot, of course, solve all this. But, following Furman (2020), it can address one strand of this difficult epistemic situation. Although multi-causality is recognized in epidemiology and biomedicine, we still lack clear conceptual frameworks to unambiguously relate different explanations. This murkiness can

[41] Very briefly: HIV, the virus, attacks immune cells (T cells), which in turn leaves an infected person vulnerable to a range of fatal infections as well as forms of cancer.

easily default to conflict between different explanatory models. Furman proposes a way to unambiguously relate biomedical and social models of AIDS, without subsuming one other to the other. Her solution is unification via "nesting and layering" (119). But, as she notes, however, this invokes the philosophical concept of "levels." That concept has been cogently critiqued, not least because it reinforces assumptions of hierarchy and explanatory priority. Those assumptions help perpetuate epistemic divides troubling the AIDS/HIV debate—especially where claims of epistemic authority are hotly contested and efforts at decolonization are ongoing. The IDE framework shows how the social explanation increases our understanding, without taking away explanatory power from HIV models or reinscribing long-held hierarchical assumptions. On this approach, each explanatory model is taken on its own terms, without presuming a neutral overall model—there isn't one. But the two-way connection between distinct models can show how each relates to the other on its own terms, with the linkage stable on both sides. IDE is a good framework for considering hard, socially-fraught cases like AIDS multi-causality.

A third insight from this case bears on philosophy of immunology.[42] A major debate in this area concerns immunology's central theory, metaphor, or unifying idea. Traditionally, this idea is thought to involve "self versus non-self." That is, the immune system defines and defends "the self" against invading pathogens; a martial metaphor that associates immunity with biological individuality and organismality. However, that traditional characterization has been challenged by immunologists, philosophers, and those working at the interdisciplinary seam between the two. Alternatives conceptualize immunity as a response to danger (Matzinger 1994), abrupt molecular discontinuity (Pradeu 2012), or a perturbation from equilibrium for the entire system (Eberl and Pradeu 2018). Motivating this debate is the background conviction that theorists need to cover all significant immune phenomena with a single metaphor, function, or concept, including tolerance and autoimmunity as well responses to infectious disease. Debate about the content of this unifying concept or theory has long dominated philosophy of immunology. Yet, immunology in practice, at least for now, exhibits no overall characterization or unifying idea beyond the thin notion of "the immune system." Explanatory particularism directly engages this aspect of immunology's practice. This suggests a deflationary resolution to the philosophical debate: there's no single answer; no single overarching theory that covers all aspects and phenomena of the immune system. Instead, pluralism in this regard is appropriate, given the nature of the immune system as we understand it. A thin notion of the immune system serves

[42] Influential work in this area of philosophy includes Tauber (1994), Pradeu (2012), Tauber (2017), and Eberl and Pradeu (2018). See Pradeu (2019) and Swiatczak and Tauber (2020) for overviews. Löwy (1992) also discusses the self/non-self distinction as a loose, "fuzzy" boundary concept crucial for immunology's origin and progress as a discipline mediating between "bench" biology and "bedside" medicine.

to loosely unify the field, with detailed substantive models and concepts mapping onto diverse aspects. The philosophical challenge is transformed, from seeking the single best overall theory/concept, to examining how different conceptions of immunity connect or clash with one another.[43]

Fourth, this case corroborates and extends recent insights from philosophy of population health. In the collection of epistemic communities concerned with public health, epidemiology sometimes assumes a position of epistemic superiority. It's "commonly referred to as the foundation of public health" (Merrill 2019, 2), and some epidemiologists characterize their field as distinctively integrative, connecting different levels of causality and of reality (e.g., Porta and Alvarez-Dardet 1998). These practitioners aspire to integrating "levels of causality," connecting micro (biological) levels and macro (social) levels. In this way, epidemiologists arrogate a bridging role to themselves, positioning their field as epistemically 'above' others concerned with population health. Other commentators push back on this, advocating more sociological analysis, methodological pluralism, integrative approaches, and less "medical domination" in epidemiology (Susser 1973). Sean Valles' *Philosophy of Population Health* (2018) synthesizes and develops critiques along these lines, arguing that "Epidemiology has long enjoyed pride of place in the interdisciplinary fields of public health science … That elevated position is inappropriate" (190). He envisages instead a collaborative "interdiscipline" combining efforts of diverse epistemic communities and social sectors to benefit population health (8). Valles' approach shares much with explanatory particularism and IDE. Because full understanding of population health exceeds the capacities of any individual person or perspective, he argues, "[n]o one has a firm grasp on the full scope of knowledge about a given population health case" (185). Accordingly, "we each need to recognize our limitations as knowers, and moving forward in population health science requires humble and non-hierarchical collaborative relationships" (2).

The appropriate attitude for making progress in population health science, Valles argues, is "epistemic humility."[44] For Valles, this a guiding norm, "the thread that ties together the disparate strands of population health research and practice" (18). Epistemic humility so understood demands rejection of any hierarchy of disciplines (e.g., epidemiology as "the foundation of public health") as well as "dominance" by any single sector of society (e.g., government agencies). No single sector or discipline should be elevated over others, in projects aimed at

[43] Ilana Löwy (1992) anticipates this pluralist response, although she too views "the immune self" as central for immunology historically (see Section 7.1). More recently, Zach and Greslehner (2023) propose a general framework for understanding immunity that is congenial to particularism, emphasizing contextuality, regulation, and trade-offs. However, they are still committed to a general theory covering all immunology, and to discovering general principles of immunity.

[44] Valles' account builds on work by Anita Ho (2011) on the importance of epistemic humility for mutual collaboration and trust between patients and medical professionals. See also Buchman and Ho (2014) on epistemic humility as a remedy for epistemic injustice in medical treatment of pain.

understanding and improving population health. Instead, each sector/discipline, and individuals working within them, should "embrace humility" in working with others on population health. All this accords with my account of EP and IDE. But my account also goes further. Valles and other advocates of epistemic humility in population health do not (to my knowledge) drill down into ways of implementing that norm. The account proposed here offers such an analysis: a finer-grained study of factors that encourage and can help institute epistemic humility in the way Valles and others envisage. IDE offers a model for population health science; an epistemic form for this emerging community.

A final insight from this case bears on relations between science and the public. "The public" of course is comprised of many, partly overlapping and complexly organized, epistemic communities (as is "science"). This case involves two scientific fields that are, in different ways, highly specialized and largely closed to the general public. However, epidemiology is more open to outsiders in at least one respect. Though technically very sophisticated, epidemiology models represent disease transmission and a vaccine's reduction thereof in ways accessible to commonsense. Immunologists can look over the bridge and, without too much effort, get a sketchy, non-expert understanding of how vaccines work in a population. And so can most anyone else: the basic ideas are accessible to non-experts. But the reverse isn't true. Immunology's understanding of vaccines is conveyed via very detailed mechanistic models that are inaccessible even to most biologists. That's unfortunate, because physiological understanding of how a vaccine works is relevant for anyone contemplating a shot in the arm. But there's no way to explain the immune response and a vaccine's effect thereon, as experts understand it, in an economical way. So a physician can't sit down with a patient and tell them clearly how a vaccine works. Even if they could, most listeners would quickly be lost in the welter of molecular and cellular detail. Even most biologists never learn how the immune system works. Grounds for distrust are solidly in place. There is of course more to anti-vaccine attitudes and conspiracy theories than this. But immunology's epistemic opacity doesn't help.

So a counterpart to the simple epidemiological picture of vaccines working in a population is desirable: a broadly accessible model of the immune system and vaccines.[45] This case study offers some guidance toward such a model. The non-expert far-side idea, that causal processes of immunity somehow operate within an individual organism, is a starting point. There's nothing technical or specialized about that notion of the immune system. What simple additions could be made to it, capturing key features of the immune system's working, and a vaccine's effect

[45] There are of course popular statements about the immune system. But, as Zach and Greslehner (2023) persuasively argue, these characterizations tend to misrepresent the immune in important ways. It's common to hear about "strong vs. weak" immune responses or systems—as though immunity were analogous to muscle-flexing. That is a bad analogy.

thereon? One thought is that a person's immune system is deeply individualized and personal; central to their life and experience as an individual organism. It is "theirs" in a way personalized medicine cannot replicate. Vaccines interact with that system—but in a very minimal way. That's why vaccines are so effective, as interventions on a complex system that's only partly understood even by experts. Vaccines are effective because they actually do very little. They don't alter or interfere with an organism's immune system. Rather, a vaccine is an experimentally-calibrated nudge that directs that system's ongoing activity in a particular way, such that a vaccinated organism doesn't experience the disease toward which the vaccine is directed. It does this (in mammals) by shifting the time-frame of an organism's adaptive immune response to a specific pathogen. The desired outcome is that if that individual organism encounters the pathogen later, it doesn't experience disease symptoms. This simplified characterization omits immunology's molecular and cellular details, to convey a basic picture of the immune system's working that's relevant to vaccines. Immune responses are specific, targeted to particular pathogens. Vaccines leverage this "exquisite specificity"—but they don't manipulate the immune system itself. They just make for a faster, more effective response to a specific pathogen; that's it. They spur one's own immune system to respond more effectively to a particular pathogen—doing what it always does, just faster and better.

7.5 Conclusion

This chapter illustrates the 'bridging models' account from Part II, for the case of immunology and epidemiology during the COVID-19 pandemic. I'll here summarize the argument and main results. I began by applying the EP/EP_M framework to each field separately to reveal the social epistemic organization of each. Immunology researchers are members of a wider, fragmented community, working for the most part in smaller sub-fields focused on understanding one or a few aspects of immunity or response to a particular disease. Epidemiology researchers often work in small interdisciplinary teams aimed at understanding and responding to public health challenges. That field has strong ties to other epistemic communities and social sectors, while lacking a single 'internal' conceptual framework analogous to the immune system. Unity for epidemiology is loosely accomplished by shared methodology and pedagogy. Both forms of social epistemic organization foster openness to IDR projects, although in different ways. The sense of one's own explanatory contribution as part of a bigger, more comprehensive picture is grounded in each field's social epistemic organization. Members of both tend to be aware of the limitations, incompleteness, and partiality of the explanatory models they construct. This is further illustrated in detail with an explanatory model from each field, which gives the flavor of each community's explanatory style and key contrasts between them. Although not full-fledged EP-explanations in Chapter 1's sense, these models exemplify the sort of explanatory work-in-progress well-suited to low-effort interdisciplinarity—that is, late-stage low-degree connections between distinct explanatory models from different fields. Applying the IDE account to these models shows how bridging requirements are (plausibly) satisfied in this case. I focus on one point of contact: the individual human organism, a common peripheral component of each model. This is just one connection between the two fields; many others are conceivable. But the bridge between concepts of the individual organism has several interesting features.

To adjust each model so as to connect with the other via a narrow but sturdy conceptual bridge, complementary additions are made to each. Each model is extended by some adjustment to an element, which coheres with the near-side specialization in an expert, nuanced way, and with the far-side in a non-expert, commonsense way. The shared new element in this case is an organism–environment boundary with causal implications. Briefly, this boundary is 'drawn into' each model, accompanied by recognition that what is on the other side of it can include phenomena relevant to that model. For the immunology model, this means allowing for phenomena outside the organism to affect an organism's immune response. For the epidemiology model, this means allowing for phenomena internal to an organism that affect population health. These complementary adjustments to each model are coherent on the near-side for each community, and extend in a commonsense way to connect with the other. Middle coherence requires that there is no inconsistency in crossing from both directions—that is, the reciprocal adjustments to each model are consistent. The non-expert, commonsense character of the adjustment as it connects on the far-side is important. Epidemiologists accept that 'something's happening inside each node/person;' immunologists have the complementary hand-wavy idea that 'something's happening out there' when organisms live their lives. Nothing more elaborate is needed. Indeed, non-specialized, commonsense, or naïve causal connections are important for middle coherence; reciprocal 'inputs' between adjusted models. This application of the IDE account yields a number of new insights, showing explanatory particularism's fruitfulness. IDE bridges are, plausibly, widespread in life and medical sciences. This case study corroborates and extends ideas from Valles and colleagues on the social epistemology of population health sciences, as well as contributing to philosophy of immunology. The case also demonstrates EP's heuristic role for philosophical theories of (causal) explanation, and motivates a new project of broader social concern: a concept of vaccines accessible to non-experts.

8
Systems Biology: Impediments to Integration

The second case is not a success. It concerns "systems biology," a relatively new field that aims to use mathematical modeling guided by principles from physics and engineering to explain biological phenomena (Kitano 2002, Klipp et al. 2009). Of course, mathematical models are widely used in many areas of biology. What distinguishes systems biology is its community identity as a proposed successor to reductionistic life science, typified by molecular biology. Systems biology is thus inherently interdisciplinary, combining epistemic resources from physics, mathematics, engineering, and computer science with experimental practices of "benchtop biology" focused on molecules, cells, and tissues. As there are many modes of interdisciplinary integration, there are many strands and styles of systems biology. One strand aims to replace reductionistic explanations based on single molecules with data-rich, mathematically rigorous "system-level" explanations. This explanatorily ambitious strand of systems biology came to prominence around the turn of the millennium, with the explicit goal of transforming explanation in large swathes of biology.

After more than two decades, this has not happened. Of course, systems-theoretic transformation of biological explanation might happen in the future. It's still worth examining our current situation. Why haven't systems biology's explanatory ambitions been realized? The IDE account offers a clear, constructive answer. As in the previous chapter, I'll begin by sketching the social epistemic organization of each community using EP_M's schema. Next, I'll examine an explanatory model from each community in more detail, then work through IDE requirements for a bridging connection between them. In this case, several requirements are unsatisfied. I use this result to diagnose the failure of systems biology's explanatory ambitions, and to propose how they might be realized in future. First, it's important to note that "the failure" in this case is very specific and of limited scope. I am not arguing that systems biology overall is a failure, nor that it lacks examples of success. Many research programs today integrate mathematical and computational resources with those of molecular and cellular biology, and can point to important successes. For example, Boogerd (2017) lists six varieties of significant advance in systems biology, including innovation of new fields (e.g., evolutionary systems biology), building research infrastructure, and applications of systems medicine. O'Malley and Soyer (2012) discuss an exemplary case of

Explanatory Particularism in Scientific Practice. Melinda Bonnie Fagan, Oxford University Press.
 DOI: 10.1093/9780191998874.003.0009

systems explanation, using a physics-based mathematical model to study vesicle transport in fungi. Carusi (2017) highlights the epistemic fruitfulness of new statistical, mathematical, and computational techniques, and related advances in "socio-technical infrastructure." Nersessian (2022) gives a fine-grained account of several research programs in the field, noting various aspects of success for these groups. Other such examples are plentiful in the literature. Notably, techniques of large-scale data collection and analysis have had a large, arguably transformative impact on many life science communities (Leonelli 2016).

All this is to note that the failure in this case is limited. But the project of broadly transforming explanatory practices in the various fields of laboratory-based biology has stalled, at best. Methods, concepts, and explanatory resources of molecular biology still predominate in these areas of life science. Methods and models of systems biology have neither replaced nor (for the most part) been added alongside them. Contrast this with the striking transformation of large swathes of life science via molecular concepts, methods, and explanatory style. "Molecularization" swept across many fields in the late-twentieth century, transforming explanatory practices in just a few decades. At its most ambitious, systems biology aspires to be molecular biology's successor; to replicate its transformative explanatory impact. Discipline-building reviews and commentaries in the early 2000s attest to this goal. For example, systems biologists Hans Westerhoff and Bernard Palsson title their 2004 *Nature Biotechnology* historical reflection "The evolution of molecular biology into systems biology." In an influential *Nature Review* that same year, systems biologist Hiroaki Kitano boldly asserts that "discovery of fundamental, systems-level principles that underlie complex biological systems is a prime scientific goal in systems biology," proposing robustness as one of several "fundamental and ubiquitously observed systems-level phenomena that cannot be understood by looking at the individual components" (2004, 826). A major task is discovering "effective guiding principles for understanding many biological phenomena, and for therapy design" (Kitano 2004, 836). But after twenty years, such principles have not emerged—at least, none that broadly impact explanations across bench biology. That's the failure examined in this chapter.

To do so, I apply the IDE account from Part II. IDE bridge building constraints apply to cases involving two (or more) explanatory models from distinct fields. Systems biology is a particular specialized scientific community, like others I've examined. The EP/EP$_M$ framework sketches its overall social epistemic organization, just as in the preceding case study. But what of the other side of the bridge? Systems biology's transformative explanatory ambitions aren't limited to just one other specialized life science community, but whole swathes of them—any that have adopted methods, concepts, and aspects of explanation originating in specialized field of molecular biology. So there's a scale discrepancy in this case, at least prima facie. Systems biology is one specialized epistemic community, with explanatory ambitions that cover a much coarser sociological grain. The IDE account can apply to this case, but only piecemeal; in terms of systems biology's relation to another specialized field. This actually brings the case closer to scientific practice, as model-construction projects in systems biology often do share targets (or other explanatory resources) with one or a few particular specialized life science communities. For the other side of the bridge in this case, I've selected stem cell biology, a field that I have experience studying.[1] Earlier chapters have presented some of its major explanatory models and their interrelations. Like any particularist study, the results of this case can't be assumed to generalize to explanations in other specialized scientific communities. But one must start somewhere. The results of this case offer a promising working hypothesis for social epistemic features of systems biology more broadly.

8.1 Social epistemic organization

I'll begin, as previously, by sketching each field's overall social epistemic organization in terms of community S, model(s) M, target X, and purpose P. Systems and stem cell biology are both particular epistemic communities at more or less the same sociological grain. Each is specialized, with an explanatory style distinct from other epistemic communities, although both exhibit complex ties to other fields as well.

Systems biology

Systems biology is a relatively new field, distinguished by methods and epistemic goals.[2] Its methods combine mathematical and computational modeling with experiments; its goals are more various.[3] Most broadly, systems biology aims to understand complex, dynamic living systems at all scales. This epistemic aim is often conjoined with the field's distinctive methodology; e.g., "account for the complexity that characterizes biological systems by capitalizing on new technologies for collection and integration of data, computational modeling, and interdisciplinary collaboration" (Mekios 2017, 193). Like epidemiology, then, systems biology is partly defined by methods that originate in other specialized fields: physics, computer science, engineering. Unlike other scientific communities discussed so far, systems biology is not defined even in part by distinctive subject matter; it's

[1] E.g., Fagan (2013, 2016b, 2020a, 2021).

[2] Philosophical and scientific perspectives on systems biology include Kitano (2001, 2002), O'Malley and Dupré (2005), Keller (2002), Boogerd et al. (2007), Bechtel (2011), O'Malley and Soyer (2012), Knuuttila and Loettgers (2013), MacLeod and Nersessian (2013a, 2013b, 2014, 2015, 2016), Braillard and Malaterre (2015), Green (2017, 2022). Nersessian's work (2017, 2022) on collaboration and cognition in integrative systems biology is a complement and fellow-traveler to this case study.

[3] For more on systems biology's breadth and heterogeneity, see Calvert and Fujimura (2011).

"not easily demarcated by a focus on a specific subject area or level of organization" (Green 2022). Instead, the field is defined mainly by its relations to others. Several influential practitioners position it at the convergence of two lines of twentieth-century research: "mainstream molecular biology, with its emphasis on individual macromolecules," and formal theoretical studies that "aimed to discover general principles rather than just being descriptive . . . the quest for the principles underlying systems, rather than just for their appearances" (Westerhoff and Palsson 2004, 1250). The first book on philosophy of systems biology follows this framing, introducing the field as "in many ways a successor to molecular biology and genomics on the one hand and mathematical biology and biophysics on the other" (Boogerd et al. 2007, 3).

The roots of present-day systems biology projects are even more various, with some strands grounded more in mathematical biology, others in physics or computer science.[4] A rich theoretical tradition on dynamical systems applied to biology has continued from the 1960s into present-day systems biology, most prominently in European contexts (e.g., Mesarović's Mathematical General Systems Theory, Savageau's Biochemical Systems Theory). Other roots are in developmental biology and physiology (e.g., Weiss, Waddington, Bernard). The community's current epistemic organization coalesced around the turn of the millennium, following completion of the Human Genome Project (HGP) in the United States. The HGP aimed to sequence the entire species-typical human genome, alongside genome-sequencing projects in a range of other species. Although successful, the HGP revealed limitations of gene-focused research and spurred interest in other, systems-oriented approaches. Briefly, knowing the entire nucleotide sequence on an organism's chromosomes doesn't directly translate into knowledge about that organism's traits—including traits relevant to health and disease. To make effective use of the massive amounts of genome sequence data accrued during the HGP and thereafter, new computational modeling approaches came to the fore. Alongside innovations in data processing and analysis came systems-theoretic ideas and explanatory ambitions, with particular interest in explaining dynamic gene regulation and organism level effects.

The systems biology community crystallized around these ideas and advances in the early 2000s, marked by the usual sociological indicators. In the US, "centers of excellence" funded by the NSF and NIH sprouted on East and West coasts, and training programs were carefully planned and implemented at technology-forward universities. The same pattern was seen internationally, as "Systems Biology" became a familiar subsection in general science journals like *Science* and *Nature*. Specialized journals such as *Molecular Systems Biology* (2005–present) and *BMC Systems Biology* (2007–2019) appeared. The International

[4] For mathematical biology antecedents, see Keller (2002). Green (2017) provides a short, reference-rich overview.

Conference on Systems Biology is currently preparing its twenty-third annual meeting, with smaller events popping up more intermittently across a lively network of institutes and centers of excellence. In little more than two decades, systems biology has sociologically flourished, effectively corralling resources (funding and otherwise) and establishing research centers. A number of high-profile international projects extend computational biology to whole systems of cells, organs, and organisms (e.g., the Virtual Cell Project, Virtual Liver Project, Virtual Physiological Human, and the Physiome Project). Its recent, rapid institutional rise is counted among the field's successes by some practitioners (see essays in Green 2017). Systems biology clearly fits the profile of a particular scientific community S.

Social epistemic organization within that community is fairly loose. There is no canonical division of systems biology into sub-fields or branches. An influential review by philosophers Maureen O'Malley and John Dupré (2005) distinguishes two strands, one focused on computational methods for managing large, diverse datasets, the other aimed at integrating mathematical systems theory with experimental biology. In practice, however, there's more a continuum of emphasis than a binary distinction. Different research groups within the systems biology community prioritize data analysis and systems principles to different degrees. Because the community in its present organization is fairly new, it's rather small in purely demographic terms. Distinct sub-communities within it are accordingly smaller-scale, and don't individually exhibit all the markers of more established scientific communities. Current research foci, as indicated by the categories at the 2023 International Conference on Systems Biology, are: systems immunology, multi-omics, microbiome and microbial systems ecology, signaling and mechanobiology, single-cell analyses, dynamics of space and time, systems medicine, network biology, gene regulatory networks (GRNs), synthetic biology, multi-scale biology, cancer systems biology, computational methods, machine learning, and systems pharmacology. Many of these subcategories, like those of epidemiology, reference larger, more established scientific communities.[5]

Because of its recent, speedy rise, systems biology is also replete with discipline-building commentaries, many of which include statements of purpose. From these, an overall idea of the field's purpose P can be gleaned. As sketched above, systems biology's aim is jointly epistemic and methodological: advance understanding of complex living systems using explanatory resources derived from physics, computer science, engineering, or some combination thereof. More traditional experiments on living things and their parts furnish data, to which mathematical modeling tools and theoretical resources, as well as computational

[5] In this respect, systems biology resembles philosophy of science: ties to particular specialized scientific communities, relatively small numbers of practitioners compared to those communities, distinct methodological foci and techniques, and a different explanatory agenda.

methods, are applied to construct new, better explanations of biological phenomena. So overall purpose P involves methods, explanatory aspirations, and IDR. The goal is to transform methods and explanations together, reconfiguring experimental studies within a new approach centered on mathematical models. This purpose is often described as "integrative"; for example, "integrate and explain global DNA, RNA, protein and metabolite data by combining mathematical modelling and extensive computational analysis with large-scale experimental techniques" (O'Malley and Dupré 2005, 1271). Epistemic resources from physics, computer science, mathematics, and engineering are to combine, somehow, with laboratory-based molecular, cellular, and developmental biology.[6] Broad, overall statements leave open what integration or combination looks like in practice. Systems biology's loftiest ambition, though, is to replace explanations associated with molecular biology with explanations grounded in mathematical modeling and physical theories. Integration in this case is thus a means to transforming a large swathe of explanations in biology. The scope of hoped-for transformation corresponds roughly to that of "molecularization"; i.e., the range of fields that have adopted methods, concepts, and aspects of explanation originating in specialized field of molecular biology. In many commentaries and programmatic statements, systems biologists position their field as a successor to, or by contrast with, molecular biology. This is a powerful unifying, community-building idea.

Accordingly, target X associated with this purpose is open-ended. In principle, anything that can be construed as a "complex living system" is a target for systems biology explanation. No consensus definition of "system" or "complexity" delimits systems biology projects. Rather, thematically related interpretations abound. So, unlike immunology, systems biology lacks a unifying explanatory target *and* a distinctive subject matter. In this respect, it more resembles epidemiology: a new explanatory and methodological toolkit applied to biological phenomena of interest. Again like epidemiology, those phenomena are also of interest to other epistemic communities. Overall, systems biology's explanatory target roughly coincides with that of experimental biology broadly construed: living organisms and their parts.[7] That is, the field aims to explain the same targets as other specialized life science communities. Where epidemiology's explanatory targets are shared with various social groups, including policy-makers and the wider public as well scientific and medical communities, systems biology shares targets primarily with specialized "bench biology" communities.[8] The motivation for this

[6] See, e.g., Boogerd et al. (2007), Green (2017), Kitano (2001).

[7] Parallels or connections between systems biology and studies of ecosystems in ecology would be a ripe area to study (e.g., Clark, under review).

[8] This may be typical of our cluttered present-day epistemic landscape; specializations form at the junctures or via splitting of pre-existing specializations.

fairly straightforward: systems biology is a response to limitations of molecular approaches to studying organisms, their behaviors, and their internal parts. But those limitations are often exaggerated by proponents of systems biology. Explanations grounded in molecular biology are critiqued as reductionistic, committed to simplistic views of causality and studying molecules or pathways in isolation.

Systems biology's explanatory style emerges from the community's central, unifying purpose, target, and method. Explanatory models M are constructed by integrating experimental data, mathematical modeling, and computational analysis, with the resulting explanations contrasting with those of molecular biology in desired ways: a shift to whole systems, abstract models, general principles or constraints, independence from particular contexts and molecular details. The epistemic goal, to be achieved by a new integrative method, is a new kind of understanding of biological systems: more fundamental, comprehensive, and systematic than that provided by traditional molecular approaches to living things. This new understanding is usually characterized in terms of a shift from studies of individual genes and molecules to the structure and dynamics of an entire living system (cell or organism). Emphasis on the "organizational structure of the system as a whole" is a hallmark of explanatory models in systems biology (Green 2022). For example, an influential early statement advocates "[s]ystem-level understanding ... [which] requires a shift in our notion of "what to look for" in biology. While an understanding of genes and proteins continues to be important, the focus is on understanding a system's structure and dynamics" (Kitano 2002, 1662). An important popular treatment associates "seeing the cell as a whole—rather than a bag of independent chemicals" with understanding "systems-level properties" of living things (Baribási 2002, 196). Tools from physics, engineering, and computer science are needed to construct these models, it is thought, because the complexity of living things is such that traditional biological modeling approaches will be overwhelmed and made intractable by the details. The sought-after models are more abstract explanations that capture the essential features (or principles, or fundamental parts) of living things, achieving "an abstract, system-level understanding of basic functional schemes [for living things] without a detailed understanding of specific mechanisms" (Green 2017, 10). An explanatory style contrasting with that attributed to molecular biology is one of systems biology's central defining features. Molecular biology is the foil for defining systems biology's explanatory style, its historical predecessor, and—in its transformative 'reach' to other areas of biology—its exemplar.

In influential essays and commentaries, systems biology's integration is conceived methodologically, as a feedback loop connecting biological knowledge, mathematical modeling and computer simulation, 'wet' experimentation and data analysis, returning to the start- and end-point of biological knowledge (e.g.,

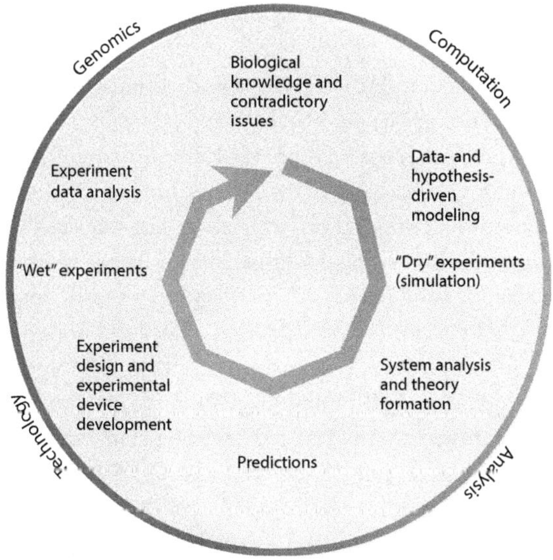

Figure 8.1 Systems biology methodology, visualized as a "research cycle"
Source: H. Kitano (2002) "Systems biology: a brief introduction." *Science* 295: 1663. Reprinted with permission from AAAS.

Kitano 2002; Figure 8.1).[9] The core idea visualized in this "research cycle" is that simulation results from a mathematical model are compared with experimental data, and any discrepancies between them spur modification of the model—and the cycle continues. This image of knowledge-production by methodological integration mirrors a prominent kind of explanatory model in systems biology: a positive feedback loop. In these programmatic images, iterative cycles of mathematical modeling and experimental test generate knowledge, analogous to the ways genetic regulatory circuits generate properties of living things in (some) systems models. In this way, broad, programmatic community-building statements indicate a definite, coherent purpose and method for systems biology. Yet the field's social epistemic organization in practice is loose, distributed, and heterogeneous. Many diverse research projects fall under this relatively new disciplinary umbrella. Many have explanatory aims, and resulting models (in EP's inclusive sense) are very heterogeneous.[10]

[9] See also Mesarovic et al.'s (2004) "triad" of biological phenomena (represented in mathematical models), computer algorithms generating data and hypotheses, and systems science as a framework for understanding; similarly, Kroh's "cycle of inquiry" (in Boogerd et al. 2007).
[10] On explanations in systems biology, see Braillard (2010), Huneman (2010), Jones (2014), Fagan (2012b, 2016c).

Stem cell biology

Stem cell biology is also an interdisciplinary field, although solidly within the life sciences.[11] Like immunology, it spans biology and clinical medicine. Indeed, many of its contributing fields overlap with areas of immunology; e.g., developmental biology, cancer biology, hematology (the study of blood). These diverse strands converge on a shared purpose and target to form the stem cell biology community. Again like immunology's, the overall shared purpose is two-fold: to understand stem cell capacities and harness them for human health benefits. The target, eponymously, is "the stem cell," its products and effects. Although this explanatory target may seem quite narrow, it covers a broad range of biological phenomena. Complicating matters, stem cell biology lacks a unifying theoretical framework analogous to the immune system. The concept of a stem cell is rather loose and equivocal. Methods in stem cell biology are also extremely diverse. Instead of unifying theories, shared principles, or focal concepts, stem cell biology is organized around a few key experimental methods and their products. The field is driven and organized primarily by experiment. External social forces have also helped forge and solidify this community's identity.

That community identity, and its social epistemic organization, is rather recent. Stem cell biology's origins are in the nineteenth century, building on explanatory achievements of Cell Theory and the phylogenetic Tree of Life.[12] Ambitious general theories of organismal development in terms of cell lineages fell by the wayside in the twentieth century. The stem cell concept then found a home in specialized medical studies of blood cells; a niche area within rapidly-growing fields of life science and medicine. After World War II, concerns about effects of radiation on humans spurred more research on blood cell reproduction, development, and radiation sensitivity (Kraft 2009). This branch of radiation research used mice in lieu of human subjects, innovating quantitative methods to measure blood stem cells from an organism, clinical methods for transplanting these stem cells (found in bone marrow) after radiation treatment, and "lines" of propagating cancer stem cells used to model of embryonic development. These experimental, clinical, and conceptual achievements spurred formation of a specialized epistemic community with ties to hematology, immunology, radiation research, cancer research and therapy, and embryology. Advances in culturing cells and tissues outside multicellular organisms led to creation of more and more artificial cell lines (Landecker 2007). The latter include stem cells derived from early embryos from various species, grown continuously in nutrient-rich media, eternally reproducing and (in the right environmental conditions) developing into more specialized cell types. In

[11] The following sketch of stem cell biology draws on my previous work; e.g., Fagan (2007, 2013, 2016b, 2021). Further references can be found there.

[12] See Coleman (1977), Dröscher (2014), Reynolds (2018).

1998, these experimental advances culminated in a method for producing human embryonic stem cells (hESC; Thomson et al. 1998). That experimental success, and ensuing scientific and public response, re-organized stem cell biology into its current community form. From a loosely-affiliated biomedical specialty centered on blood cells, it's grown into a broader community encompassing "adult" and "embryonic" branches, with a strong emphasis on internal standards, proactive ethical and policy stances, and a marked entrepreneurial bent encouraging ties with biotechnology and private industry. This entrepreneurial character and spur to private industry partnerships arose in part because of barriers to public/government funding for research using human embryos in many national contexts.[13]

Currently, stem cell biology exhibits all the usual sociological markers of a scientific community S. Within five years of the hESC innovation, The International Society for Stem Cell Research (ISSCR) formed, a dedicated journal (*Cell Stem Cell*) was inaugurated, training programs and academic divisions were instituted, and efforts to navigate ethical controversy about human embryo use began. Intense bioethical and policy debate about the latter, with attendant public interest and scrutiny, raised concerns about the legality of and funding for embryonic stem cell research. Rather than split into two specialized fields, focusing on adult and embryonic stem cells respectively, community leaders emphasized ties between them, with many research centers (even individual laboratories) pursuing both. The community was also geographically inclusive from the outset, with leading centers in Europe, both US coasts, Israel and parts of the Middle East, Australia, Japan, Korea, Singapore, and more. This international character persists. A recent editorial on 'Integrating international perspectives on stem cell challenges and potential,' appearing in the ISSCR's dedicated journal, *Stem Cell Reports*, was authored by researchers from twenty different institutes/academic units in eighteen different countries (two centers in the UK and South Africa contributing; Fears et al. 2021). The community's emphasis on commonality across kinds of stem cells as well as national borders, is partly a reaction to its origins in the crucible of bioethics controversy. Political pressure and public scrutiny have made for a strong internal sense of community, and long-standing interest in expanding through collaboration as well as pedagogy. As a result, the community has grown quickly. The International Society of Stem Cell Research (ISSCR) is the field's largest professional society; its annual meeting in 2023 had nearly 3,000 attendees, with an additional "nearly 600 participating virtually."[14] Many other societies for stem cell research exist worldwide. At a smaller sociological grain are institutes and centers, numbering in the hundreds at least.

This sprawling international enterprise is unified by a shared purpose and target. As noted, stem cell biology's overall purpose P has epistemic and practical

[13] Gottweis et al. (2009), Hauskeller et al. (2019).
[14] ISSCR annual meeting archives <https://www.isscr.org/the-archives>

aspects. The epistemic, explanatory goal is to understand stem cell phenomena: reproductive and developmental capacities of stem cells, and products of exercising those capacities in multicellular organisms (and, increasingly, in artificial "in vitro" cell culture). But this isn't understanding for its own sake. Therapeutic aspirations shape the kind of understanding at issue. The practical side of purpose P is to harness stem cells' regenerative capacities for new clinical therapies, and thereby improve human health. Advances in understanding should not only satisfy explanatory norms (more on those shortly) but also potentially yield new clinical interventions. The ideal is a therapeutically useful explanatory model of some aspect of cell, tissue, organ, or organism development. Of course, not all stem cell research projects are clinically oriented. As in immunology, potential health benefits shape the field's explanatory norms. That therapeutic aim also links many otherwise disparate projects as a common purpose. This both unifies and directs the community, shaping its explanatory norms. Although a few stem cell therapies are in use (notably those involved in bone marrow transplantation), the field's clinical goals are largely hoped-for rather than realized, at least for now.

The epistemic side of purpose P invokes stem cell biology's overall target X. As in the immunology community, X remains somewhat inchoate. The concept of a stem cell is variously understood—indeed, as I've argued elsewhere, context-dependence is its signature feature (Fagan 2013, 2022). Most generically, a stem cell is defined as a cell capable of both self-renewal and differentiation.[15] The former is production of offspring cells similar to the parent; i.e., cell reproduction. Stem cells make more stem cells. This occurs by cell division, aka mitosis. So a self-renewing stem cell gives rise to a continuous lineage of stem cells—a stem cell line—that can be represented as a branching tree. Differentiation is the process by which unspecialized cells of an early embryo give rise to more specialized cell types of a multicellular organism—neurons, muscle cells, white blood cells, and many more. A "minimal model" of a stem cell is then a cell that divides to produce one stem cell and another, more differentiated, specialized cell. But the overall target X of stem cell biology is not an individual cell, nor one cell type. It is much broader; loose and ambiguous. The meaning of "stem cell" is perennially, deeply unsettled. Stem cell phenomena are various, encompassing processes of cell reproduction and development within (and outside) multicellular organisms. Stem cell capacities involve both reproduction and development, enacted within (and forming) the various organs and tissues of a multicellular organism. Alongside organismal development, stem cell phenomena also include behaviors of cancer cells and tumors, and organisms' regenerative powers. So the range of explanatory targets is very wide indeed. Currently, no theoretical principles cover that range. As a community, stem cell biology is not organized around principles of

[15] E.g., "Stem cells are functionally defined as having the capacity to self-renew and the ability to generate differentiated cells" (Melton 2013, 7).

development, nor of cell identity, nor of organismal form. Instead, it is organized around experiments and their cellular products.

Experiments in stem cell biology are very heterogeneous; e.g., clinical trials, ex vivo cell culture, functional genomics, cell reprogramming, fluorescence-activated cell sorting, and more. The list is long and open-ended, as new technologies bring forth continual additions and older methods fall out of use. Although much about stem cell phenomena remains unknown, significant advances have been made in understanding them. These achievements are associated with concrete experimental success, such as embryonic stem cell lines (Thomson et al. 1998), adult cells "reprogrammed" into an embryo-like state (Takahashi and Yamanaka 2006), and simplified model organs (e.g., "mini-brains"; Lancaster et al. 2014). Overall, explanatory models M in stem cell biology are tied to specific experimental methods—primarily but not exclusively involving concrete experimental systems (cells and tissues) in controlled artificial cell culture environments. Many such models include detailed depictions of interacting molecules, cells, and tissues, constructed via interplay of diverse abstract and concrete models. Stem cell researchers work back and forth between concrete experiments and more abstract representations of the molecular and cellular processes therein, to construct explanatory models of stem cell phenomena. As new methods for characterizing and tracking cells are innovated, concepts and details in those models change. In this way, scientists' characterizations of stem cells are relative to experimental contexts. Inter-model relations generate a tightly yet variously connected web of experimental insights and explanatory models, with notable experimental achievements as hubs (see Chapter 2). I am inclined to think this social epistemic organization is not limited to stem cell biology, but extends across a wide swathe of life science: a tangled bank of diverse specialized communities, multiply interconnected via ties between explanatory models, alongside points of contact in methods, technologies, applications, background assumptions, and more.

Sub-areas of stem cell research cluster around a paradigmatic method of generating a particular kind of cell, tissue, or organism. The 2024 ISSCR conference, like those preceding it, has six tracks, or themes: new technologies; pluripotency and development; somatic stem cells, cancer, and regeneration; disease modeling and drug discovery; clinical applications; and ethics, policy, and standards.[16] This breakdown reflects the way the field as a whole (at least, its leadership as represented in the community's largest professional organization) parses and prioritizes areas of inquiry. Yet in practice, many laboratories and institutes pursue a combination of these; the six themes don't limn the community's sociological organization. The primary social epistemic "organizer" for stem cell biology is experimental success. Accordingly, stem cell researchers are ecumenical and

[16] ISSCR 'Overview.' 2024 meeting program. <https://www.isscr2024.org/overview>

opportunistic in seeking new avenues for these achievements. Stem cell biology is a quite open community, opportunistic in forging collaborative ties, affording considerable diversity in concepts and methods, and eager to use and adapt new technologies. In some ways, it seems a perfect match for systems biology's integrative aims. And yet ... Explanatory IDR connections between these two fields are elusive.

Summary

The two specialized communities sketched above exhibit very different social epistemic organization. Systems biology is organized primarily by its relations to other communities; stem cell biology by experimental achievements. To caricature somewhat: systems biology is a conduit between communities that innovate mathematical and computational models, and life science communities that focus on "wet" experiments using molecules, cells, and tissues. Stem cell biology is one such community. Although ecumenical in methods and approaches, borrowing profligately from other fields, stem cell biology is organized mainly around a few exemplary, fruitful experimental systems; hESC, reprogramming, CRISPR-editing, organoids, and more. These exemplar experiments produce new living entities (cell lines, tissues, miniature organs, models of whole embryos) alongside explanatory models grounded thereon. The research communities that coalesce around and radiate from these exemplary models are international, even global in scale. In contrast, systems biology models are often constructed by a handful of researchers, a single team or even one individual. These innovations are more local and individualized. This social epistemic contrast is rooted in each community's signature methods. Computer simulation and mathematical modeling can be performed in a single lab, while experiments on cells and tissues are distributed across dozens, if not more. Concrete experiments on living things and their parts are more socially extended. So there's an asymmetry of scale between these two communities—the inverse, so to speak, of the asymmetry that introduces this chapter. The systems biology community is much smaller, demographically. The 2023 International Conference of Systems Biology included a total of 157 presentations; 98 talks (including 5 keynotes) and 59 posters.[17] The 2023 meeting of the International Society for Stem Cell Research had >370 talks and >1150 posters, plus 7 additional plenary sessions. This is just one point of comparison, of course, but indicative of demographic discrepancy.

[17] ISCB23 conference website <https://icsb2023.bioscience-ct.net/index.html >

8.2 Explanatory models

Having set out the overall social epistemic organization of the two scientific communities in this case, and introduced their respective explanatory styles, I next present an explanatory model from each in more detail. Of course, neither is representative of either field in totality—no single explanatory achievement could be. But each exhibits important features of its community's style. Appropriately for systems biology, their model is intended for an IDR project. The stem cell model, while not explicitly so intended, does afford multiple opportunities for IDR connection, in keeping with that community's openness. Yet IDE conditions are not met in this case. This result is emblematic of the fate of systems biology's broader explanatory ambitions.

Explanatory model 1: dynamical systems of cell development

As noted, systems biology explanations share targets with explanations from other life science fields—notably molecular biology. But the explanatory models directed at those targets are very different. One major contrast is the role of mathematical and computational models: central and prominent in systems biology, largely absent in molecular biology and related fields.[18] There are of course many different mathematical theories and techniques at work across systems biology as a whole. One prominent theoretical framework is dynamical systems theory (DST). DST is a branch of mathematics that defines "systems" as sets of variable quantities changing over time (Strogatz 2000). Changes are represented by equations specifying dependency relations among variables. Parameters in the equations represent the form and strength of dependency relations. The set of such equations for a given system is a dynamical model of that system: a DS model. The system as a whole is represented by the state variable(s): a quantity (or quantities) reflecting the value of each variable in the system at a given time (or interval). Any change in the value of any component variable of a system is reflected in a change in the value of the state variable. Because patterns of dependency among variables of a system are complex and nonlinear, it's usually impractical to solve a DS model's equations analytically. Instead, systems biology investigation with DST proceeds by translating the initial model into a qualitative geometric framework termed "the state space." The state space is defined by all possible values of the state variable, which encompass all possible combinations of values for each variable in the DS model. At any given time, a system is located at some point in its state space. Changes in values of variables are represented as a path between points in state

[18] These models are parts of explanatory models in EP's sense, not (always) coextensive with them.

space. The path taken is determined by parameter values and other formal features of the DS model's equations. Different possible combinations of parameter values define another abstract multidimensional framework: "parameter space." The latter affords a comprehensive view of patterns of change to system parameters, represented as paths in parameter space.

Explanations in a DST framework focus on features of those patterns and paths. Understanding of a system's behavior is advanced in those terms. The most straightforward such pattern to detect is a "steady-state:" a point in state space such that, if a system reaches it, values of its variables remain constant (absent any perturbations to the system). Such points can often be identified for a set of equations in the absence of general solutions. Patterns of change in a system from points near a steady-state are approximated via another theoretical device, linear stability analysis.[19] If those patterns tend to converge toward the steady-state, that steady-state is "stable"; an "attractor." If the system's paths tend to radiate away from the steady-state when nearby, that steady-state is "unstable." These dynamical features of a system are taken to represent how a system responds to perturbations—roughly, external things that impinge on a system, or internal fluctuations within the system itself ("noise") that change the value of one or more of its variables. Steady-states that are stable for some perturbations but unstable for others are mixed cases, or "saddle points."[20] The term "saddle" brings in a visual aspect of DST models: the "landscape" of a system's state space. A landscape, in DST, is a three-dimensional representation of regions and boundaries defined by a system's behavior around its steady-states.[21] These regions are delineated by "basins of attraction" around stable steady-states. A basin of attraction encompasses all the points in state space representing combinations of the values of variables that, if the system is initially assigned that combination of values, then (absent any perturbations that fling it out of the basin) the system will change so as to reach that steady-state. Unstable steady-states define peaks on a system's dynamical landscape; the obverse of basins. Saddle-points are associated with both; a tarn or mountain lake with a nearby drop, so to speak. Boundaries of basins divide a system's state space into regions. In this way, topographic features of a system's dynamical "landscape" are determined by dependency relations codified in equations of a DS model.

Explanatory models in a DST framework characterize a system's stability behaviors, represented as a landscape topography showing where a system will 'move' from a given starting point in state space. Sometimes a path can be calculated analytically by solving the equations for a given initiating point, but more

[19] For details, see Newman (2010, 679–686), Strogatz (2000, 24–26, 150–155).

[20] Other special cases involve complex values of matrix vectors, such that the steady-state is not a fixed point, but an oscillation or spiral.

[21] C.H. Waddington's pioneering "epigenetic landscape" diagram (1957) is viewed by some systems biologists (reasonably, I think) as an ancestor of systems biology models today (Fagan 2012b). Sewall Wright's evolutionary landscape emerges from a different theoretical tradition.

often paths are approximated via linear stability analysis, inferred from simulation results, or estimated using a combination of mathematical and computational techniques. A general representation of paths emanating from each point in a system's state space yields a "flow diagram" for that system. Changes to parameters in the set of equations are visualized as changes in landscape topography (for state space) or as paths in parameter space.[22] None of this theorizing, obviously, references living things. DST explanations are abstract, using mathematical techniques to describe and predict patterns of change (and patterns in those patterns) in variables of an antecedently-characterized system. Any system whatsoever that satisfies the equations is covered by—i.e., falls within the explanatory scope of—that DS model. This generality is a key explanatory value for systems biologists; alongside mathematical rigor, it's vital to DS models' explanatory power.[23] Biological targets are specified by interpreting the variables and dependency relations among them as components of a biological network: genetic, metabolic, or biochemical. Most biological networks represented this way are within a single cell, but inter-cellular networks can be modeled as well. For simplicity, I'll refer to the interpreted assignment for variables in a DS model as "molecular components"; genes, RNA, proteins, metabolites, or some combination thereof.[24] The value of a variable refers to the amount of that molecular component in the system at a particular time. Changes to these values are represented as paths in a DS landscape, as described above.

DS models make predictions about what happens to a system over time, given initial amounts of each molecular component. Does the system reach a steady-state, oscillate, or behave unpredictably? Outcomes are in terms of the state variable: amounts of each molecular component. "The system" is sometimes identified as a single cell. But it's important to recognize that this is reductionist, effectively defining the cell as a whole as a subset of its molecular components.[25] In some areas of biology, this reductive assumption is accepted—indeed, it's central for explanations in some areas of genetics. Many systems biologists endorse the reductive assumption for gene regulatory networks (GRNs). In this variety of DS model, variables are identified as "genes" and their values represent the amount of each gene product (mRNA or protein). If the overall system is defined as a single

[22] Regions in parameter space are delineated by features such as number and relative position of attractors (Klipp et al. 2009, Jaeger et al. 2012, Jaeger and Sharpe 2014). A transition from one region of parameter space to another means a sudden change in a system's dynamical properties, its stability behavior—termed a "bifurcation."

[23] In the 2000s, there was a major explanatory achievement concerning "small-world networks," originating in mathematics but applied to systems that can be represented as network diagrams in engineering, sociology, biology, and more (see, e.g., Baribási 2002).

[24] Amounts can be conceptualized as concentrations, number of molecules, etc. ODEs are the most commonly used formal framework for cellular systems models, both in systems biology generally, and for stem cell phenomena in particular. Alternative frameworks include directed graphs, Bayesian and Boolean networks (see Fagan 2013, ch. 9 for more detail).

[25] Fagan (2013) discusses this issue more fully.

cell, then the state of that cell at time *t* is defined as a specific pattern of gene expression; i.e., a "fingerprint" of the amounts of each gene product in the network. Different cell types of a multicellular organism can then be identified in terms of such patterns; a different gene expression pattern for skin cells, neurons, liver cells, muscle, etc. This is the basic idea of Stuart Kauffman's pioneering theory of cell development (1969).[26] Kauffman's theory identifies cell types as attractors. If one takes a DST view of cell development, stem cells are a special case: precursors of "mature" cell types; i.e., stable attractors. Just as multicellular organisms pass through a sequences of developmental stages from zygote to early embryo to late-stage embryo and beyond, organisms' component cells pass through a sequence of stages, each defined by a distinctive pattern of gene expression. Only the last, however, is a stable attractor—a final, mature cell state. Stem cells are mixed, metastable attractors. Kauffman's theory suggests an explanatory model of cell differentiation.

Kauffman's theoretical approach also includes general principles of self-organization in complex systems. Two key principles are in tension: stochastic variation and robustness. In the 2010s, building on Kauffman's ideas, two groups of systems biologists proposed DS models of stem cells. One was constructed by Sui Huang, a collaborator of Kauffman's from the University of Calgary; the other by Chikara Furusawa and Kunihiko Kaneko, from (respectively) RIKEN and the University of Tokyo (Huang 2009, Furusawa and Kaneko 2012). Although not identical, the two models are similar in structure, content, aims, and explanatory features. So for my purposes they can be treated as one explanatory achievement for systems biology. (It is not, however, a *major* explanatory achievement, for reasons discussed shortly.) The model advances understanding of stem cells' distinctive capacities of reproduction and development, according to explanatory desiderata and values of the systems biology community. Principles of stochastic variation and robustness further enrich these modelers' approach (Furusawa and Kaneko 2001, Huang 2009, Huang 2011, Kaneko 2011). Both apply to cells conceptualized as dynamical systems. Stochastic variation in cell state follows from thermodynamic principles applied to processes of binding and diffusion for small numbers of molecules. Robustness refers here to a system's ability to maintain its state over time despite small perturbations. A cell-system, in Kauffman's theoretical framework, must balance stochastically-driven plasticity with robustness.

In this framework, stem cells embody the central tension. With respect to reproduction, they are stable—continuously dividing to make more stem cells. But

[26] Stuart Kauffman's relation to the systems biology community is interesting. Kauffman has been an influential theorist for decades, and played an important role in establishing the Santa Fe Institute as an epicenter for theoretical biology in the US. His contributions to studies of complex living systems and self-organization are extensive, dating back to the 1960s and continuing to present-day. Yet Kauffman appears less involved in systems biology's recent community- and institution-building efforts (Section 8.1). It would be interesting to delve further into his role, and that of the Sante Fe Institute, in the systems biology community today.

they are also "plastic," giving rise to more developed, differentiated cell types. So stem cells exemplify both principles at once. In DST terms, a stem cell is at once at an attractor, and a state from which different attractors (more mature cell states) are accessible (e.g., Furusawa and Kaneko 2001, 415). The explanatory DS model derives these and other characteristic features of stem cells from a simple GRN consisting of two genes that each affect their own and one another's expression. Each gene encodes a product that is self-activating and inhibits the other; a simple regulatory circuit. This simple system is represented by a set of coupled ordinary differential equations with parameters for rates of activation, inhibition, and degradation for each gene product. Additional parameters and terms are added to represent cell-cell interaction, stochastic variation in molecular concentrations, and cell division. So this mathematical model represents a population of cells, each identified as a system comprised of a simple two-node network. Dynamics of this cellular system are determined by computer simulation.[27] Each run of a simulation describes a path in state space from a given initial state. Many repetitions using different initial states reveal landscape topography for this multicellular system: the pattern of basins and attractors.[28]

Systems biologists use computer simulation to perform thousands of repeated "runs" with systematic variation of GRN features to infer general "design principles" for the systems so represented. In this case, Huang and Kaneko (working separately) sought to replicate stem cells' defining abilities with a simple GRN model (plus elaborations, which differed somewhat between the two groups). Computer simulations showed that the GRN's landscape has two stable attractors—one for each gene being highly-expressed, the other's expression correspondingly low. These are interpreted as different developmental pathways. Between these stable states is a saddle point—metastable attractor straddling the two basins of attraction (Figure 8.2). A cell in that state can enter either pathway. Stochastic fluctuations in gene expression determine which pathway is taken, in a given simulation run. Over many simulation runs, the potential for development to either stable state is shown. The crucial explanatory feature is that the model's dynamical behavior replicates stem cells' characteristic abilities to stably maintain (and reproduce in) the stem state, and to develop along multiple pathways to more specialized, mature cell types. For example, this simple two-node model "recapitulates" stem cells' potential to develop along multiple pathways (Kaneko 2011, 408). More elaborate variants of the simple model can represent other features of biological stem cells, such as broader differentiation potential, low levels of "promiscuous" gene

[27] This requires the mathematical model be "translated" into programming language—yielding a computational model. The explanatory achievement in this case involves all these formal models as well as their results and theoretical interpretations thereof.

[28] Computer simulation also allows for variation in network structure; e.g., changes in the number of nodes or parameter values. Effects of these changes too can be inferred from thousands of repeated simulation runs.

expression, long-term maintenance of a stem cell population, the developmental distinction between stem cells and cells "committed" to a particular pathway (Kaneko 2011, 408), and even the potential for cell reprogramming (see next section). In virtue of its similarities with experimentally-characterized stem cells, the DS stem cell model is recognized (by systems modelers) as an explanation of stem cell capacities in living systems.

Experiments on stem cells in living tissues or artificial cell culture don't figure in this explanation, except as a basis for comparison with stem cells in the model or to suggest values of variables and parameters in elaborations of the basic two-node model. (Many more combinations of values are explored via simulation, however—the DST model is not constrained by values of variables and parameters found in living cells.) Abstraction from experimental details is a major feature of this explanatory style. Generic behaviors of GRNs are derived from simulation studies, and then compared with observed developmental phenomena such as existence of discrete cell types, homeostatic stability of cell types, and branching pathways of cell differentiation ("macroscopic observables" of cell development; Kauffman 1993, 409). Insofar as they match, generic GRN properties explain observed features of biological development. Huang and Kaneko make these explanatory claims explicit; e.g., "[u]sing a dynamic-systems approach to explain the differentiation of stem cells, we have described here how fluctuating and oscillatory gene expressions underlie the essence of stemness" (Furusawa and Kaneko 2012, 217). Generic network properties, revealed by canvassing the full range of

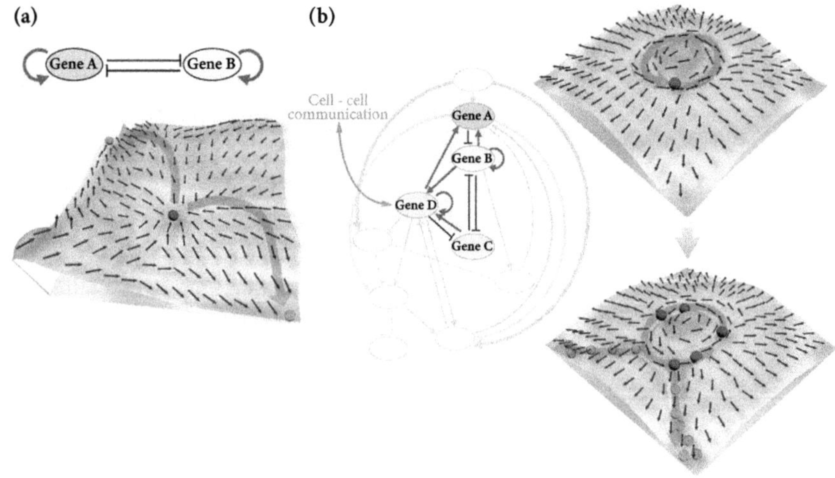

Figure 8.2 Systems model of stem cells, showing the basic GRN and associated landscape with attractors.

Source: C. Furusawa and K. Kaneko (2012), "A dynamical-systems view of stem cell biology." *Science* 338: 215–217 (figure 2a, p. 216). Reprinted with permission from AAAS.

possible network structures via simulation, are fundamental basic principles that entail key features of cell development.[29]

We can see a number of features of systems biology's explanatory style in the DS stem cell model. Explanatory power is primarily localized to the mathematical model and its grounding in DST. Although general principles of stochasticity and robustness enhance that explanatory power, it's the mathematical formulation that's essential. Kauffman (1993) aims "to account for" known facts of development by showing they are "explicable as natural consequences of the framework" (520). Alongside rigor, an advantage of DS models is that they offer general, universal predictions for cells' dynamical behavior. If conditions for a DS model are satisfied, then the "course of differentiation" it predicts "appears universally" (Kaneko 2011, 403).[30] These remarks indicate the epistemic values of DS modelers—the kind of explanation and understanding they are seeking. Crucially, the biological features of stem cells are explananda to be explained by the DS model. Simulation results from the model show that GRNs with specific structures reproduce (or "recapitulate") dynamic traits and processes of biological stem cells.[31] The systems modelers interpret this mapping as having one-way explanatory significance. The DS model is not associated with concrete biological systems, except via the model-target relation; no explanatory power is attributed to the biological aspects of the model. In the GRN, the latter are abstracted away. The modelers want to specify typical dynamical behavior for all GRNs that satisfy certain constraints "no matter what complicated processes are involved" (Kaneko and Yomo 1999, 254). Generic characteristics of *possible* networks (studied via simulation) are taken to reveal essential features of *actual* biological networks—bypassing the complex and varied details of cell lineage-, tissue- or species-specific molecular pathways, and nearly all molecular, cellular, and organismal details.[32]

Traditional explanatory unification is implicated as well. The DS model subsumes stem cell phenomena to a wider domain, 'covering' it along with phenomena of normal development, cancer, and evolution. For DS modelers, this is a significant advance in understanding; "a unifying formal framework that connects [diverse developmental phenomena], and thereby, explains the seemingly

[29] This explanatory style conforms well to Hempel's classic D-N account (see Chapter 1; Fagan 2016c).

[30] For example: "[w]e believe that our results are universal in a class of dynamical systems satisfying minimal requirements of the developmental process" (Furusawa and Kaneko 2001, 412); "[r]obustness and plasticity are essential features of all biological systems" (Kaneko 2011, 403); and selective adaptation for orderly networks "may be an ahistorical universal in biology" (Kauffman 1993, 409).

[31] In this case, computer simulation seems more a tool than a source of explanatory power, although further examination is needed to confirm this. The explanatory power and values accorded to computer simulation may be an incisive way to distinguish different strands or subcommunities of systems biology.

[32] For more on abstraction and general principles in biological (specifically mechanistic) explanations, see Braillard (2010), Levy and Bechtel (2013), Green et al. (2015), Breidenmoser and Wolkenauer (2015).

spontaneous order and patterns . . . that so robustly produces the diversity of cell types during metazoan development" (Huang et al. 2009, 870). Experiment, as noted, plays a minor, supporting role in model-construction. The "underlying" GRN structure is dictated by theoretical considerations, constrained only somewhat by experimental data. The DS stem cell model is itself a human construct, albeit incorporating background knowledge acquired from traditional experiments. To fix parameters of the mathematical model of this network, the modelers used simulations. The main results showing stem-like features were also from simulations, with no experimental counterpart or sequel. These methods are harmoniously suited to the explanatory values—and aspirations—of the ambitious, systems-theoretic strand of systems biology (see Section 8.1).[33]

Explanatory model 2: reprogramming

As noted, the stem cell research community is organized around focal experimental methods and their products. Explanatory models in that field are tightly associated with at least one such experimental system. Examples in earlier chapters (e.g., embryo models) show a profusion of closely-related models, grounded on variations of one focal experimental system (see Chapter 2, Chapter 4). The most famous—or infamous—such experimental system is the method of producing human embryonic stem cells (hESC). These cells, if grown in nutrient-rich artificial media, can retain abilities normally restricted to the cells of very early embryos: unlimited division and the potential to develop into any kind of cell found in the human body (e.g., muscle, neurons, skin, liver). Because they continue to divide, hESC are effectively immortal "cell lines"—as long as they stay in the right cell culture conditions, without getting too "overgrown." Unlike the various kinds of specialized cells making up human bodies (and other multicellular organisms), hESC are rather nondescript under the microscope. Their distinguishing morphological features are flat round shape, large nuclei, and prominent nucleoli; because they divide often, they appear in clumped colonies when grown on a flat surface. Gene expression patterns are more lively; hESC characteristically show high expression of certain genes. However, details of their gene expression profile change in response to changes in technology and our knowledge base—so although there's a lot of information about hESC's "genetic signature," not much of it is robust or stable. But hESC's reproductive and developmental abilities are both. They are a powerful generative cellular tool, able to produce seemingly limitless human cells of any type or developmental stage.

[33] For example, Kitano (2004) anticipates a general theory of biological robustness ("there are specific architectural features required to make organisms robust and . . . they might be universal to any robust and evolvable complex system," 835).

As discussed in Chapter 2, hESC are one of many interconnected embryo models, which collectively advance understanding of early human development. Of all these interrelated models, hESC are the most striking achievement. Their explanatory power is proportionate to, and to an extent derives from, their generative powers. hESC are fecund and versatile; a uniquely "generative" experimental system (Fagan 2016b). Their explanatory contributions are inseparable from these cells' concrete biological power. And the latter resides in the experimental method for producing and maintaining hESC. The achievement, in large part, is the experimental method itself. That method was innovated by a group led by James Thomson at the University of Wisconsin, as a modification of earlier stem cell culture experiments in non-human primates (Thomson et al. 1998). Producing hESC begins with an early human embryo (about five days) in tissue culture, acquired after IVF donation. Part of the embryo's inner layer of cells is removed, and the "explant" placed in artificial media to grow. In this new environment, some of the explanted cells divide to produce colonies with the appearance and gene expression pattern of hESC. Cells with these features are selected for continued culture, continuously dividing to generate new colonies. Samples from an "immortal" hESC cell line develop into other cell types if placed in a different artificial culture environment. That is, by altering cell culture conditions, experimenters can direct hESC to develop along one or another pathway, to become any kind of mature specialized cell found in the human body (including reproductive cells). More elaborate 3D cell culture "scaffolds" can spur development of miniature, simplified organs—"organoids" that model liver, brain, intestine, heart, etc.[34]

Human ESC were immediately controversial, because they are made by removing cells from an early human embryo (although not a candidate for normal human development; "spare" embryos donated after IVF procedures are used). This raised ethical and political controversy in many countries (Gottweis et al. 2009, Hauskeller et al. 2019). Consequently, many stem cell researchers sought ways to avoid controversy, by making cells with hESC's generative abilities *without* using human embryos. Shinya Yamanaka thought to do this starting from regular differentiated mammalian cells—skin, neurons, etc. His approach built on three lines of earlier experimental work: hESC, "cloning" experiments on amphibians[35], and intriguing experiments on mature cells showing that one molecular factor (MyoD) can switch skin cells to a muscle phenotype. The molecular factor is expressed within cells—a protein encoded by a gene. Yamanaka's idea was to find molecular factors like MyoD that can switch a cell's developmental state, from some kind of differentiated cell to an embryo-like state—that of ESC. The task was

[34] Experimenters cannot, however, direct hESC to produce an entire human organism—just "sheets" of cells, spherical "embryoid bodies," more complex "gastruloids," or simple "organoids" (see Chapter 2).

[35] These "somatic cell nuclear transfer" experiments showed that an unfertilized egg can induce the nucleus of a mature cell to 'restart' embryonic development.

to find a few molecular factors that can do the same job as the egg cytoplasm—an enormous simplification of the problem of development. So, the task was to build a simple model to control a cell's phenotype: making a new developmental pathway that effectively reverses the normal process of development.

Based on this idea, Yamanaka's research team at Kyoto University developed the method of cell reprogramming, which yielded induced pluripotent stem cells (iPSC).[36] Briefly: differentiated cells (mouse or human) are placed in artificial culture, and multiple genes added to their nuclei. After several weeks, in successful experiments, a few cells (usually ≤0.05 percent of the total culture) show signs of pluripotency, and these are selected for further growth and proliferation. Cells that divide and morphologically resemble ESC under these conditions initiate a line of induced pluripotent stem cells (iPSC). The key to the method is exactly what combination of genes is added to specialized cells, to change them to an embryo-like state—the "recipe" is very specific. Yamanaka's team developed their reprogramming method through a sophisticated sequence of experiments, that allowed them to discover a small set of effective "reprogramming" factors. hESC provided the motivation, background, and test baseline for the entire series of experiments. The experimenters wanted to discover a powerful network of interacting genetic factors without prior knowledge of that network's structure—no mathematical modeling required. Instead, the concrete experimental system did the programming and reprogramming work. The only part of the experiment involving anything like a human-made program was the very first step, in which Yamanaka's team canvassed earlier stem cell culture experiments to identify twenty-four "candidate factors"—genes that, based on published results, might induce a pluripotent cell state (Takahashi and Yamanaka 2006, 663). A range of experiments were considered, on early embryos, cancers, and embryonic stem cell lines.

Once Yamanaka's team identified their twenty-four candidate factors, they added all of them to cultured mouse skin cells, and used a special cell culture medium to select only cells with ESC abilities. The selective cell culture was indispensable for the whole series of experiments. The team first showed that all twenty-four factors together can induce the embryo-like cell state (in a very few cultured cells). The experimenters then removed each candidate individually. If the other twenty-three factors together couldn't induce an ESC-like state, then the missing factor was identified as necessary for producing that state. Ten candidate factors were identified in this way. Combined, they induced pluripotency *more efficiently* than the set of twenty-four. "Individual subtraction" experiments

[36] A few features of the original iPSC experiments, later found unnecessary, are here omitted for simplicity. Reprogramming was first accomplished in mouse cells; the method was adapted for human cells a year later.

were then performed for each of the ten. That round of experiments showed that four molecular factors out of ten are necessary to induce the embryo-like state. Yamanaka's group tried those four together, and no others, and got about the same efficiency as with the ten. They identified those four factors as the "core set" of factors sufficient to induce pluripotency in differentiated mammalian cells (mice, then humans): Oct3/4, Sox2, Klf4, and c-Myc. All these are "transcription factors"—genes encoding proteins that influence expression/activation of other genes. Expression relations between these genes can be represented by an interacting gene network, a GRN.

Producing iPSC via reprogramming was a major achievement for Yamanaka's team and for stem cell biology more broadly (Takahashi and Yamanaka 2006). The method was lauded as *Science* magazine's 2008 Breakthrough of the Year, garnering a Nobel Prize for Yamanaka in 2012. But there was more to the explanatory achievement of iPSC. Once the factors were identified, Yamanaka's findings were rapidly replicated in many laboratories across the world. This led to both refinement of the method and generation of many variants on the original procedure. Different combinations of specific reprogramming factors, starting cell types, culture conditions (and more) produce subtly different iPSC populations with specific developmental capacities, such as the ability to produce neurons, blood, or cardiac muscle. So instead of a single standardized procedure, reprogramming methods are a family of variants on a general schema, the fruits of a globally-distributed experimental process.[37] The upshot is an extraordinarily dense and richly-connected family of models of development, each grounded in a concrete experimental system. One major explanatory achievement that emerged from this thicket of iPSC experiments and models is "the core pluripotency network" (Figure 8.3). This network model isn't identical to the results of Yamanaka's experiments, nor is it fully general or fixed in the community. It is, however, influential and important in stem cell biology—a significant achievement, building on the iPSC triumph. The model depicts three factors (Oct4, Sox2, and Nanog) that together comprise a "developmental switch" determining whether cells are pluripotent or differentiate. That is, depending on the balance of gene expression activity among these three components, a cell can stay in (or enter) a stem state, or differentiate along one pathway or another.

In some ways, this simple GRN resembles the DS model of stemness—the network diagram aspect, anyway. But as explanations, the two are very different. Most obviously, the stem cell biology community's model is very closely tied to experiments. One influential review depicts specific molecular interactions implicated in cells' entering or leaving a "stem state" as "[t]he accumulated

[37] Reviewed in Maherali and Hochedlinger (2008), Hochedlinger and Plath (2009), Stadtfeld and Hochedlinger (2010).

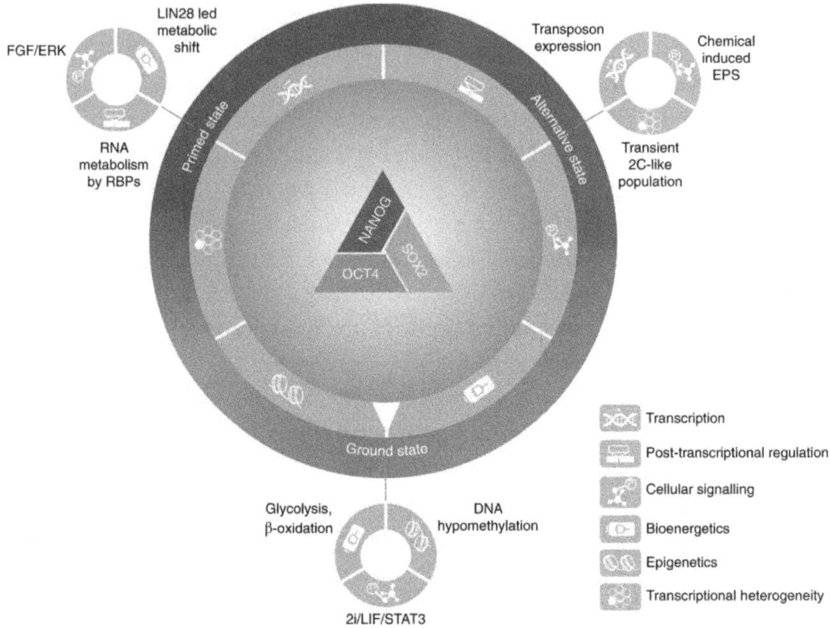

Figure 8.3 The core pluripotency network model

Source: used with permission of Springer Nature, from "Deconstructing the pluripotency gene regulatory network," by M Li and JC Izpisua Belmonte, *Nature Cell Biology* 20(4), 2018; figure 1, p. 383.

knowledge of the molecular circuitry that underpins pluripotency" (Hackett and Surani 2014, 417). That "accumulated knowledge" is from myriad experiments on reprogramming and manipulation of mouse and human ESCs. A second, related contrast is that the simple three-factor network doesn't hold all explanatory power for the phenomena of interest. The three genes (Oct4, Sox2, and Nanog) "together with a suite of interconnected regulators, underpin pluripotency" (Hackett and Surani 2014, 417). The three genes alone are not represented as dominating or controlling pluripotency, but as part of a much more complex regulatory system. Their significance as a "core set" is due to their exceptionally broad and robust experimental support, rather than a privileged causal role. More detailed models expand the description of regulatory pathways beyond the core set, embedding this minimal network within a more inclusive signaling network within and without the cell. The three-factor network "hooks into" a cell's molecular context in ways not fully understood, but which produce changes in cell phenotype that are represented as different developmental stages. The next task is to work out how this simple network hooks into other cellular pathways. The core set is thought to play two key roles: repress genes associated with differentiation, and activate ESC-specific genes (Stadtfeld and Hochedlinger 2010, 2249). The full picture is an intricately complex mechanism, interactions of which hinge on molecular specificity (see Chapter 7). Working out the details is an ongoing explanatory project in stem cell biology.[38]

Though a work-in-progress, this experimentally-grounded model of pluripotency indicates important features of explanatory models in this community. The target to be understood is early-stage human development, inaccessible to direct experimental study for practical and ethical reasons. Cultured embryonic stem cell lines are concrete models that, as part of a network of interrelated experimental products, advance understanding of that process. An extrapolative inference from concrete model systems to unmanipulated human (or mammalian) development is crucial, for this style of explanation to succeed. Debates about whether such inferences are justified are therefore quite high-stakes, in the many life science communities that use model organisms or experimental systems.

The GRN model of pluripotency summarizes results of myriad experiments. The components represented in the network represent real molecular networks underlying stem cell behaviors—not in every detail, but with respect to certain features of the molecules and other structures involved, and the relative time and location of interactions. More detailed "mechanistic" models include the character of those interactions (e.g., binding, breaking, pushing, dividing, transmitting); these are abstracted away in the simple GRN. That model "covers" both embryo-derived and induced pluripotent stem cells, and so is sometimes hailed as a "general" pluripotent stem cell model. Many stem cell biologists consider the pluripotent state to be *fundamental* for subsequent developmental stages—so understanding this state at the molecular/biochemical level is the starting point, or foundation, for understanding the rest. But this understanding is not vouchsafed by derivation or mathematical modeling. Instead, the achievement rests on showing how an organism proceeds from one developmental stage to the next, articulated in terms of components of the organism and its environment that are themselves characterized and accessible to manipulation. Similarly, although Hackett and Surani describe recent results in terms of "principles," the explanatory model cannot be expressed as a concise set of statements or patterns. The more complete pluripotency network is a complex, detailed account of interactions between particular molecules, biochemical, and biophysical factors. "Principles" in this experimental context refer to what philosophers of biology term "mechanisms," or parts thereof.

The term "programming" also has a somewhat counterintuitive meaning in reprogramming experiments.[39] Yamanaka's subtractive experiments selected for interactions among a small set of factors, but did not dictate how they are related. The whole series of experiments was designed to reveal GRN organization,

[38] More precisely targeted gene editing using CRISPR is now routinely combined with iPSC methods, generating an even more intricate profusion of experimental methods and cellular products, which fit into the explanatory array.

[39] See also Brandt (2010).

without pre-supposing anything about that organization. The "program" in these experiments is not a product of human design, but of biological processes internal to the cells. It is the molecular circuitry thought to control development, which experimenters discover by combining results of many experiments (still a work-in-progress). "Reprogramming" refers to a manipulation of this circuitry; forcing a change by changing amounts of specific genes. There's no counterpart to computer simulation, mathematical modeling, or design of hypothetical networks, as in systems biology.

Summary: contrasts

Both models advance understanding of stem cells, in their respective communities. They both do so by showing how simple molecular networks (GRNs) can give rise to dynamic cell phenotypes. But their means of showing this are very different: cell culture experiments on the one hand, computer simulations on the other. The two models are constructed via different processes, drawing on divergent epistemic and technical resources. Reprogramming rests on a background of prior experiments/models, DST models on mathematical theories and computational methods. Both explanatory achievements bear on biological principles about the nature of development, but in diametrically-opposed ways. The systems biology model posits essential principles for biological development that converge on "the stem cell" as the foundation for subsequent cell development. Such principles are central for this style of explanation. Reprogramming, in contrast, contradicts deeply-entrenched principles of irreversible, internally-directed development. This overturning of a principle is seen as a positive result—an advance in understanding by means of experiment. This contrast epitomizes the stark difference in explanatory style, for these two GRN models.

8.3 Bridging to IDE

The relevance of these two models to one another is obvious. They share a target: stem cells and their developmental sequelae. IDR projects involving both are to be expected on both sides, albeit for different reasons. On the stem cell biology side, the reprogramming model is part of a major community achievement. Its significance is such that Yamanaka was awarded a Nobel Prize in 2012 (the other co-recipient was John Gurdon, who pioneered the method of "reversing" development of a cell nucleus with egg cytoplasm). This broader recognition motivates members of other communities to connect their efforts to this success. Central to that success was the reprogramming model (and method) being endorsed, refined, and taken up by many small-scale research groups (labs, teams, etc.) in the stem cell community. Because that community is markedly interdisciplinary, one would expect the reprogramming model to connect with those of the many other fields that feed into stem cell biology. Systems biology is one such—the first annual Systems Biology of Stem Cells meeting was held in 2010 (at UC Irvine), attended by Huang and a number of other DS modelers. And, obviously, connections are to be expected on the systems biology side. That community's explicit integrative aim provides clear motivation for systems biologists to engage in IDR projects with experimental biologists. In the case of the DS stem cell model, that integrative aim is part of the modelers' explanatory efforts, as attested in their publications. Huang, Kaneko and colleagues explicitly positioned their DS model as a contribution to stem cell biology—an explanation of stem cell capacities and their role in cell development. Although they didn't discuss the reprogramming model specifically, these DS modelers sought uptake and engagement from the stem cell research community more broadly.

Yet, despite features that would seem to encourage IDE in this case, none ensued. The DS stem cell model achieved respectable recognition in theoretical biology circles, but had no appreciable uptake by stem cell biologists. To my knowledge, stem cell biologists have not made any response to it whatsoever. Papers discussing these theoretical models do not appear in stem cell journals, are not cited by stem cell biologists, and are not included on the program of the largest annual meeting for stem cell researchers.[40] One of the DS modelers, in a contribution to *BioEssays*, hypothesized that editors and reviewers are subject to the Dunning-Kruger effect, thus blocking hoped-for uptake of, e.g., gene network models (Huang 2013).[41] Absence of IDE in this case is all the more puzzling, because other ideas and methods from systems biology have been eagerly adopted by members of the stem cell community: gene network models, landscape representations of cell development, and more (Fagan 2012b, 2013). These 'transfers' from systems biology are especially common in studies of pluripotency—the phenomenon highlighted by reprogramming, central to that model. Indeed, GRNs figure in both explanatory models discussed. Altogether this case seems very ripe for IDE, given features of the particular models and social epistemic organization of the communities involved. Why was there a failure to connect, in this case? The IDE account from Part II offers an answer, and furthermore suggests how the obstacle can be overcome.

To briefly recap, my IDE account focuses on "low-effort interdisciplinarity," offering minimal conditions for connecting two EP-explanations from different

[40] Citations to Furusawa and Kaneko (2012) as of 7/1/2014: Web of Science (23) and GoogleScholar (38). None of these citations are in specialized stem cell biology journals. Programs of the International Society for Stem Cell Research annual meeting assessed: 2011–2014, 2018.

[41] The lack of connection is even more puzzling for Huang's broader oeuvre, which adopts and adapts Waddington's epigenetic landscape—a figure also adopted and adapted by the cell reprogramming community at roughly the same time (late 2000s–early 2010s). See Fagan (2012b) for a detailed examination of Waddington's landscape model and its use by both communities.

communities. Unlike higher-effort, resource-intensive IDR projects, in low-effort IDE only a small conceptual link between disparate models is sought; a "narrow, sturdy bridge." More ambitious IDR projects seek to establish what amounts to a new "transdisciplinary" community with a distinctive explanatory style, negotiated by participants from their diverse community backgrounds. Low-effort IDE is far less demanding, well-suited to cases where explanatory models are well underway as achievements in their respective communities (see Chapter 4). The reprogramming model, also still a work-in-progress, is evidently a major explanatory achievement for stem cell research. The DS stem cell model, although more modest, is recognized and credited as a contribution to theoretical biology, and comfortably fits that community's explanatory style. As the product of a couple of research groups rather than a wider community, the DS model is typical of a hoped-for explanatory contribution to an IDR project. In this case though, there's no stem cell biology counterpart—no ongoing IDR project at all. Working through the IDE requirements, minimal though they are, reveals why.

Point of contact

The most obvious way to connect two distinct models is to identify a point of contact between them, involving similarity, difference, or both. In this case, there are many areas of overlap. The most obvious is the shared term "stem cell," differently conceptualized in each model. In the systems biology model, a stem cell is a mathematical construct, conceptualized in terms of DST and the principles of stochastic variation and robustness. More precisely, a stem cell is a metastable attractor, embodying both principles as a cell state that is stable under some perturbations, unstable under others. A cell in the stem state can enter other developmental pathways, giving rise to more specialized cell types. But it also can remain a stem cell, depending on stochastic fluctuations in the concentration of relevant gene products. Crucially for this model, these features of the stem state are derived from a simple two-node GRN: two gene products that inhibit one another and self-induce. For stem cell biology, 'the stem cell' is the field's unifying explanatory target. It's conceptualized somewhat differently across specific research projects, but a rough characterization in terms of self-renewal and differentiation is fairly robust. In the reprogramming model, stem cells of interest are pluripotent (PSC for short); i.e., able to develop into any cell type of a mature organism (in practice, a mouse or human). Most other kinds of stem cell have more limited developmental potential, while specialized, mature cells can only make more of the same cell type. Induced PSC are constructed from mature mammalian cell, by increasing the expression of a few specific genes: "the core set." Their self-renewal and differentiation abilities are shown in cell culture experiments. Crucially, in this model, iPSC are derived from mature cells in culture, via manipulation of gene expression. A three-element gene network underlies the pluripotent cell state.

Obviously, 'the stem cell' is a central element of both models. Its abilities are derived from a simple gene network, demonstrated by computer simulation (DST) or cell culture experiments (reprogramming). There are many points of similarity that could serve as sites for bridging these two explanatory models. However, as a central concept and target of explanation for both models, 'the stem cell' isn't well-suited to IDE. It's not such as to gesture beyond a model's limits (as the next section will show). A more promising point of contact is the concept of a molecular network; a central circuit controlling development in each model. The systems GRN has two nodes, representing interacting gene products. It's a mathematical construct, based on a theoretically-grounded set of equations. Nothing about the two nodes' biochemical character is included; just their amounts and interactive effects. The latter are specified in rigorous detail by the DS model's equations. The reprogramming GRN is an abstract representation of three molecules that interact within cells. It's based on experiments, closely tied to molecular and biochemical details of intra-cellular processes. The specifics of interactions between the core set of molecules are unknown, although reliably elicited in experiments. Although the two GRNs don't coincide (having different numbers of nodes), they do complement each other, in aspects rigorously characterized: molecular interactions and biochemical structure. Both are clearly important for understanding stem cell behaviors. So this point of contact looks more promising. But it too is a central element of each model. It's difficult to identify a point of contact between these two models that has the peripheral, liminal character best-suited to low-effort bridging connections. This hints at problems to come.

Near-side coherence

IDE connections between models require some effort, albeit less than transdisciplinary projects. Some adjustment is needed on both sides, satisfying the minimal coherence requirements set out in Chapter 5. An advantage of IDE is the adjustment is minor—indeed, the more modest and peripheral the better. But in this case, the obvious points of contact between the two models are major components, implicating nearly the entire model. This sets up more of a barrier to IDE than one would like, in fostering interdisciplinary ties. Even so, I'll work through coherence requirements for an IDE for this case. On the near-side, the adjustment made to each model must cohere with that field's explanatory style while gesturing beyond the model's limits. Many such adjustments are possible in principle; only those that introduce a mystery, non-sequitur, or direct explanatory challenge to the community are ruled out. For the two models considered here, these minimal conditions are hard to satisfy.

Consider first 'the stem cell' as the point of contact for the two models. On the systems biology side, 'the stem cell' is the derivational result of a simple circuit comprised of a pair of self-enhancing, mutually-inhibiting, molecular species. Simulations using that network display 'the stem cell's' developmental role. So in a way the entire model represents 'the stem cell.' The "essence of stemness" appears in multiple guises: coupled ordinary differential equations, a network diagram, and computer code. Each of these forms is remote from biological material—constructed using mathematical theory, general ideas about biological development, computer programming techniques, and pieces of physical theory. These modeling resources are diverse, but all are distinct from biological material; living cells and tissues in organisms or artificial cell culture. In this sense, 'the stem cell' in the DS model is abstract. So a simple way to extend and adjust that model is to make that abstraction explicit; e.g., "There are concrete, biologically active stem cells in living organisms and experimental systems, to which the abstract, mathematical stem cell is somehow related." The nature of that relation is unspecified. Minimally, we can posit an abstract-to-concrete continuum, mediating between the DS stem cell model and other, more biologically concrete, models.[42] This is a straightforward near-side adjustment, appealing to a relation familiar to many scientific modelers. Yet the near-side coherence condition is not met, minimal though it is.

A near-side adjustment to one field's model indicates that model's limits, gestures beyond them, *and* comports with that community's explanatory style. The last clause is the near-side coherence constraint, ruling out anomalies, non sequiturs, and direct explananda. Stem cells in the material world, affecting development in organisms and experimental set-ups, aren't anomalous or non sequiturs for the DST model. There's nothing mysterious or irrelevant about indicating a connection toward living, concrete stem cells. But the latter are, for systems biologists, explananda for their model. The DST modelers make this plain in their publications (see Section 8.2). So it would seem that systems biologists in this case conceive the relation between mathematical and material stem cells as *itself explanatory*. Therefore, the proposed adjustment to the DST model doesn't meet the near-side coherence requirement. To meet that requirement, the abstract-to-concrete relation must be drained of explanatory relevance; systems biologists must conceptualize it in some other way. But it's difficult to imagine how such an abstract-to-concrete relation could comport with systems biology's explanatory style. So bridging connections for IDE in this case fall at the first hurdle, so to speak. This obstacle also shows up in participants' attitudes in IDE as collaborative activity—more on that shortly.

[42] In Part I's framework, this makes explicit an indirect difference relation between two models.

The stem cell concept isn't the only possible point of contact between these two models, of course, just the most obvious. The simple two-node circuit—the GRN—is a more promising site for near-side adjustment. Although this regulatory network can be considered an aspect of 'the stem cell' in the DS model, it can also be parsed as an element on its own. Try the latter interpretation as a site for adjustment; just the two-node network of self-inducing and mutually-repressing elements. The abstract-to-concrete relation between this GRN and networks instantiated in concrete biological context is, plausibly, less loaded with assumptions about explanation than the stem cell concept. So a promising near-side adjustment to this element of the DS model is: 'There are concrete biological systems whose molecules interact in this abstract network pattern,' or 'The abstract GRN is instantiated in some biological material.' This more focused adjustment to one element of the DST model indicates its limits, gesturing to the concrete material world. There's no anomaly or non sequitur involved, because the point of a systems biology model is to integrate mathematical and computational abstractions with experimental studies of concrete living things. Because instantiated GRNs in living cells aren't the explanatory target of this DS model, nor systems biology more broadly, this near-side adjustment doesn't introduce any explananda. Molecules that interact so as to instantiate the two-node GRN are just material counterparts of that abstract element of the DS model. But, if the abstract-to-concrete relation is imbued with explanatory force, such that, for example, an abstract GRN "covers" its multiple material realizations so as to explain them, then the near-side coherence requirement isn't met. The bridging connection has to be such as to indicate the explanatory limits of each model, not show that some phenomenon falls within those limits. The corresponding attitude of IDE participants is openness. There is more to the target phenomenon than what the DS model explains (non-exclusivity), allowing for other explanations to connect.

This requirement, for models and associated attitudes, holds on both sides. Consider first stem cells in the reprogramming explanation. Like most models in stem cell biology, it's not about stem cells in general, but a particular kind of stem cell. Here the kind is 'the pluripotent stem cell' of a particular species (usually mouse or human, in practice). These cells (PSC) are produced in cell culture by extraction from early embryos (ESC) or reprogramming more specialized cells (iPSC). Both are experimental constructs produced in artificial cell culture, from cells removed from living organisms. Pluripotent stem cells are thought to exist in unmanipulated living organisms for brief developmental intervals; i.e., very early embryonic stages. These PSC are included in stem cell biology's explanatory target: the pluripotent stem cell of a mammalian species (mouse or human). Reprogramming experiments produce cells of this type; that's the main achievement in this case. The explanatory model, with its associated advance in understanding, is grounded on that experimental triumph. Details of reprogramming experiments reveal components of "the pluripotency network:" three interacting transcription factors that, when highly expressed in a cultured cell, induce a stable embryo-like state; i.e., PSCs. So a PSC, in the reprogramming model, is the product of a reprogramming experiment (and the model's explanatory target). Because of this, it's hard to imagine an adjustment to the PSC concept that doesn't introduce an explanandum for the model. There is of course the background idea of multiple kinds of stem cell; many stem cells aren't PSCs, and are the focus of different experiments. Those connections between different experimentally-grounded models are, as I've discussed in this and previous chapters, the main social epistemic integument of this community.

However, systems biology doesn't fit comfortably into that community—at least, not the strand with high explanatory ambitions. So the adjustment to the PSC concept in this case, to connect DST and reprogramming models, needs to mirror the abstract-to-concrete link; concrete-to-abstract. We need an extension of the PSC concept that indicates explanatory territory in more abstract domains. Reasoning along these lines might proceed as follows:

> Reprogramming experiments show there are multiple paths in cell development, leading to a pluripotent state. We could think more abstractly about cell development, as pathways linking states, with PSC as the most "primitive" (the "ground state"). Theoretical study of abstract patterns of these pathways could yield new insights about cell development and stem cells' role therein.

This is a more elaborate line of reasoning than one hopes for, in low-effort IDE. But it does connect the reprogramming PSC concept to a more abstract domain, acknowledging that work in that domain could advance understanding of the explanatory target of interest. The problem is that experimental work on PSC across species shows that regulatory networks for pluripotency are species-specific. The molecular network for mice isn't the same as for humans. Reprogramming factors for PSC are different, depending on the organism. Similarly, the starting cell type and developmental stage, even within a species, appears to subtly bias the developmental abilities of iPSC from reprogramming. The cells used, their history in organismal evolution and development, make a difference to reprogramming. More abstract approaches, such as systems biology theories and analysis, group stem cell phenomena in more general categories and derive insights thereby. Lumping and grouping in this way, for reprogrammers, is the wrong move to advance understanding. So the same problem arises for the near-side adjustment to the stem cell model, as for the DS model. Adjusting the stem cell concept so as to connect with the other either introduces an explanandum or conflicts with the community's explanatory style. Prospects for IDE with this point of contact are dim.

So, again, consider the GRN on its own. In the reprogramming model, the GRN is "the core set" of three molecular factors that interact with one another, and other components of intra- and inter-cellular networks, to induce PSCs in reprogramming experiments. Those interactions aren't fully understood, though experimental results are robust and the individual genes and proteins of the core set are well-characterized. This GRN is a direct, though schematic, representation of concrete stuff—physical, chemical, and biological. So a promising adjustment to this element of the reprogramming model is to take it as a point of departure for abstract theoretical studies, whatever those might be: 'The concrete GRN can be studied abstractly.' This bridging statement makes the concrete-to-abstract connection explicit, indicates that the abstract side is beyond the scope of reprogramming experiments, and (because the tie is minimal and modest) avoids direct clashes in explanatory style. So it doesn't introduce a mystery, non-sequitur, or direct explanatory challenge for cell reprogrammers. Openness to explanatory activities on the abstract side is at least possible. By narrowing the point of contact between the two models, we can imagine bridging connections that are near-side coherent. Both sides of this abstract/concrete connection acknowledge the limits of their field's model, indicating what is outside its boundaries.

Far-side coherence

Far-side coherence is more relaxed, so (usually) easier to satisfy. The adjustment to each field's model should coherently connect with a non-expert's idea of what the other field is up to. That non-expert idea is vague, hand-wavy, and largely ignorant of the specialized explanatory practices on the other side. This requirement rules out only obvious incompatibilities between the bridging idea and the other model. This corresponds to the second-layer relation of collective intention; members of one field attempting to see an activity from the other side. For participants in an IDE project, this amounts to recognizing the other field's model as making an explanatory contribution relevant to one's own. This is articulated as some commonsense statement linking the bridging extension to the other field's explanatory contribution. From the systems biology side, far side coherence is satisfied if DS modelers can articulate some explanatory benefit emanating from the stem cell side. Consider the more promising of the near-side adjustments discussed above: the abstract GRN is instantiated in concrete biological material. Advancing understanding of concrete GRNs in stem cells could lead to further insights about their abstract counterpart. The DS stem cell model is idealized in omitting other molecular and cellular networks. In concrete biological systems, no GRN is fully isolated as a distinct module; molecular pathways are deeply interactive, and incompletely understood. Interconnections between a concrete counterpart of the DS GRN and other molecular/cellular networks could lead to novel features, that could be explored in elaborated versions of the DS model. Such results don't directly impact the DS model, or detract from its explanatory contribution. But understanding how it's instantiated/embodied could suggest improved or refined

DS models of the stem cell GRN and its practical significance. This far-side connection can be stated as: 'Understanding the concrete GRN in biological cells and tissues offers insights for refining the DS stem cell model.'

Mild though it is, DS modelers might reject this statement. Systems biology is defined as a community by its leaders and institution-builders, largely by contrast with molecular biology, and specifically by contrast with explanations that hinge on molecular details. There's a deeply-ingrained tendency to think of those details as non-explanatory; as data to be explained via mathematical models in systems biology rather than explanatory contributions in a different style. For example, Huang (2013) contrasts "the profound concepts that Conrad Waddington had in mind when he coined the term [epigenetic landscape]"—precursors to DST explanations in biology—with "the concrete, little molecular modifications of DNA and chromatin that their lab is studying" (415). The two aren't considered to be on the same epistemic level, so to speak. And this asymmetry, on the systems biology side, undercuts "other-regarding" attitudes associated with collaborative IDE-construction. From their perspective, on the other side of the bridge is not another (albeit different) explanation that could bear on their own, but data about a target their model explains. This isn't a quirky belief of a few systems biologists; rather, it's a central feature of the field's social epistemic organization. Systems biology has established itself as a distinct field largely by contrast with molecular biology's explanations. This strategy for effective community-building undercuts systems biology's integrative explanatory aims. To be clear: this doesn't undercut other IDR connections between systems and stem cell biology, nor systems biology and other life science communities whose explanatory style prioritizes molecular details. There are in fact many such connections, mainly in the realm of data collection and processing. It's constructing IDEs that's difficult.

Relatedly, there's a tendency in systems biology to resist acknowledging lack of expertise about biological phenomena. This is also a problem for far-side coherence, because the bridging connection on this side should be openly, honestly non-expert. This is how the norm of humility for collaborative activity plays out in IDE. Systems biology textbooks and training programs often treat biological subject matter in one or two chapters, or as a short unit of a course. There is often the suggestion, in my experience, that systems biologists feel they can acquire specialized knowledge about living cells, organisms, and tissues very quickly and efficiently, so as to become expert enough to engage with those areas of life science as the equals of biologists. And the reverse is not assumed; biologists working on experimental systems of cells and molecules are thought ignorant of DS models. (Most are! Though of course there are exceptions—on both sides of the bridge.) If systems biologists think they are experts in the areas of biology relevant to their projects, the attitudes they bring to IDE construction won't include the humility implied by social action norms.

The main takeaway here is that features of systems biology's social epistemic organization mitigate against far-side coherence and associated collaborative attitudes requisite for IDE. This is not unrelated to the problems for near-side coherence. The abstract-to-concrete relation between elements of a systems model and molecular networks embodied in living cells is, for many systems biologists, imbued with explanatory power. If molecular details are conceived as part of an explanatory target rather than as a different kind of explanatory contribution, then neither coherence nor collaborative attitude requirements for an IDE are met. This hinders IDE construction, and so frustrates integrative explanatory ambitions for systems biology. (Of course, that doesn't detract from the explanatory achievement of the DS model on its home turf, so to speak. It's the connection to stem cell biology, in this case, that's obstructed.)

Similarly for the reprogramming model's far-side connection, tying concrete experimental studies to abstract investigation using mathematical modeling tools. Most stem cell biologists have little understanding of those specialized methods; far-side coherence is designed to accommodate this. The requirement demands only some intelligible commonsense statement of how systems biology studies might inform the reprogrammers' model of PSC and their abilities. The most promising point of contact is the network "circuit" of molecular factors that effect reprogramming of other cells into PSC. So far-side connection can be articulated as: 'Studying the pluripotency network abstractly can offer insights for reprogramming experiments, advancing understanding of the pluripotency network in living cells.' One way this might happen is seeing the abstract counterpart of this three-element GRN as part of a broader, more inclusive class of networks, identifying properties of those networks, and looking to experiments to determine if those properties are found in PSC. Even negative results could be informative, leading to refinement of reprogramming models. One could imagine that systems modeling and simulation could be another domain of trial-and-error exploration of possible developmental pathways for cultured mammalian cells—but yielding results that group PSC development into broader classes of phenomena. Seeing reprogrammed cells and their developmental lineages as members of more general classes could lead to new insights and ideas for experiments. The emphasis on experiment makes this a plausible connection from the reprogramming side, and only a vague, hand-wavy notion of what mathematical modeling and simulation involves is needed. (Of course, a number of stem cell biologists are familiar with those methods—far-side coherence just doesn't demand that expertise.)

I don't think there's any major obstacle to reprogrammers accepting that systems modeling could offer insights for reprogramming experiments to reveal more about the pluripotency network. But there's no impetus to adopt it either, for the stem cell research community. There is an ethos in the field to combine technologies and adopt new methods—stem cell biology being interdisciplinary and highly opportunistic—but this doesn't motivate connections to systems modeling specifically. However, it would be reasonable to expect a few reprogrammers to follow this path. Indeed (in my view) that's how events have played out since 2013. Some reprogrammers studying pluripotency today make use of systems biology concepts and mathematical modeling tools.[43] These projects haven't transformed the community or its explanatory style in a major way, but they are present. My view is that these forays toward systems biology absorb those methods and concepts into the prevailing explanatory style, rather than constructing an IDE or a more deeply integrated explanatory model. But these are quick reflections. It will be interesting to see how these systems-infused explanatory achievements fare in the stem cell community going forward. For comparison, since 2013 there have been several new experimental techniques that have had major impact on the stem cell community: CRISPR/Cas9 and organoids. These experimental techniques fit smoothly with the prevalent explanatory style in stem cell biology.

The takeaway is that one can imagine, for both epistemic communities in this case, contributions to their model from the far-side. Explanatory activities in the other field, vaguely-conceived as they are, might have some explanatory relevance to one's own model. But on the systems biology side, there are reasons to think these posited far-side connections wouldn't be accepted. On the stem cell side, there are no major obstacles but no specific attractions either. Because IDE bridging connections are mutual, the far-side coherence requirement isn't satisfied in this case.

Middle coherence

Middle coherence demands that reciprocal bridging adjustments to-and-from each model don't undermine one another; i.e., there's no direct conflict between them, so a participant in IDE could accept both at once. In this case, GRN-focused connections are compatible in this way. From the systems biology side: 'The abstract GRN is instantiated in some biological material, and understanding this concrete GRN offers insights for refining the DS stem cell model.' From the stem cell side: 'The concrete GRN can be studied abstractly, and these studies can offer insights for reprogramming experiments to reveal the pluripotency network in living cells.' The question is whether systems and stem cell biologists on either side accept these bridging statements. As detailed above, features of both communities' social epistemic organization tend to discourage (or at least do not encourage) their acceptance. IDE is possible in principle, but not promising in practice. Attitudes toward the bridge in each community are the deciding factor. Norms of openness and non-exclusivity, equality and humility, and respect for contributions

[43] See, e.g., Rackham et al. (2021), Ho and Morsut (2021), Schaffer and Ideker (2021).

emanating from the other field, would be instrumental in instilling those attitudes. But in their absence, failure to construct an IDE is to be expected.

8.4 Diagnosis and further results

At first glance, it may seem surprising that the two explanatory models in this case fail to connect. An IDE is manifestly possible, in principle. A GRN mediated by an abstract–concrete tie is not the only potential point of contact. However, working through this example shows why failure to connect is unsurprising in this case. Given each community's explanatory style and social epistemic organization, even IDE's minimal bridging requirements are unlikely to be met. Failure to meet these requirements is understandable on both sides. For stem cell researchers, there's scant reason to connect their reprogramming model with the DS stem cell model, because the latter doesn't offer any details about the molecular network controlling pluripotency. According to systems biology's own explanatory norms, of course, that's an explanatory virtue. But from an experimental reprogrammer's viewpoint, nothing is gained from the concrete-to-abstract connection. The DS model captures key features of "stemness"; e.g., reprogramming is possible but rare, stem cells give rise to multiple cell types. Reproducing these features is a triumph for the DS model on its own terms. However, for experimental researchers this amounts to restating, in the language of state spaces and attractors, results that experiments have already achieved.

Systems biologists do have reason to connect their model to stem cell research: integration of mathematical modeling, computer simulation, and experiment is the distinctive goal (and method) of their scientific community. But their community's explanatory norms, grounded in a coherent package of values, methods, commitments, and aspirations, make it difficult for systems biologists to recognize stem cell biologists' own achievements as explanatory. To the DS modelers, detailed molecular mechanisms and networks are mere "hand waving" or "ad hoc metaphors" (Huang 2011, 2249). Absence of general principles and mathematical rigor means, to them, that an explanation is lacking. Showing that familiar stem cell features are derivable from general principles of development, via mathematical models and computer simulation, is just what is wanted. But the explanatory norms of experiment driven stem cell research are different. So although systems biologists aim at integration in this case, their explanatory style undercuts those same efforts. The most obvious points of contact between the two models are linked by relations imbued with explanatory significance on the systems biology side. As a result, requirements for IDE aren't satisfied. So, despite systems biology's avowed interdisciplinary aim, and the two fields' shared interest in understanding stem cells, the two models in this case remain isolated from one another. Importantly, both research groups have good reason to respond as they do. Stem cell researchers (including reprogrammers) ignore the systems models rather than engaging them. DS modelers dismiss reprogrammers' explanatory models as "mere hand-waving." It's hard to see how to get around the limitations of each community's own specialized view of explanation, so as to collaboratively construct an IDE.

However, there are a few positive takeaways from this case. First, the IDE account offers a better diagnosis of this case than others on offer, which tend to be one-sided. One such is "ignorance" on the part of stem cell researchers. Some DS modelers blame experimental biologists' ignorance of mathematical modeling techniques and commitment to outdated reductionist assumptions (e.g., Huang 2013). Philosophers of science tend to sympathize with this "epistemic deficit" view. If only stem cell researchers understood more about mathematical modeling and other systems biology techniques, then explanatory integration would proceed smoothly—so this line of thinking goes. But ignorance is not an obstacle to IDE. Indeed, recognizing one's lack of expertise in another specialized area is important for the attitudes associated with collaborative activity: openness, humility, respect for different explanatory contributions to the bridging effort. Researchers in specialized fields tend to be ignorant of others; that's true of stem cell research, but also of systems biology. Indeed, in science today no one's expertise extends very far. Furthermore, given stem cell biology's opportunistic ethos, ignorance would not stop researchers in this community from seizing on any new technique that might benefit their experiments. Stem cell biologists eagerly adopt and adapt ideas and tools from many different fields. The problem is not that stem cell biologists don't understand mathematical models (etc.) but rather than they see no benefit to using them. (And from their point of view, that's not a problem!) Diagnosing the situation as due to "ignorant biologists" is beside the point.

The IDE account also improves on my own earlier view of this case (Fagan 2016c). In that work, I argued that the two specializations fail to communicate in this case, because each is committed to a different account of explanation: covering-law or mechanistic, respectively. The systems modelers in this case emphasize their explanation's generality, mathematical rigor, the role of principles, broad scope, and the virtue of showing that well-evidenced stem cell behaviors are unsurprising when interpreted in terms of their model. All these features suggest an explanatory style that conforms to the classic covering-law theory (see Chapter 1). In contrast, the reprogramming model fits well with recent philosophical accounts of mechanistic explanation (Machamer et al. 2000, Craver 2007, Bechtel 2011, Craver and Darden 2013). The latter describe mechanisms, succeeding insofar as they accurately represent the organized entities and activities that produce or underlie the phenomenon of interest. The explanandum is how some overall phenomenon (e.g., stem cell pluripotency) occurs. The explanans describes how different parts constituting the mechanism for that phenomenon work together. For the pluripotency model, specific DNA, RNA, and protein molecules interact with one another, with other small molecules, and with features of cells. Mechanistic explanations are traditionally contrasted with law-based explanations, emphasizing causally-relevant details of particular systems (or classes of systems) rather than a wide range of phenomena or mathematical rigor. Improvements are made by adding further detail to a lower-level description, or further precision in characterizing the upper-level phenomenon of interest. Rather than an abstract unifying framework, mechanistic explanations integrate a wide array of experimental results, which "contribute piecemeal to the construction of a complex and evidentially robust mechanistic explanation" (Craver 2007, 19).

This diagnosis is OK, as far as it goes. Explanatory models on both sides, in the context of each field's social epistemic organization, are positioned to "clash" in ways captured by two different philosophical theories of explanation. Stem cell biologists seek molecules and interactions underlying stem cell capacities, such as transcription factors Oct4 and Nanog (e.g., Takahashi and Yamanaka 2006). On the other, DST models abstract from molecular details, defining a stem cell as a metastable local attractor on a rugged epigenetic landscape. The stalemate is set. Divergent explanatory commitments are central to the failure to communicate in this case. The two philosophical theories of explanation highlight key contrasts between the two explanatory models. DS models are relatively simple, lend themselves to theoretical interpretation, are amenable to formal analysis, and offer the prospect of discovering general biological principles. In systems biology, these features are associated with explanatory virtues of simplicity, coherence with established theory, formal expression and rigor, generality. Stem cell biology doesn't prioritize these explanatory virtues, however. Explanatory models in that field are not simple, do not appeal to theories (there are no theories, in the traditional sense, in stem cell biology), include little to no formalism, and do not (at this time) aspire to generality. These two clusters of virtues do correspond fairly well to different philosophical accounts of explanation.

But contrasting philosophical theories don't reveal the actual grounds of the failure in this case. The reason for failure is not in philosophy, but in those particular communities' own social epistemic organization and aims. Nor does framing the problem in terms of different philosophical theories indicate a positive way forward. The IDE account does better in both respects. The reason for failure in this case is not that one community (or both) is using a flawed philosophical theory of explanation. It is rather that each community's positive commitments to a particular style of explanation are such that IDE conditions are unlikely to be met in this case. There's no direct conflict—no inconsistency—between the two explanatory models, because they fail to connect at all. Stem cell biologists, as noted, do not respond to the DS stem cell model, suggesting that they don't recognize it as explanatory in their sense. The DS modelers return this attitude, some stating explicitly that experimental stem cell research does not yield satisfactory explanations of stem cell capacities, but only "hand-waving or ad hoc metaphors" (Huang 2009,

557).[44] In 2016, I argued that the way forward in this case is for systems biologists to adopt a mechanistic framework and adapt their models to it. But this is a one-sided solution, assuming a hierarchy of explanatory styles that privileges the mechanistic model. IDE offers a better answer. Neither community should be expected to give up deep explanatory commitments and values, nor the hard-won explanatory achievements they bring to the table. The problem is not systems biologists' explanatory style, nor stem cell biologists' stubborn ignorance of mathematical modeling, but rather than the two fields don't see the other's model as bearing on their own work so as to satisfy IDE norms.

Philosophical theories of explanation can do more for this case than rationalize the failure to communicate. They can be deployed as heuristics, to conceive of more productive bridging connections—ways IDE norms could be satisfied in this case. Those norms pinpoint the attitudes and associated bridging relations needed to accomplish this. The failure to connect is rooted in features of both communities, which tend to discourage meeting IDE requirements. But both communities have features that tend to encourage IDE as well. Systems biology's integrative aims are vital for that community, while stem cell biology has a tradition of methodological openness and inclusivity. Theories of explanation can indicate how these IDE-conducive commitments can be leveraged, so as to satisfy IDE requirements. Many philosophers of science today argue that all explanation is causal explanation. Is there a causal interpretation of DS models that could show an advantage to reprogrammers? Alternatively, is there a non-causal interpretation of stem cell models that DS modelers might recognize as explanatory? Substantive philosophical theories of explanation, shorn of monist commitments, are valuable tools and probes for articulating bridging connections that can satisfy IDE requirements. In this way, the IDE account both diagnoses the problem in this case, and points the way to a cure.[45]

8.5 Conclusion

This chapter illustrates the IDE/bridging account from Part II, for the case of systems biology and stem cell research. Systems biology is by definition integrative or interdisciplinary. The basic idea is to apply concepts and tools from physics, computer science and engineering (especially electronic and control engineering) to

[44] Huang introduces his DST framework to a general scientific audience with the aim is "to present here a conceptual framework that has long evaded the attention of experimental biologists, but will offer a different perspective in which both reprogramming and the 'ground state' character of pluripotency will appear natural and expected" (2009, 546).

[45] There's a related discussion in philosophy of biology about the scope and limits of mechanistic explanation, hinging on whether systems biology explanations are of this type or not (e.g., Braillard 2010, Levy and Bechtel 2013).

biological questions. At its most ambitious, systems biology aims to be a successor to molecular biology, replacing its reductionist explanatory style with something more holistic, interactive, dynamic, and mathematically-expressed. This is not, of course, the ambition of every systems biologist—just one strand of a constellation of research programs united under one banner. The aim of transforming biology in the twenty-first century, as molecular biology did in the twentieth, is a powerful social unifier. But that lofty explanatory goal hasn't been realized. The IDE account not only shows why, but also clarifies the integrative successes that have emerged, and indicates how these could be extended.

9
Conclusion: Further Work

I introduced explanatory particularism in terms of three themes: explanation as collaborative process and product, diversity as epistemic resource, and engaging scientific practice. These guiding themes shape this book's argument, main claims, and objectives. I began with the idea that explanation is a form of collaborative activity, taking the social epistemic organization of science into specialized communities as basic. Alongside this community focus is commitment to the idea of understanding through difference; epistemic benefits of connecting (without unifying in any traditional sense) different explanatory products. That commitment also shapes my overall argument. Although this approach departs from philosophical tradition, my goal is not refutation. The book's argument is mainly positive. Critiques of explanatory monism motivate a particularist approach. Shorn of monist assumptions, other philosophical theories of explanation are valuable resources for particularist study of explanation.

The book's objective is three-fold. Its modest goal is to clarify, articulate, and explore implications of the idea that explanation is a form of collaborative activity, inherently pro-social. Parts I–III accomplish this in stages. Applying that idea yields some key results, which in turn satisfy the book's substantive goal. The latter is to articulate epistemic benefits of different scientific explanations. This result shows diversity as a resource for advancing understanding within and across particular scientific communities. The key idea here is relational understanding, extended and elaborated to abstract ideals of unity of science and particularist objectivity (Chapter 6). Importantly, these epistemic advantages and ideals don't supplant or replace the core thesis of explanatory particularism and its implications (Part I). Results from applying EP/EP_M to interdisciplinary contexts (Part II) go alongside my earlier results and theses. And similarly for my view of scientific explanation and other philosophical accounts of same. This bears on the book's third, dialectical goal: reconfiguring the relation between philosophical theories and scientific practice. This commitment informs the style of argument used within and across each chapter. Explanatory particularism offers a more inclusive, practice-oriented way for philosophers to study scientific explanation. This approach takes diverse explanatory practices in scientific communities as the starting-point. It's an intervention on the current debate; a departure but not wholesale rejection of the established philosophical tradition of studying explanation. Explanatory particularism is not opposed to any other specific philosophical theory of explanation. Rather, I seek to reframe

Explanatory Particularism in Scientific Practice. Melinda Bonnie Fagan, Oxford University Press.
 DOI: 10.1093/9780191998874.003.0010

those theories into a particularist approach to studying explanation, which allows philosophers to engage the full range of diversity of styles of explanation in scientific practice.

9.1 Major takeaways

Here I briefly review the main claims and takeaways of my version of explanatory particularism. As noted, the latter is both an approach to studying explanation, and a (thin) general account of explanation as collaborative process and product. This new approach is designed for our current epistemic situation: awash in information and different varieties of knowledge, often lacking a way to put things together—to see a wider picture. This epistemic situation applies to many communities that aren't scientific specializations. Although this book focuses on relations between the latter, nothing about explanatory particularism restricts its application to scientific groups. Rather, my account applies straightforwardly to cases in which scientists interact with members of other, non-scientific epistemic communities. EP defines explanations as major achievements of understanding, which may be more difficult to identify for communities traditionally considered non-scientific. But that's no reason to exclude those communities. Explanation and understanding may not be as significant to those communities, nor as central to their members' identities. (But then again, they might be!) The way I conceptualize explanations, as products of particular communities, does away with any strict, univocal, or principled division between science and society, at least as regards explanatory models. There are just communities constructing explanations, the latter being the fullest expression of a community's epistemic values in that context (i.e., all the technological, cultural background, etc.). As long as there are different communities with interests in understanding some target, and epistemic products (explanatory models) produced in light of that purpose, then minimal IDE constraints and collaborative norms apply. Communities with very different values, goals, and epistemic standards can be connected via these minimal yet substantive ties—irrespective of the "science–society boundary." Indeed, explanatory particularism rejects any such boundary for science in general—it is, in this sense, anti-demarcationist.

Exploration of science–society cases is a task for future work, by a community of scholars. The preceding chapters include takeaways in their own right, too. One such is the general framework for analyzing pairwise relations between distinct explanatory models, adapted from philosophy of models/modeling (Chapter 2). This framework yields a taxonomy of model–model relations, and is not restricted to explanatory models. So it may be fruitful for case studies of scientific modeling more generally. The framework consists of two cross-cutting distinctions and two continuous axes. The axes represent a project's stage in the process of model-construction and degree-of-connection between models. Orthogonally, relations between elements of models may be direct or indirect, and involve similarity or difference. Norms for collaborative social activity adapted from social action theory are another general result (Chapter 3). The three-layer interactive structure of collective intention in collaborative activity builds on ideas from philosophy of social action, notably those of Michael Bratman. Participants in a collaborative activity each regard the shared goal and their individual role therein; the corresponding position of others as attending to the shared goal and their individual role therein; and the corresponding attitude of others as analogously other-regarding. This reciprocal, intercalated intentional structure implies norms for collaborating agents: non-exclusivity, humility, transparency, and respect. Although not surprising results, these norms apply straightforwardly to explanatory model-construction within a particular community. The social action framework also distinguishes several aspects of participants' understanding involved in this collaborative process: understanding of oneself as a participant, of other community members as collaborative partners, and of the model (distinct from its target). This strand of the project connects with recent work in social epistemology of science, developing ideas from Longino's critical constructive empiricism.[1] The general EP account, developed by ideas about models/modeling and social action, should be a useful framework for a wide range of case-based projects in philosophy of science. (For a detailed summary of this framework, see the conclusion of Part I at the end of Chapter 3.)

The main results of this book, however, concern interdisciplinary research (IDR). The general framework for analyzing model–model relations, so applied, delivers a broad framework for classifying and characterizing IDR projects (Chapter 4). This limns a new category, "low-effort interdisciplinarity." Another result of the IDR application is the insight that scientists' diverse ideas about explanation can pose an important challenge or obstacle for IDR. The IDE account (Chapter 5) answers that challenge, by applying social action norms to the special case of connecting distinct explanatory models (in the "low-effort" mode). The result is a set of minimal constraints on such "bridging connections," with associated norms for participants in IDR projects. This account is illustrated by an extended example (hemoglobin research) and two detailed case studies (Part III). Explanatory practice in accordance with IDE norms is more pluralistic, equitable, tolerant of diversity, and should be better able to engage non-expert communities. Chapter 6 further operationalizes the IDE norms for participants into a set of guidelines for IDE bridge-building. More abstract takeaways concern aspects of scientific understanding and unity of science. IDE connections, I propose,

[1] See Fagan (under review) for more on this connection.

advance understanding in a distinctive way, via "relational understanding." Understanding of self, other participants, and models as such are all interrelated in the process of advancing relational understanding via IDE. An idealized extension of this idea is "patchwork unity of science," a non-hierarchical form of unification that preserves diversity across explanatory communities (Chapter 6). This pro-social, normative image of scientific explanation and understanding has important implications for concepts of scientific objectivity, science–society relations, and other issues in socially-engaged philosophy of science today.

9.2 Ideas for future work

A full, substantive particularist account of scientific explanation is beyond the capacity of any one scholar of even an ordinary-sized team of scholars. Such a full substantive account requires a wide range of studies of explanation in many distinct scientific communities, at different historical periods. History and philosophy of science already includes a rich literature of such cases, as well as studies relegated to so-called "philosophy of the special sciences." Explanatory particularism showcases these case studies as contributions to philosophical study of explanation, and (with added focus on interdisciplinarity) suggests fruitful new ones. An open-ended list of such cases is a fitting way to conclude this monograph. In no particular order:

- Distributed cognition is, roughly, the view that cognition and other mental phenomena are localized in part beyond individual organisms, extending into what's usually categorized as the environment. The basic idea is applicable at many scales, and, according to some philosophers, points us to general principles of memory, language, problem-solving, and related phenomena. A key idea is that boundaries of cognitive systems are flexible, such that different cognitive systems can be identified and studied, depending on the scale/phenomenon of interest. Many proponents of this approach argue that cognitive systems that cut across (traditional) biological boundaries are of great explanatory value (reviewed in Theiner 2018). But other theorists argue that cognitive systems coincide with (parts of) biological systems, such that cognitive boundaries cutting across biological boundaries are deficient as explanations. Explanatory particularism offers a way forward in this debate, which stems from different ideas about the role of biological units and systems in cognitive science explanations. We can separate these two schools of thought, consider each style of explanation on its own terms, and then use IDE criteria to connect these different explanatory contributions. Letting go of the idea that each specialized field can have only one style of explanation in play at a given time helps deflate the debate as well.

• Machine learning (ML) and AI projects raise distinctive, socially-relevant questions about explanation for those research communities, and for communities impacted by their products (i.e., most of us). A defining goal of such research is to construct artificial epistemic agents, through processes that are (plausibly, in most or all cases) opaque, inscrutable, and nonintuitive. What must (or should) members of ML communities understand? How does "interpretability" bear on explanation/understanding in these communities? What form/aspects of understanding are requisite for safe and ethical applications of AI research? ML research is currently roiled by debates about these questions. Some researchers propose that "explainable AI" is requisite, while others are open to ceding some epistemic tasks to machines. Whether the latter are epistemic agents on par with (or superior to) humans in any respect is hotly debated. At issue are values that inform the style of explanation for computer science and its applications. EP/IDE offers a fruitful way to approach these debates. Could artificial epistemic agents participate in collaborative social activities, including explanation? If so, what constraints would that place on such agents, and how do those constraints impact practices in ML fields?

• In many ways, explanatory particularism builds on ideas from Helen Longino's influential social epistemology of science (Longino 1990, 2002; Fagan, under review). One such idea is pluralism about explanation. Longino's articulation of diverse approaches to studying (human) behavior is a major part of her philosophical oeuvre. Some of these approaches, she argues, "parse causal space" in ways that are mutually incompatible, producing explanatory models of behavior that cannot be integrated or consistently combined (Longino 2013). The account presented here offers a related but distinct framework for examining distinct styles of explaining behavior. One thought is that Longino's insights about mutually incompatible explanatory models could be reconceptualized as bridging connections satisfying IDE norms. Failure to satisfy these norms, on the other hand, could indicate hierarchical attitudes about explanation pertaining to the biology-psychology divide. It would be interesting to use EP/IDE to examine the contested explanatory terrain among life, behavioral, and social sciences, and compare results with Longino's pluralistic insights.

• Models in different sub-fields of economics are interrelated in complex ways. Caterina Marchionni's work on this topic is especially relevant to explanatory particularism (e.g., Marchionni 2008, Kuorikoski and Marchionni 2014). Marchionni argues that integrating diverse models is important for progress in the social sciences. She further proposes that explanatory norms can be "frictions" for integration, and that ideas of model-unification (in some sense) are important in both economics and sociology. Explanatory particularism—especially IDE norms—could build on the cases Marchionni considers, updating her diagnoses of situations in which models appear incompatible or incoherent with one another.

• Feminist philosophers of science pioneered many ideas and theses that inform explanatory particularism. The Gender and Science Lab, established in 2018, is one research program in this area.[2] Led by Sarah Richardson, this group is committed to an interdisciplinary, collaborative approach to advancing understanding of sex/gender issues. The group began as a reading group in 2010, and expanded from this starting point to develop a contextualist approach that "brings together gender scholars and biomedical scientists to build and validate new theory and methods for the intersectional study of gender/sex in human populations" (GenderSciLab 2020). This interdisciplinary research center has an explicit methodology for fostering collaboration toward this end, influenced by ideas from Helen Longino and other feminist philosophers. Explanatory particularism and IDE norms offer a promising framework for analyzing the GenderSci Lab as an interdisciplinary research project—especially ways explanations are constructed, evaluated, and communicated to a wider epistemic community.

• History of science offers numerous cases of combining distinct explanatory models.[3] For example, the founders of theoretical population genetics calculated mathematical consequences of Darwinian evolution under Mendelian inheritance, thereby synthesizing Mendelism, Darwinism, and biometry (Provine 1971). Initially conceived as incompatible alternatives (gradual vs. discontinuous evolutionary change), Mendelian and Darwinian models were viewed as complementary by 1918, providing the foundation for the new field of theoretical population genetics. How was the relation between these explanatory models transformed, exactly? Part of the answer is technical (formulation of Hardy–Weinberg equilibrium, Yule's demonstrations)—but other aspects involve social epistemic organization. The biometricians were a distinct scientific community from other Darwinians of the day, defined by their commitment to using quantitative measurement and statistical methods to study evolution. Conceptual changes required for crucial results by Yule, Hardy, and Weinberg involve multiple wider communities as well, including breeding experiments and early agricultural science. Re-examining the Modern Synthesis through the lens of explanatory particularism could reveal how inter-community relations were profoundly altered, offering a fresh take on this well-studied case. This community-focused treatment could yield new insights about (for example) the exclusion of embryology/developmental biology from the Synthesis, with implications for present-day evolutionary developmental biology as well as recent debates over the explanatory role of Hamilton's Rule (Birch 2014). Other well-studied HPS cases that could be similarly re-examined in this way include the Copernican revolution, alchemy and atomism, and controversies over phylogenetic inference.

[2] For more information, see https://www.genderscilab.org.
[3] Thanks to Jim Lennox for this suggestion.

- The concept of the chemical bond is central to chemistry, a lynchpin of the chemistry–physics relation, and importantly implicated in many biological explanations. An early twentieth-century concept of covalent electron-sharing (canonically articulated by G.N. Lewis) has been important to explanations of molecular structure in many fields within the chemistry–biology axis. But reconciling this core structural idea with insights from quantum physics is difficult. "The chemical bond" has undergone significant conceptual change over the past century, and its interface with quantum physics remains vexed and problematic. One proposal is the quantum theory of atoms in molecules (QTAIM), which models the bond as interactive. Thematic resonance with explanatory particularism makes this an intriguing example. It would be interesting to examine the specifics of how features of electron density distribution are linked to chemical concepts in QTAIM. Do these linkages satisfy minimal bridging requirements for IDE? If so, do these ties go beyond IDE's minimal conditions? If not, how does this disconnect map onto social epistemic organization in chemistry and QTAIM's status as an explanatory achievement? This case builds on scholarship from philosophy of chemistry, too-long sidelined in general philosophical studies of explanation.

- The Climate and Global Dynamics Laboratory, part of the National Center for Atmospheric Research, hosts an NSF-funded Growing Convergence Research project that is aimed at "designing scenarios of combined climate interventions for application to Earth System models to better understand the efficacy and safety of these responses to the climate crisis" (CGDL 2024).[4] Different methodologies for constructing climate models give different perspectives on the complex phenomenon. These designed scenarios could be analyzed in terms of IDE bridging connections, and the project itself via associated norms for participant attitudes. The typology of model–model relations from Part I could also apply to this research project, as well as other projects that involve diversely-constructed climate models. The particularist approach may be helpful in clarifying social epistemic relations between model-building and policymaking communities. Monica Morrison, currently working on this Convergence project, used an earlier version of Part I's modeling framework in Part I to examine relations between models included in the Community Earth System Model Large Ensemble project (CESM-LE).[5] She attends to both similarities and differences among contributed models, as well as patterns of shared history for specific model components. Different pieces of the complex earth system are combined via social epistemic organization among different modeling groups. More broadly, climate change science is a highly interdisciplinary research community with complex, substantive ties to many

[4] See https://www.cgd.ucar.edu/.
[5] Personal communication, 9/12/2019.

other epistemic communities. Failures of communication could be diagnosed in terms of attitudes requisite for IDE—gaps between members of physical and social science communities might be of particular interest.

- Case studies of interdisciplinary research projects by Nancy Nersessian, Hanne Andersen, and their various collaborators overlap somewhat with my particularist account.[6] But, as with Longino's work, there are important differences as well. Notably, these scholars use a cognitive (Nersessian) or argument-focused approach (Andersen), rather than the social action norms of explanatory particularism. The latter offers a framework that highlights some of these scholars' key insights concerning explanation (alongside other aspects of scientific practice). Lines of descent linking research projects, epistemic dependence relations within collaborative groups, and ways of managing tensions between systems modeling and "wet" labwork are among these. A closer look at connections and complementarity between EP and these other approaches to IDR studies would be interesting to pursue.

- Cognitive neuroscience aims to link biological with cognitive/psychological explanations. The former are in terms of cells, molecular factors, and electrochemical signals; explanatory modeling uses both experiment-based and mathematical modeling tools. In systems biology, integrating these can be a challenge (see Chapter 8). So examining how similar methods are combined (if they are) in another epistemic community could be illuminating. But the main explanatory challenge for cognitive neuroscience concerns how explanatory resources from life and mind sciences are related. There is tendency to partition explanatory power unevenly, with models of biological parts and interactions given greater weight. The interdisciplinary, diverse cognitive neuroscience community is currently divided on this issue. A number of philosophers have examined these challenges (e.g., Bechtel 1986), while connections with other epistemic communities continue to accrete: artificial intelligence research, linguistics, data science, animal ethology, and more. The particularist image of science as a mesh of complexly-related explanatory models could be a fruitful approach to examining current debates in the field.

- Theoretical biology and philosophy converge on the task of analyzing the nature of organisms and their organization. Biological theories in this area draw on numerous other epistemic communities—including ideas consonant with explanatory particularism. For example, Bechtel and Bich (2021) propose a conceptual bridge between, on the one hand, mechanistic models of organismal phenomena, and, on the other theories of organismal autonomy and self-organization. This

[6] See discussions in Chapters 4 and 8 for details and references.

proposed bridge is very much in the spirit of IDE, building on three conceptual resources: "(1) a framework of work and constraints, (2) a conception of function grounded in the organization of an autonomous system, and (3) a focus on control" (Bechtel and Bich 2021, 1). This and other examples attest to insights incorporated into explanatory particularism being present in the wider philosophy of science literature, yet unrecognized as contributions to philosophical studies of explanation. Relatedly, some influential views of mechanistic explanation could be fruitfully compared and contrasted with my particularist approach (e.g., Bechtel 2011).

- Biogeophysics is a new field at the intersection of microbiology, biogeochemistry, geophysics—these parent strands all originating as interfield convergences too. This example showcases the intricacy of interfield connections in science today. The inciting discovery by Estella Atekwana was in geophysics, investigating environmental effects of military fuel (Atekwana and Slater 2009, Atekwana 2021). Surprising results at several sites led Atekwana and colleagues to hypothesize that microbes were breaking down hydrocarbon contaminants. This spurred the geophysicists to become acquainted with some microbiology literature, eventually producing an explanatory achievement recognized by their wider epistemic community: geophysics methods show that microbes can have large-scale effects on the abiotic environment. The group's result catalyzed a larger research program, spanning many scales, data-acquisition methods, and explanatory modeling approaches. A deeper dive into this initially one-sided link between earth and life science communities is needed, to see if this emerging specialization conforms to IDE constraints or shows stronger two-way ties. A key part of this method is dovetailing physical methods and measurements with features of microbes gleaned from life science explanations. Characterizing the bridging or stronger links between explanatory models from such disparate scientific communities would be informative and fruitful. Closer ties and understanding between explanatory achievements in life and earth sciences would enrich environmental sciences more broadly. Explanatory particularism is well-suited to examine this constellation of fields.

- Explanations of injustice and related phenomena are another promising site for applications of explanatory particularism. Such explanations are important for multiple epistemic communities, spanning social science, psychology, and the humanities. Within and across those fields, explanations in terms of individuals' attitudes and preferences are often seen as competing alternatives to explanations in terms of social–structural patterns. A forthcoming monograph (Erin Beeghly's *What's Wrong with Stereotyping?*) uses some of EP's ideas to re-conceptualize relations between different explanatory resources, thereby not only integrating multiple lines of evidence but also prioritizing the voices of those impacted in those theories; "an explanation that itself enacts justice" (MS, 8). Applied social philosophy projects of this sort further articulate the pro-social normative aspect of my EP/IDE account.

- Archaeology and its diverse evidence base (ancient DNA, present-day genomics, history, anthropology, chemistry, earth sciences, art, and more) is another promising example to consider. Alison Wylie's pioneering work on pluralism and standpoint approaches in archaeology has many thematic resonances with explanatory particularism—unsurprising, as Wylie's work has been influential for my own. Wylie and colleagues apply her standpoint approach to interactions between scientific and Indigenous epistemic communities (1999, 2003, 2015). Mutual epistemic respect is a cornerstone of this approach: recognizing distinct epistemic communities as agents with their own evidential assessments, not passive consumers of knowledge from other scientific fields. Wylie also proposes that archaeology is a "trading zone" (after Galison), fusing together different streams of evidence from many different specialized fields. This suggests a specialized method that harmonizes and balances diverse epistemic resources into a model that advances understanding about some aspect of our human past. Wylie refers to the process of combining these diverse epistemic resources as "bridging," here meaning a complex chain of ampliative inferences, ranging from concrete material to general psychological/moral claims. My social action framework and associated norms of explanatory particularism could also apply to these cases, complementing and extending projects in philosophy of archaeology.

- Another idea to explore via EP/IDE is that models of zoonotic disease have "social and political lives"—i.e., bear on policy dynamically, carrying social and political values that can be made explicit. This proposal was made by social scientists at the University of Sussex about seven years before the COVID-19 pandemic. Claims such as "if unchecked, many zoonoses threaten global pandemics" and "Complex interactions of epidemiological, ecological, social and technological processes shape zoonotic disease emergence, transmission, risks and vulnerabilities, influenced by wider socioeconomic and environmental drivers" now have the weight of hard-won experience (Leach and Schoones 2013, 10). This approach to modeling disease requires explanatory contributions from epidemiology, ecology, policy, a range of social sciences, and (arguably) the humanities. Leach and Schoones advocate adding "participatory" modelling based on anthropological, ethnographic, and participatory approaches to more established epidemiological methods (2013, 11). On their view, different contributing models need not be tightly integrated—yet it seems advisable for some connections among them to be articulated. The IDE account is well-suited to that task. Leach and Schoones also remark on hierarchical power asymmetries in relations between relevant epistemic communities studying zoonotic outbreaks—including modeling efforts guided by local participants "on the ground"—and especially the epistemic prominence of quantitative modeling, raised over other explanatory resources in a number of fields (Leach and Schoones 2013, 15). These scholars recommend detailed "triangulation" between multiple models, making explicit the limits as well as advantages of each. Many insights of EP/IDE are already present in their analysis; more could be gleaned from applying the normative account developed here.

• Last but not least: explanatory models in molecular biology are cited by philosophers as exemplars of mechanistic explanation (e.g., protein synthesis, see Darden 2006). Yet some classic explanatory achievements of this epistemic community have features associated with other philosophical accounts of explanation: structural, law-like, or mathematical. Karina Alleva and colleagues (2017) examine the case of the Monod–Wyman–Changeux model of allostery, arguing that different philosophical theories of explanation capture different aspects of this explanatory achievement. The operon model (Fagan 2016a) is another such case. Historical and philosophical studies of these and other classic explanatory achievements indicate the value of repurposing philosophical theories as heuristics for studying explanations in scientific practice. In this context-dependent way, philosophical accounts of explanation and explanatory virtues can play a role in forging and clarifying bridges between explanatory models. Conceptions of cause, structure, derivation, part-whole relations, etc., are tools to try, given two or more different specialized models. No one of these will work for all cases. Yet some general insights may well emerge from a range of cases. One thought is that diverse explanatory aspects in a single model track the latter's interfield origins—a plausible idea for classic molecular biology achievements. Relatedly, a complex, 'dense' explanatory profile could make a model well-suited to different kinds of future use—adding to the image of unity of science proposed in Chapter 6. Explanatory particularism leaves open many avenues for further articulation.

References

Aach, J, Lunshof, J, Iyer, E, and Church, GM (2017) Addressing the ethical issues raised by synthetic human entities with embryo-like features. *elife* 6: e20674

Abbas, AK, Lichtman, AH, and Pillai, S (2021) *Cellular and Molecular Immunology*, 10th edition. Amsterdam: Elsevier

Achinstein, P (1983) *The Nature of Explanation*. Oxford: Oxford University Press

Achinstein, P (1985) The pragmatic character of explanation. *Proceedings of the Biennial Meeting of the Philosophy of Science Association, Volume Two: Symposia and Invited Papers* (1984), 275–292

Alleva, K, Díez, J, and Federico, L (2017) Models, theory structure and mechanisms in biochemistry: The case of allosterism. *Studies in History and Philosophy of Biological and Biomedical Sciences* 63: 1–14

Alonso, FM (2018) Reductive theories of collective intention. In: Jankovic, M and Ludwig, K (eds.) *Routledge Handbook of Collective Intentionality*. New York: Routledge/Taylor & Francis, 34–44

American Academy of Arts and Sciences (2016) Humanities indicators, 2015

Amsterdamska, O (2005) Demarcating epidemiology. *Science, Technology, & Human Values* 30: 17–51

Andersen, H (2010) Joint acceptance and scientific change: a case study. *Episteme* 7: 248–265

Andersen, H (2016) Collaboration, interdisciplinarity, and the epistemology of contemporary science. *Studies in History and Philosophy of Science* 56: 1–10

Andersen, H, and Wagenknecht, S (2013) Epistemic dependence in interdisciplinary groups. *Synthese* 190: 1881–1898

Anderson, RM, and May, RM (1985) Vaccination and herd immunity to infectious diseases. *Nature* 318: 323–329

Araz, OM, Ramirez-Nafarrate, A, Jehn, M, and Wilson, FA (2020) The importance of widespread testing for COVID-19 pandemic. *Health Systems* 9: 119–123

Atekwana, EA (2021) Biogeophysics: geophysical imaging of microbial processes, searching for biosignatures in geophysical data. Atkinson Distinguished Lecture Series, University of Utah, February 18, 2021

Atekwana, EA, and Slater, LD (2009) Biogeophysics: a new frontier in earth science research. *Review of Geophysics* 47: RG4001, doi:10.1029/2009RG000285

Bangu, S (2017) Scientific explanation and understanding: unificationism reconsidered. *European Journal for Philosophy of Science* 7: 103–126

Baribási, A-L (2002) *Linked: The New Science of Networks*. New York: Perseus Books

Batterman, R (2002) *The Devil in the Details*. Oxford: Oxford University Press

Batterman, R (2010) On the explanatory role of mathematics in empirical science. *British Journal for the Philosophy of Science* 61: 1–25

Batterman, R, and Rice, C (2014) Minimal model explanations. *Philosophy of Science* 81: 349–376

Beatty, J (1997) Why do biologists argue like they do? *Philosophy of Science* 64: S432–S443

Beaver, D, and Rosen, R (1978) Studies in scientific collaboration, Part I. *Scientometrics* 1: 65–84

Bechtel, W (1986) The nature of scientific integration. In: Bechtel, W (ed.), *Integrating Scientific Disciplines*. Dordrecht: Springer, 3–52

Bechtel, W (2006) *Discovering Cell Mechanisms: The Creation of Modern Cell Biology*. Cambridge: Cambridge University Press

Bechtel, W (2011) Mechanism and biological explanation. *Philosophy of Science* 78: 533–557

Bechtel, W, and Abrahamson, A (2005) Explanation: a mechanist alternative. *Studies in History and Philosophy of Biological and Biomedical Sciences* 36: 421–441
Bechtel, W, and Bich, L (2021) Mechanism, autonomy and biological explanation. *Biology and Philosophy* 36: 1–27
Bechtel, W, and Richardson, R (1993) *Discovering Complexity*. Princeton, NJ: Princeton University Press
Beeghly, E (forthcoming) *What's Wrong with Stereotyping?* Oxford: Oxford University Press
Bich, L, Mossio, M, Ruiz-Mirazo, K, and Moreno, A (2016) Biological regulation: controlling the system from within. *Biology and Philosophy* 31: 237–265
Birch, J (2014) Hamilton's Rule and its discontents. *British Journal for the Philosophy of Science* 65: 381–411
Blues Brothers, The (1980) Universal Pictures
Boesch, B (2019) The means-end account of scientific, representational actions. *Synthese* 196: 2305–2322
Bogen, J (2005) Regularities and causality. *Studies in History and Philosophy of Biological and Biomedical Sciences* 36: 397–420
Boix Mansilla, V (2006) Assessing expert interdisciplinary work at the frontier: an empirical exploration. *Research Evaluation* 15: 17–29
Bokulich, A (2008) Can classical structures explain quantum phenomena? *British Journal for the Philosophy of Science* 59: 217–235
Bokulich, A (2011) How scientific models can explain. *Synthese* 180: 33–45
Bokulich, A (2018) Representing and explaining: the eikonic conception of scientific explanation. *Philosophy of Science* 85: 793–805
BonJour, L (1985) *The Structure of Empirical Knowledge*. Cambridge, MA: Harvard University Press
Boogerd, F (2017) Systems biology in the broad sense. In: Green, S (ed.), *Philosophy of Systems Biology*. Dordrecht: Springer, 49–58
Boogerd, F, Bruggeman, F, Hofmeyr, J-H, and Westerhoff, H (2007) Towards philosophical foundations of Systems Biology: introduction. In: Boogerd, F, et al. (eds.) *Systems Biology: Philosophical Foundations*. Amsterdam: Elsevier, 3–19
Boon, M, and Van Baalen, S (2019) Epistemology for interdisciplinary research—shifting philosophical paradigms of science. *European Journal for Philosophy of Science* 9: 16
Boyd, K (2021) Group understanding. *Synthese* 198: 6837–6858
Braillard, P-A (2010) Systems Biology and the mechanistic framework. *History and Philosophy of the Life Sciences* 32: 43–62
Braillard, P-A, and Malaterre, C (eds.) (2015) *Explanation in Biology: An Enquiry into the Diversity of Explanatory Patterns in the Life Sciences*. Dordrecht: Springer
Brandt, C (2010) The metaphor of 'nuclear reprogramming.' In: Barahona, A, Suarez-Díaz, E, and Rheinberger, H-J (eds.) *The Hereditary Hourglass: Genetics and Epigenetics, 1868–2000*. Berlin: Max Planck Institute for History of Science, 85–95
Bratman, ME (1999) Shared intention. In: *Faces of Intention: Selected Essays on Intention and Agency*. Cambridge: Cambridge University Press, 109–129
Bratman, M (2014) *Shared Agency*. Oxford: Oxford University Press
Bredekamp, H, Dünkel, V, and Schneider, B (eds.) (2015) *The Technical Image: A History of Styles in Scientific Imagery*. Chicago: University of Chicago Press
Breidenmoser, T, and Wolkenauer, O (2015) Explanation and organizing principles in systems biology. In: Braillard, P-A, and Malaterre, C (eds) *Explanation in Biology*, 249–264
Brigandt, I (2010) Beyond reduction and pluralism: toward an epistemology of explanatory integration in biology. *Erkenntnis* 73: 295–311
Brigandt, I (2013) Integration in biology. *Studies in History and Philosophy of Biological and Biomedical Sciences* 44: 461–465
Brint, SG, Turk-Bicakci, L, Proctor, K, and Murphy, SP (2009) Expanding the social frame of knowledge: interdisciplinary, degree-granting fields in American colleges and universities, 1975–2000. *Review of Higher Education* 32: 155–183

Brivanlou, A (2018) Human embryo research: what we know, how we know it and what more we can learn [public presentation, Rice University, Houston TX, 5/23/2018). Workshop: Drawing the line: ethical, policy and scientific perspectives on US embryo research]

Broadbent, A (2013) *Philosophy of Epidemiology*. London: Palgrave Macmillan

Bubar, KM, Reinholt, K, Kissler, SM, Lipsitch, M, Cobey, S, Grad, YH, and Larremore, DB (2021) Model-informed COVID-19 vaccine prioritization strategies by age and serostatus. *Science* 371: 916–921

Buchman, DZ, and Ho, A (2014) What's trust got to do with it? Revisiting opioid contracts. *Journal of Medical Ethics* 40: 673–677

Calvert, J, and Fujimura, JH (2011) Calculating life? Duelling discourses in interdisciplinary systems biology. *Studies in History and Philosophy of Biological and Biomedical Sciences* 42: 155–163

Campbell, DT (1969). Ethnocentrism of disciplines and the fish-scale model of omniscience. In: Sherif, M, and Sherif, CW (eds.) *Interdisciplinary Relationships in the Social Sciences*. Chicago: Aldine, 328–34

Carusi, A (2017) Enactments of systems biology. In: Green, S (ed.) *Philosophy of Systems Biology*. Dordrecht: Springer, 59–67

Cartwright N (1999) *The Dappled World*. Cambridge: Cambridge University Press

Cartwright, N, Pemberton, J, and Wieten, S (2020) Mechanisms, laws and explanation. *European Journal for Philosophy of Science* 10: 25–35

Castro, VF, and Pacherie, E (2021) Joint actions, commitments and the need to belong. *Synthese* 198: 7597–7626

Cat, J, and Tuboly, A (eds.) (2019) *Neurath Reconsidered*. Dordrecht: Springer

Chakravartty, A (2010) Perspectivism, inconsistent models, and contrastive explanation. *Studies in History and Philosophy of Science* Part A 41: 405–412

Chang, H (2012) *Is Water H_2O?* Dordrecht: Springer

Chant, S (2018) Collective action and agency. In: Jankovic, M and Ludwig, K (eds.) *Routledge Handbook of Collective Intentionality*. New York: Routledge/Taylor & Francis, 13–24

Chirimuuta, M (2018) The development and application of efficient coding explanation in neuroscience. In: Reutlinger, A and Saatsi, J (eds.) *Explanation Beyond Causation*. Oxford University Press, 164–184

Clark, C (under review) Geography's role in causal explanations of ecological complexity

Climate and Global Dynamics Laboratory (2024) Understanding the Earth System and its human relationships. <https://www.cgd.ucar.edu> (accessed 6/2/24)

Coleman, W (1977) *Biology in the Nineteenth Century*. Cambridge: Cambridge University Press

Corlett, JA (1996) *Analyzing Social Knowledge*. Totowa: Rowman & Littlefield Publishers

Craver, C (2007) *Explaining the Brain: Mechanisms and the Mosaic Unity of Neuroscience*. Oxford: Oxford University Press

Craver, C (2014) The ontic account of scientific explanation. In: Kaiser, MI, Scholz, OR, Plenge, D, and Hüttemann, A (eds.) *Explanation in the Special Sciences: The Case of Biology and History*. Dordrecht: Springer, 27–52

Craver, C, and Darden, L (2013) *In Search of Mechanisms*. Chicago: University of Chicago Press

Crick, F (1970) Central dogma of molecular biology. *Nature* 227: 561–563

Crombie, AC (1994) Styles of scientific thinking in the European tradition: Vol. 2. Bloomsbury Academic

Crone, K (2021) Foundations of a we-perspective. *Synthese* 198: 11815–11832

Curd, JA, Cover, M, and Pincock, C (2013) (eds.) *Philosophy of Science: The Central Issues*. WW Norton & Company

Dang, H (2019) Do collaborators in science need to agree? *Philosophy of Science* 86: 1029–1040

Darden, L (2006) *Reasoning in Biological Discoveries*. Cambridge: Cambridge University Press

Darden, L, and Maull, N (1977) Interfield theories. *Philosophy of Science* 44: 43–64

de Regt, H (2015) Scientific understanding: truth or dare? *Synthese* 192: 3781–3797

de Regt, H (2017) *Understanding Scientific Understanding*. Oxford: Oxford University Press de Regt, H (2019) From explanation to understanding: normativity lost? *Journal for General Philosophy of Science* 50: 327–343

de Regt, H, and Dieks, D (2005) A contextual approach to scientific understanding. *Synthese* 144: 137–170

de Regt, H, and Gijsbers, V (2016) How false theories can yield genuine understanding. In Grimm, SR, Baumberger, C, and Ammon, S (eds.) *Explaining Understanding.* New York: Routledge/ Taylor & Francis, 50–75

de Regt, H, Leonelli, S, and Eigner, K (2009) Focusing on scientific understanding. In: De Regt, H, Leonelli, S, and Eigner, K (eds.) *Scientific Understanding: Philosophical Perspectives.* University of Pittsburgh Press, 1–17

de Ridder, J (2014) Epistemic dependence and collective scientific knowledge. *Synthese* 191: 37–53

Deutscheland Forschungsgemeinschaft (1996) German Research Foundation <https://www.dfg.de/en >

DiMarco, M (2022) Wishful intelligibility, black boxes, and epidemiological explanation. *Philosophy of Science* 88: 824–834

Dowe, P (2000) *Physical Causation.* Cambridge: Cambridge University Press

Downes, S (1992) The importance of models in theorizing: A deflationary semantic view. *PSA: Proceedings of the Biennial Meeting of the Philosophy of Science Association* 1992, 142–153

Downes, S (2020) *Models and Modeling in the Sciences: A Philosophical Introduction.* New York: Routledge

Doyle, Y, Egan, S, Graham, N, and Khalifa, K (2019) Non-factive understanding: a statement and defense. *Journal for General Philosophy of Science* 50: 345–365

Dragos, C (2021) Epistemic autonomy and group knowledge. *Synthese* 198: 6259–6279

Dröscher, A (2014) Images of cell trees, cell lines, and cell fates. *History and Philosophy of the Life Sciences* 36: 157–186

Eberl G, and Pradeu T (2018) Towards a general theory of immunity? *Trends in Immunology* 39: 261–263

Eckert, R, Randall, DJ, and Augustine, G (1988) *Animal Physiology: Mechanisms and Adaptations.* WH Freeman

Eigenbrode, SD, O'Rourke, M, Wulfhorst, JD, Althoff, DM, Goldberg, CS, Merrill, K, Morse, W, Nielsen-Pincus, M, Stephens, J, Winowiecki, L, and Bosque-Pérez, NA (2007) Employing philosophical dialogue in collaborative science. *BioScience* 57: 55–64

Elgin, C (2007) Understanding and the facts. *Philosophical Studies* 132: 33–42

Elgin (2017) *True Enough.* Cambridge, MA: MIT Press

Epstein, B (2016) *The Ant Trap.* Oxford: Oxford University Press

Ernst, Z, and Chant, S (2007) Collective action as individual choice. *Studia Logica* 86: 415–434

European Commission (2004) Report of activities (2001-2004), European Union Research Advisory Board: Interdisciplinarity in research. < https://op.europa.eu/en/publication-detail/-/publication/f0058dce-d869-4655-b67a-d91e7e84e6a4/language-en >

European Commission (2017) Work Programme for Horizon 2020. <https://commission.europa.eu/research-and-innovation_en >

Fagan, MB (2007) The search for the hematopoietic stem cell. *Studies in History and Philosophy of Biological and Biomedical Sciences* 38: 217–237

Fagan, MB (2011) Is there collective scientific knowledge? *Philosophical Quarterly* 61: 247–269

Fagan, MB (2012a) Collective scientific knowledge. *Philosophy Compass* 7: 821–831

Fagan, MB (2012b) Waddington redux: models and explanation in stem cell and systems biology. *Biology and Philosophy* 27: 179–213

Fagan, MB (2013) *Philosophy of Stem Cell Biology.* Palgrave Macmillan

Fagan, MB (2014) Do groups have scientific knowledge? In: Chant, S, Hindriks, F, Preyer, G (eds.) *From Individual to Collective Intentionality: New Essays.* Oxford: Oxford University Press, 163–186

Fagan, MB (2016a) Interventionist omissions: a critical study of mechanistic explanation in biology. *Philosophy of Science* 83: 1082–1097

Fagan MB (2016b) Generative models: human embryonic stem cells and multiple modeling relations. *Studies in History and Philosophy of Science* 56: 122–134
Fagan, MB (2016c) Stem cells and systems models: clashing views of explanation. *Synthese* 193: 873–907
Fagan, MB (2017a) Stem cell lineages: between cell and organism. *Philosophy and Theory in Biology* 9, Special Issue: The Ontologies of Living Beings. <http://www.philosophyandtheoryinbiology.org/>
Fagan, MB (2017b) Explanation, multiple perspectives, and unification. *Balkan Journal of Philosophy* 9: 19–34, Invited paper, Special Issue 'Explanation and Understanding'
Fagan, MB (2020a) Organoide: Ein wesentliches Element in einem generativen Modellgefüge (trans: Anja Pichl), in: *Organoide.* Ihre Bedeutung für Forschung, Medizin und Gesellschaft, edited by Sina Bartfeld, Hannah Schickl, Cantas Alev, Bon-Kyoung Koo, Anja Pichl, Angela Osterheider, Lilian Marx-Stölting, Berlin-Brandenburg Academy of Sciences and Humanities, 149–170
Fagan, MB (2020b) Interdisciplinarity, perspectives, and explanation. In: Massimi, M, and McCoy, CD (eds.) *Understanding perspectivism: Scientific Challenges and Methodological Prospects.* New York: Taylor & Francis, 28–49
Fagan, MB (2021) *Stem Cells: Elements.* Cambridge: Cambridge University Press
Fagan, MB (2022) Biology's uncertainty principle. *Institute of Arts and Ideas News Magazine* <https://iai.tv/articles/biology-uncertainty-principle-auid-2170>
Fagan, MB (2023) Immunology and population health: collaboration without convergence. *Philosophy of Science* 90: 1265–1275
Fagan, MB (in press) Causal explanation in context. To appear in: Russo, F, and Illari, P (eds.) *The Routledge Handbook of Causality and Causal Methods.* New York: Routledge
Fagan, MB (under review) Socializing explanation: Longino and explanatory particularism
Fagan, MB, and Maienschein, J (2022) Theories of biological development. *Stanford Encyclopedia of Philosophy* (Summer 2022 Edition), Edward N Zalta (ed.), <https://plato.stanford.edu/archives/sum2022/entries/theories-biological-development/>
Faye, J (2014) *The Nature of Scientific Thinking.* London: Palgrave Macmillan
Fears, R, Akustu, H, Alentajan-Aleta, LT, Caicedo, A, de Carvalho, ACC, et al. (2021) Inclusivity and diversity: Integrating international perspectives on stem cell challenges and potential. *Stem Cell Reports* 16: 1847–1852
Ferreira, A, Marguti, I, Bechmann, I, Jeney, V, Chora, Â, Palha, NR, Rebelo, S, Henri, A, Beuzard, Y, and Soares, MP (2011). Sickle hemoglobin confers tolerance to *Plasmodium* infection. *Cell* 145: 398–409
Fleck, L ([1935] 1979) *Genesis and Development of a Scientific Fact.* (Bradley, F, and Trenn, TJ, trans.; Trenn, TJ, and Merton, RK, eds.) Chicago: University of Chicago Press
Franklin, A (2018) On the renormalization group explanation of universality. *Philosophy of Science* 85: 225–248
Friedman, M (1974) Explanation and scientific understanding. *Journal of Philosophy* 71: 5–19
Frigg, R, and Nguyen, J (2018) Scientific representation. *Stanford Encyclopedia of Philosophy* (Winter 2018 Edition), EN Zalta (ed.), <https://plato.stanford.edu/archives/win2018/entries/scientific-representation/>
Frodeman, R, Klein, JT, and Pacheco, RCS (eds.) (2017) *The Oxford Handbook of Interdisciplinarity*, 2nd edition. Oxford: Oxford University Press
Frost, WH (1928) *Epidemiology.* In: Maxcy, K (ed.) *Papers of Wade Hampton Frost.* New York: Commonwealth Fund
Furman, K (2020) Mono-causal and multi-causal theories of disease: how to think virally and socially about the aetiology of AIDS. *Journal of Medical Humanities* 41: 107–121
Furusawa, C, and Kaneko, K (2001) Theory of robustness of irreversible differentiation in a stem cell system: chaos hypothesis. *Journal of Theoretical Biology* 209: 395–416
Furusawa, C, and Kaneko, K (2012) A dynamical-systems view of stem cell biology. *Science* 338: 215–217

Garfield, E (1988) Mapping the world of epidemiology. Part 1. The disease detectives. Volume 11: 281–289, *Current Comments* 35: 3–11
Gender and Science Laboratory (2020) Our mission <https://www.genderscilab.org> (accessed 3/27/24)
Gerson, EM (2013) Integration of specialties: an institutional and organizational view. *Studies in History and Philosophy of Biological and Biomedical Sciences* 44: 515–524
Gerson, EM (2021) Biological research and engineering collaboration. Biological Engineering Collaboratory seminar (online). 5 October 2021.
Giere, R (1988) *Explaining Science: A Cognitive Approach.* Chicago: Chicago University Press
Giere, R (2004) How models are used to represent reality. *Philosophy of Science* 71: 742–752
Giere, R (2010) An agent-based conception of models and scientific representation. *Synthese* 172: 269–281
Gieryn, T (1999) *Cultural Boundaries of Science.* Chicago: University of Chicago Press
Gijsbers, V (2013) Understanding, explanation, and unification. *Studies in History and Philosophy of Science* 44: 516–522
Gilbert, M (1989) *On Social Facts.* Princeton, NJ: Princeton University Press
Gilbert, M (2000) Collective belief and scientific change. In: *Sociality and Responsibility: New Essays in Plural Subject Theory.* Lanham: Rowman & Littlefield, 37–49
Glennan, S (1996) Mechanisms and the nature of causation. *Erkenntnis* 44: 49–71
Glennan, S (2002) Rethinking mechanistic explanation. *Philosophy of Science* 69: S342–S353
Glennan, S (2017) *The New Mechanical Philosophy.* Oxford: Oxford University Press
Godfrey-Smith, P (2006) The strategy of model-based science. *Biology and Philosophy* 21:725–740
Goldman, A (2001) Experts: which ones should you trust? *Philosophy and Phenomenological Research* 63: 85–110
Goodwin, W (2013) Quantum chemistry and organic theory. *Philosophy of Science* 80: 1159–1169
Gorman, M (2010) (ed.) *Trading Zones and Interactional Expertise.* Cambridge, MA: MIT Press
Gottweis, H, Salter, B, and Waldby, C (2009) *The Global Politics of Human Embryonic Stem Cell Science.* Palgrave Macmillan
Green, S (2017) Introduction to philosophy of systems biology. In: Green (ed.) *Philosophy of Systems Biology.* Dordrecht: Springer, 1–23
Green, S (2022) Systems and synthetic biology. *Stanford Encyclopedia of Philosophy* (Summer 2022 Edition), Edward N Zalta (ed.), <https://plato.stanford.edu/archives/sum2022/entries/systems-synthetic-biology/>.
Green, S, Bechtel, W, and Levy, A (2015) Design sans adaptation. *European Journal for Philosophy of Science* 5: 15–29
Green, S, Fagan, M, and Jaeger, J (2015) Explanatory integration challenges in evolutionary systems biology. *Biological Theory* 9: 18–35
Griffiths, PE, and Stotz, KC (2007) Gene. In: Hull, DL, and Ruse, M (eds.) *The Cambridge Companion to the Philosophy of Biology.* Cambridge: Cambridge University Press, 85–123
Grimm, SR (2006) Is understanding a species of knowledge? *British Journal for the Philosophy of Science* 57: 515–553
Grimm, SR (2010) The goal of explanation. *Studies in History and Philosophy of Science* 41: 337–344
Grimm, SR (2012) The value of understanding. *Philosophy Compass* 7: 103–117
Grüne-Yanoff, T (2016) Interdisciplinary success without integration. *European Journal for Philosophy of Science* 6: 343–360
Grüne-Yanoff, T, and Mäki, U (2014) Introduction: Interdisciplinary model exchanges. *Studies in History and Philosophy of Science* 48: 52–59
Gupta, A, Lutolf, MP, Hughes, AJ, Sonnen, KF (2021) Bioengineering in vitro models of embryonic development. *Stem Cell Reports* 16: 1104–1116
Hackett, JA, and Surani, MA (2014) Regulatory principles of pluripotency: from the ground state up. *Cell Stem Cell* 15: 416–430
Hacking, I (1992) The self-vindication of the laboratory sciences. In: Pickering, A (ed.) *Science as Practice and Culture.* Chicago: University of Chicago Press, 29–64

Hauskeller, C, Manzeschke, A, Pichl, A (eds) (2019) *The Matrix of Stem Cell Research*. New York: Routledge
Havstad, J (in preparation) Values and incentives in science
Hazelwood, LF (2001) *Can't Live Without It: The Story of Hemoglobin in Sickness and in Health*. Nova Science Publications
Hempel, CG (1965) *Aspects of Scientific Explanation and Other Essays in the Philosophy of Science*. New York: Free Press
Hempel, CG (1966) *Philosophy of Natural Science*. Prentice Hall
Hempel CG, Oppenheim P (1948) Studies in the logic of explanation. *Philosophy of Science* 15:135–175. Reprinted in Hempel (1965), 245–290
Herrick, JB (1910) Peculiar elongated and sickle-shaped red blood corpuscles in a case of severe anemia. *Archives of Internal Medicine* 6: 517–521
Hill, RW, Wyse, GA, and Anderson, M (2016) *Animal Physiology*, 4th edition. Sinauer Associates
Hitchcock, C, and Woodward, J (2003) Explanatory generalizations II. *Noûs* 37: 181–199
Ho, A (2011) Trusting experts and epistemic humility in disability. *International Journal of Feminist Approaches in Bioethics* 4: 102–123
Ho, C, and Morsut, L (2021) Novel synthetic biology approaches for developmental systems. Stem Cell Reports 16: 1051–1064
Hochedlinger, K, and Plath, K (2009) Epigenetic reprogramming and induced pluripotency. *Development* 136: 509–523
Hochstein, E (2017) Why one model is never enough: a defense of explanatory holism. *Biology and Philosophy* 32: 1105–1125
Høffding, S, and Satne, G (2021) Interactive expertise in solo and joint musical performance. *Synthese* 198 (Suppl 1): S427–S445
Holbrook, JB (2013) What is interdisciplinary communication? *Synthese* 190:1865–1879
Huang, S (2009) Reprogramming cell fates: reconciling rarity with robustness. *Bioessays* 31: 546–560
Huang, S (2011) Systems biology of stem cells: three useful perspectives to help overcome the paradigm of linear pathways. *Philosophical Transactions of the Royal Society, Series B* 366: 2247–2259
Huang, S (2013) When peers are not peers and don't know it: the Dunning–Kruger effect and self-fulfilling prophecy in peer-review. *BioEssays* 35: 414–416
Huang, S, Ernberg, I, and Kauffman, S (2009) Cancer attractors: a systems view of tumors from a gene network dynamics and developmental perspective. *Seminars in Cell and Developmental Biology* 20: 869–876
Hull, D (1972) Reduction in genetics—biology or philosophy? *Philosophy of Science* 39: 491–499
Humphreys, P (1989) *The Chances of Explanation: Causal Explanation in the Social, Medical, and Physical Sciences*. Princeton, NJ: Princeton University Press
Humphreys, P (2000) *Analytic versus Synthetic Understanding*. Oxford. Oxford University Press
Huneman, P (2010) Topological explanations and robustness in biological sciences. *Synthese* 177: 213–245
Huneman, P (2018) Diversifying the picture of explanations in biological sciences: ways of combining topology with mechanisms. *Synthese* 195: 115–146
Huutoniemi, K, Klein, JT, Bruun, H, Hukkinen, J (2010) Analyzing interdisciplinarity: typology and indicators. *Research Policy* 39: 79–88
Ishida, Y (2020) In dialectical tension: realist and instrumentalist attitudes in scientific practice. *Synthese* 197: 2665–2694
Jacob, F (1965) Nobel Foundation, the Nobel Prize in Physiology or Medicine 1965. http://www.nobelprize.org/nobel_prizes/medicine/ laureates/1965/ press.html
Jacob, F, and Monod, J (1961) On the regulation of gene activity. *Cold Spring Harbor Symposia in Quantitative Biology* 26: 193–211
Jacobs, JA, and Frickel, S (2009) Interdisciplinarity: a critical assessment. *Annual Review of Sociology* 35: 43–65

Jaeger, J, Irons D, and Monk N (2012) The inheritance of process: a dynamical systems approach. *Journal of Experimental Zoology Series B (Mol. Dev. Evol.)* 318B: 591–612
Jaeger, J, and Sharpe, J (2014) On the concept of mechanism in development. In: Minelli, A, and Pradeu, T (eds) *Towards a Theory of Development*. Oxford: Oxford University Press, 56–78
Jankovic, M and Ludwig, K (2018) Introduction. In: Jankovic, M and Ludwig, K (eds.) *Routledge Handbook of Collective Intentionality*. New York: Routledge/Taylor & Francis, 1–8
Johannsen, W (1911) The genotype conception of heredity. *American Naturalist* 45: 129–159
Jones, N (2014) Bowtie structures, pathway diagrams, and topological explanation. *Erkenntnis* 79: 1135–1155
Judson, HF (1979) *The Eighth Day of Creation*. Cold Spring Harbor Laboratory Press
Kaneko, K (2011) Characterization of stem cells and cancer cells on the basis of gene expression profile stability, plasticity, and robustness. *BioEssays* 33: 403–413
Kaneko, K, and Yomo, T (1999) Isologous diversifcation for robust development of cell society. *Journal of Theoretical Biology* 99: 243–256
Kaplan, D, and Craver, C (2011) The explanatory force of dynamical and mathematical models in neuroscience: a mechanistic perspective. *Philosophy of Science* 78: 601–627
Kaplan, JM, and Valles, S (2021) Reflecting on what philosophy of epidemiology is and does, as the field comes into its own. Introduction to the Special Issue on Philosophy of Epidemiology. *Synthese* 198: 2383–2392
Kauffman, S (1969) Metabolic stability and epigenetics in randomly constructed genetic nets. *Journal of Theoretical Biology* 22: 437–467
Kauffman, SA (1976) Constraints on the sociobiologists' program. *Proceedings of the Biennial Meeting of the Philosophy of Science Association, Volume Two: Symposia and Invited Papers*, 32–47
Kauffman, S (1993) *The Origins of Order*. Oxford: Oxford University Press
Keller, EF (2000) Models of and models for: theory and practice in contemporary biology. *Philosophy of Science* 67 Supplement, Part II: Symposia Papers, S72–S86
Keller, EF (2001) *The Century of the Gene*. Cambridge, MA: Harvard University Press
Keller, EF (2002) *Making Sense of Life*. Cambridge, MA: Harvard University Press
Khalifa, K (2017) *Understanding, Explanation, and Scientific Knowledge*. Cambridge: Cambridge University Press
Khalifa, K (2020) Understanding, truth, and epistemic goals. *Philosophy of Science* 87: 944–956
Khalifa, K, Millson, J, and Risjord, M (2021) Inference, explanation, and asymmetry. *Synthese* 198 (Suppl. 4) S929–S953
Kibble, TWB (2015) Spontaneous symmetry breaking in gauge theories. *Philosophical Transactions of the Royal Society* A 373: 20140033. http://dx.doi.org/10.1098/rsta.2014.0033
Kimbrough, S (1979) On the reduction of genetics to molecular biology. *Philosophy of Science* 46: 389–406
King, A (2018) Explanatory models: a framework for instrumentalism. Presentation at Models and Simulations 8, University of South Carolina, March 16, 2018
Kitano, H (2001) (ed.) *Foundations of Systems Biology*. Cambridge, MA: MIT Press
Kitano, H (2002) Systems biology: a brief overview. *Science* 295: 1662–1664
Kitano, H (2004) Biological robustness. *Nature Reviews Genetics* 5: 826–837
Kitcher, P (1981) Explanatory unification. *Philosophy of Science* 48: 507–531
Kitcher, P (1984) 1953 and all that. A tale of two sciences. *Philosophical Review* 43: 335–371
Kitcher, P (1989) Explanatory unification and the causal structure of the world. In: Kitcher, P and Salmon, W (eds.) *Scientific Explanation*, Minneapolis: University of Minnesota Press, 410–505
Kitcher, P (1991) Socializing knowledge. *Journal of Philosophy* 88: 675–676
Klein, JT (1990) *Interdisciplinarity: History, Theory, and Practice*. Wayne State University Press
Klein, JT (2005) Interdisciplinary teamwork: the dynamics of collaboration and integration. In: Derry, SJ, Schunn, CD, and Gernsbacher, MA (eds.) *Interdisciplinary Collaboration: An Emerging Cognitive Science*. Mahwah, NJ: Lawrence Erlbaum Associates, 23–50
Klipp, E, Liebermeister, W, Wierling, C, Kowald, A, Lehrach H, and Herwig, R (2009) *Systems Biology: A Textbook*. Weinheim: Wiley-VCH

Knuuttila, T, and Loettgers, A (2013) Basic science through engineering? *Studies in History and Philosophy of Biological and Biomedical Sciences* 44: 158–169
Koskinen, I, and Mäki, U (2016) Extra-academic transdisciplinarity and scientific pluralism: What might they learn from one another? *European Journal of Philosophy of Science* 6: 419–444
Kostić, D (2019) Unifying the debates: mathematical and non-causal explanations. Introduction to Special Issue, *Perspectives on Science* 27: 1–6
Kraft, A (2009) Manhattan transfer: lethal radiation, bone marrow transplantation, and the birth of stem cell biology, ca. 1942–1961. *Historical Studies in the Natural Sciences* 39: 171–218
Kuhn, TS (1962) *The Structure of Scientific Revolutions*. Cambridge, MA: Harvard University Press
Kuhn, TS (2000) *The Road Since Structure*. Chicago: University of Chicago Press
Kuorikoski, J (2021) There are no mathematical explanations. *Philosophy of Science* 88: 189–212
Kuorikoski, J, and Marchionni, C (2014) Unification and mechanistic detail as drivers of model construction. *Studies in History and Philosophy of Science* 48: 97–104
Kutz, C (2000) Acting together. *Philosophy and Phenomenological Research* 61: 1–31
Kvanvig, JL (2003) *The Value of Knowledge and the Pursuit of Understanding*. Cambridge: Cambridge University Press
Lancaster, MA, and Knoblich JA (2014) Organogenesis in a dish. *Science* 345: 1247125-1-8
Landecker, H (2007) *Culturing Life*. Cambridge, MA: Harvard University Press
Lange, M (2016) *Because without Cause: Non-Causal Explanations in Science and Mathematics*. Oxford: Oxford University Press
Langmuir, AD (1976) William Farr: founder of modern concepts of surveillance. *International Journal of Epidemiology* 5: 13–18
Last JM (2001) (ed.) *Dictionary of Epidemiology*, 4th edition. Oxford: Oxford University Press
Leach, M, and Schoones, I (2013) The social and political lives of zoonotic disease models: narratives, science and policy. *Social Science & Medicine* 88: 10–17
Leavy, P (2012) *Essentials of Transdisciplinary Research*. Walnut Creek, CA, Left Coast Press
Le Bihan, S (2016) Enlightening falsehoods: a modal view of scientific understanding. In: Grimm, SR, Baumberger, C, and Ammon, S (eds.) *Explaining Understanding*. New York: Routledge/Taylor & Francis, 127–152
Ledford, H (2015) Team science. *Nature* (Special Issue: Interdisciplinarity) 525: 308–311
Lehrer, K (1990) *Theory of Knowledge*. Boulder: Westview Press
Lenhard, J (2006) Surprised by a nanowire. *Philosophy of Science* 73: 605–616
Lenhard, J, and Winsberg, E (2010) Holism, entrenchment, and the future of climate model pluralism. *Studies in History and Philosophy of Modern Physics* 41: 253–262
León, F, Szanto, T, and Zahavi, D (2019) Emotional sharing and the extended mind. *Synthese* 196:4847–4867
Leonelli, S (2016) *Data-Centric Biology: A Philosophical Study*. Chicago: University of Chicago Press
Levy, A, and Bechtel, W (2013) Abstraction and the organization of mechanisms. *Philosophy of Science* 80: 241–261
Li, M, and Izpisua Belmonte, JC (2018) Deconstructing the pluripotency gene regulatory network. *Nature Cell Biology* 20: 382–392
Lilienfeld, AM & Lilienfeld, DE (1980) *Threads of Epidemiologic History. Foundations of Epidemiology*. Oxford: Oxford University Press
List, C and Pettit, P (2006) Group agency and supervenience. *Southern Journal of Philosophy* 44: 85–105
List, C, and Pettit, P (2011) *Group Agency*. Oxford: Oxford University Press
Lloyd, EA (1988) *The Structure and Confirmation of Evolutionary Theory*. Princeton, NJ: Princeton University Press
Lloyd, EA (1994) Normality and variation: the Human Genome Project and the ideal human type. In: Cranor, CF (ed.) *Are Genes Us?* Rutgers University Press, 99–112
Lloyd, EA, and Anderson, CG (1993) Empiricism, objectivity, and explanation. *Midwest Studies in Philosophy* 18: 121–131
Longino, HE (1990) *Science as Social Knowledge*. Princeton, NJ: Princeton University Press

Longino, HE (2002) *The Fate of Knowledge.* Princeton, NJ: Princeton University Press
Longino, HE (2013) *Studying Human Behavior.* Chicago: University of Chicago Press
Löwy, I (1992) The strength of loose concepts—boundary concepts, federative experimental strategies and disciplinary growth: the case of immunology. *History of Science* 30: 371–396
Ludwig, K (2007) Collective intentional behavior from the standpoint of semantics. *Noûs* 41: 355–393
Ludwig, K (2016) *From Individual to Plural Agency. Collective Action, Volume 1.* Oxford: Oxford University Press
McCoy, CD, and Massimi, M (2018) Simplified models: a different perspective on models as mediators. *European Journal of Philosophy of Science* 8: 99–123
Machamer, P, Darden, L, and Craver, C (2000) Thinking about mechanisms. *Philosophy of Science* 57: 1–25
MacLeod, M (2018) What makes interdisciplinarity difficult? Some consequences of domain specificity in interdisciplinary practice. *Synthese* 195: 697–720
MacLeod, M, and Nagatsu, M (2016) Model coupling in resource economics: conditions for effective interdisciplinary collaboration. *Philosophy of Science* 83: 412–433
MacLeod, M, and Nersessian, N (2013a) Coupling simulation and experiment: the bimodal strategy in integrative systems biology. *Studies in History and Philosophy of Biological and Biomedical Sciences* 44: 572–584
MacLeod, M, and Nersessian, N (2013b) Building simulations from the ground up: modeling and theory in systems biology. *Philosophy of Science* 80: 533–556
MacLeod, M, and Nersessian, N (2014) Strategies for coordinating experimentation and modeling in integrative systems biology. *Journal of Experimental Zoology (Molecular and Developmental Evolution)* 322B: 230–239
MacLeod, M, and Nersessian, N (2015) Modeling systems-level dynamics: understanding without mechanistic explanation in integrative systems biology. *Studies in History and Philosophy of Biological and Biomedical Sciences* 49: 1–11
MacLeod, M, and Nersessian, N (2016) Interdisciplinary problem-solving: emerging modes in integrative systems biology. *European Journal of Philosophy of Science* 6: 401–418
MacLeod, M, and Nersessian, N (2019) Mesoscopic modeling as a cognitive strategy for handling complex biological systems. *Studies in History and Philosophy of Biological and Biomedical Sciences* 78: 101201
Magnello, E, and Hardy, A (2004) *Statistical Methods in Epidemiology.* Berlin: Birkhauser Verlag
Maherali, N, and Hochedlinger, K (2008) Guidelines and techniques for the generation of induced pluripotent stem cells. *Cell Stem Cell* 3: 595–605
Mäki, U (2013) Scientific imperialism: difficulties in definition, identification, and assessment. *International Studies in the Philosophy of Science* 27: 325–339
Mäki, U (2016) Philosophy of interdisciplinarity. What? Why? How? *European Journal for Philosophy of Science* 6: 327–342
Mantzavinos, C (2016) *Explanatory Pluralism.* Cambridge: Cambridge University Press
Marchionni, C (2008) Explanatory pluralism and complementarity. *Philosophy of the Social Sciences* 38: 314–333
Marchionni, C (2013) Playing with networks: how economists explain. *European Journal for Philosophy of Science* 3: 331–352
Martens, J (2018) Exploring the relation between the sense of the other and the sense of us: for agency cognition, emergent coordination, and the sense of agency. *Journal of Social Philosophy* 49: 38–60
Massachusetts Institute of Technology (2016) *Convergence: The Future of Health.* Cambridge, MA
Mattila, E (2005) Interdisciplinarity in the making: modelling infectious diseases. *Perspectives on Science* 13: 531–553
Matzinger, P (1994) Tolerance, danger, and the extended family. *Annual Review of Immunology* 12: 991–1045
Mazumdar, PMH (1989) *Immunology 1930–1980: Essays on the History of Immunology.* Toronto: Wall and Thompson

Mekios, C (2017) From biological research to a philosophy of systems biology. In: Green, S (ed.), *Philosophy of Systems Biology*. Dordrecht: Springer, 193–204

Melton, D (2013) Stemness: definitions, criteria, and standards. In: Lanza, R and Atala, A (eds.) *Essentials of Stem Biology*, 3rd edition. San Diego: Academic Press, 7–17

Merrill, RM (2019) *Introduction to Epidemiology*. Jones & Bartlett Learning

Mesarovic, MD, Sreenath, SN, and Keene, JD (2004) Search for organising principles: understanding in systems biology. *Systems Biology* 1: 19–27

Millgram, E (2015) *The Great Endarkment*. Oxford: Oxford University Press

Mills, CW (2005) "Ideal theory" as ideology. *Hypatia* 20: 165–184

Mitchell, S (2002) Integrative pluralism. *Biology and Philosophy* 17: 55–70

Mitchell, S (2009) *Unsimple Truths*. Chicago: University of Chicago Press

Mitchell, S, Daston, L, Gigerenzer, G, Sesardic, N, and Sloep, PB (1997) The whys and hows of interdisciplinarity. In: Weingert, P, Mitchell, SD, Richerson, PJ, and Maasen, S (eds.) *Human by Nature: Between Biology and the Social Sciences*. Mahwah, NJ: Lawrence Erlbaum Associates, 103–150

Mitchell, S, and Gronenborn, A (2017) After fifty years, why are protein x-ray crystallographers still in business? *British Journal for the Philosophy of Science* 68: 703–723

Morabia, A (2004) (ed.) *A History of Epidemiological Methods and Concepts*. Basel: Birkhäuser Verlag

Moris, N, Alev, C, Pera, M, and Arias, AM (2021) Biomedical and societal impacts of in vitro embryo models of mammalian development. *Stem Cell Reports* 16: 1021–1030

Morrison, M (2011) One phenomenon, many models: Inconsistency and complementarity. *Studies in History and Philosophy of Science* Part A 42: 342–351

Morrison, M (2015) *Reconstructing Reality: Models, Mathematics, and Simulations*. Oxford: Oxford University Press

Morgan, MS and Morrison, M (1999). Models as mediating instruments. In: Morgan, MS, and Morrison, M (eds.) *Models as Mediators*. Cambridge: Cambridge University Press, 10–37

Morrison, MA (2021) A socio-epistemic theory of climate model development. Ph.D. thesis, Indiana University, Bloomington

Nagel, E (1961) *The Structure of Science*. New York: Harcourt, Brace & World

National Academy of Sciences (2006) Facilitating interdisciplinary research. US National Science Foundation Report

National Science Foundation (2008), Impact of transformative interdisciplinary research and graduate education on academic institutions. US National Science Foundation, Workshop Report

National Science Foundation (2016) Solicitation NSF 19-551: call for convergence research. US National Science Foundation

Nature Editorial Board (2015) Special issue: interdisciplinarity. *Nature* 525: 305

Nersessian, N (2006) The cognitive-cultural systems of the research laboratory. *Organization Studies* 27: 125–145

Nersessian, N (2017) Systems biology modeling practices: reflections of a philosopher ethnographer. In: Green, S (ed.) *Philosophy of Systems Biology*. Dordrecht: Springer, 215–225

Nersessian, N (2019) Interdisciplinarities in action: cognitive ethnography of bioengineering sciences research laboratories. *Perspectives on Science* 27: 553–581

Nersessian, N (2022) *Interdisciplinarity in the Making*. Cambridge, MA: MIT Press

Neto, C (2019) What is a lineage? *Philosophy of Science* 86: 1099–1110

Newman, M (2010) *Networks: An Introduction*. Oxford: Oxford University Press

Niiniluoto, I (2003) Science as collective knowledge. In: Sintonen, M, Ylikoski, P, and Miller, K (eds.) *Realism in Action*. Kluwer, 269–278

O'Connor, C, and Weatherall, JO (2016) Black holes, black-scholes, and prairie voles: an essay review of *Simulation and Similarity*, by Michael Weisberg. *Philosophy of Science* 83: 613–626

O'Connor, C, and Weatherall, JO (2019) *The Misinformation Age*. Yale University Press

Odenbaugh, J (2021) Models, models, models: a deflationary view. *Synthese* 198: 1–16

O'Donnell, AM, and Derry, SJ (2005) Cognitive processes in interdisciplinary groups: problems and possibilities. In: Derry, SJ, Schunn, CD, and Gernsbacher, MA (eds.) *Interdisciplinary Collaboration: An Emerging Cognitive Science*, Mahwah, NJ: Lawrence Erlbaum Associates, 51–82

Olsson, E (2017) Coherentist theories of epistemic justification. *Stanford Encyclopedia of Philosophy* (Spring 2017 Edition), Edward N Zalta (ed.), <https://plato.stanford.edu/archives/spr2017/entries/justep-coherence/>

O'Malley, M (2013) When integration fails: Prokaryote phylogeny and the tree of life. *Studies in History and Philosophy of Biological and Biomedical Sciences* 44: 551–562

O'Malley, M (2014) *Philosophy of Microbiology*. Cambridge: Cambridge University Press

O'Malley, M, and Dupré, J (2005) Fundamental issues in systems biology. *BioEssays* 27:1270–1276

O'Malley, M, and Soyer, O (2012) The roles of integration in molecular systems biology. *Studies in History and Philosophy of Biological and Biomedical Sciences* 43: 58–68

Oppenheim, P, and Putnam, H (1958) The unity of science as a working hypothesis. In: G. Maxwell, G, Feigl, H, and Scriven, M (eds.) *Concepts, Theories, and the Mind–Body Problem*. Minnesota University Press, 3–36

O'Rourke, M, and Crowley, S (2013) Philosophical intervention and cross-disciplinary science: the story of the Toolbox Project. *Synthese* 190: 1937–1954

O'Rourke, M, Crowley, S, and Gonnerman, C (2016) On the nature of cross-disciplinary integration: A philosophical framework. *Studies in History and Philosophy of Biological and Biomedical Sciences* 56: 62–70

Palermos, SO (2022) Epistemic collaborations: distributed cognition and virtue reliabilism. *Erkenntnis* 87: 1481–1500

Paneth, N, Susser, E, and Susser, M (2004) Origins and early development of the case-control study. In: Morabia, A (ed.) *A History of Epidemiological Methods and Concepts*. Basel; Birkhäuser Verlag, 291–311

Parker, W (2020) Model evaluation: an adequacy-for-purpose view. Philosophy of Science 87: 457–477

Paul, WE (1983) Preface to Volume 1. *Annual Review of Immunology* 1: vii

Paul, WE (2012) *Fundamental Immunology*, 7th edition. Wolters Kluwer

Pauling, L, Itano, HA, Singer, SJ, and Wells, IC (1949) Sickle cell anemia, a molecular disease. *Science* 110: 543–548

Peirce, CS (1878) How to make our ideas clear. *Popular Science* 12: 286–302

Pera, MF, de Wert, G, Dondorp, W, Lovell-Badge, R, Mummery, CL, Munsie, M, and Tam, PP (2015) What if stem cells turn into embryos in a dish? *Nature Methods* 12: 917–919

Petrie, HW (1976) Do you see what I see? The epistemology of interdisciplinary inquiry. *Educational Researcher* 5: 9–15

Pettit, P, and Schweikard, D (2006) Joint actions and group agents. *Philosophy of the Social Sciences* 36: 18–39

Pincock (2018) Explanatory relevance and contrastive explanation. *Philosophy of Science* 85: 806–818

Porta, M, and Alvarez-Dardet, C (1998) Epidemiology: bridges over (and across) roaring levels. *Journal of Epidemiology and Community Health* 52: 605

Porter, AL, and Rafols, I (2009) Is science becoming more interdisciplinary? Measuring and mapping six research fields over time. *Scientometrics* 81: 719–745

Posfai, E, Lanner, F, Mulas, C, and Leitch, HG (2021) All models are wrong, but some are useful: Establishing standards for stem cell-based embryo models. *Stem Cell Reports* 16: 1117–1141

Poston, T (2014) *Reason and Explanation: A Defense of Explanatory Coherentism*. New York: Palgrave Macmillan

Potochnik, A (2017) *Idealization and the Aims of Science*. Chicago: University of Chicago Press

Potochnik, A, and de Oliveira, GS (2020) Patterns in cognitive phenomena and pluralism of explanatory styles. *Topics in Cognitive Science* 12: 1306–1320

Povich, M (2018) Minimal models and the generalized ontic conception of scientific explanation. *British Journal for the Philosophy of Science* 69: 117–137
Pradeu T (2012) *The Limits of the Self: Immunology and Biological Identity* (trans. Elizabeth Vitanza). Oxford: Oxford University Press
Pradeu T (2019) *Element: Philosophy of Immunology*. Cambridge: Cambridge University Press
Provine, W (1971) *The Origin of Theoretical Population Genetics*. Chicago: University of Chicago Press
Punt, J, Stranford, S, Jones, P, and Owen, J (2018) *Kuby Immunology*, 8th edition. New York: WH Freeman
Putnam, H (1981) *Reason, Truth, and History*. Cambridge: Cambridge University Press
Quine, W and Ullian, J (1970) *The Web of Belief*. New York: Random House
Rabinow, P, and Bennett G (2012) *Designing Human Practices: An Experiment with Synthetic Biology*. Chicago: University of Chicago Press
Rackham, O, Cahah, P, Mah, N, Morris, S, Ouyang, JF, Plant, AL, Tanaka, Y, and Wells, CA (2021) Challenges for computational stem cell biology: a discussion for the field. *Stem Cell Reports* 16: 3–9
Reutlinger, A (2016) Is there a monist theory of causal and non-causal explanations? *Philosophy of Science* 83 733–745
Reutlinger, A (2018) Extending the counterfactual account of explanation. In: Reutlinger and Saatsi (eds.) *Explanation Beyond Causation*, 74–95
Reutlinger, A, and Saatsi, J (eds.) (2018) *Explanation Beyond Causation*. Oxford: Oxford University Press
Reynolds, AS (2018) *The Third Lens: Metaphor and the Creation of Modern Cell Biology*. Chicago: University of Chicago Press
Rhodes, R (2012) *The Making of the Atomic Bomb*, 25th Anniversary Edition. Simon & Schuster
Rice, C (2015) Moving beyond causes: optimality models and scientific explanation. *Noûs* 49: 589–615
Rice, C (2018) Idealized models, holistic distortion, and universality. *Synthese* 195: 2795–2819
Rice, C (2021) *Leveraging Distortions*. Cambridge, MA: MIT Press
Rice, C, and Rohwer, Y (2021) How to reconcile a unified account of explanation with explanatory diversity. *Foundations of Science* 26: 1025–1047
Richardson, HS (1997) *Practical Reasoning about Final Ends*. Cambridge: Cambridge University Press
Richardson, RC (1997) Natural and artificial complexity. *Philosophy of Science* 64: S255–S267
Rivron, N, and Fu, J (2021a) It takes a village to form embryo models. *Stem Cell Reports* 16: 1011–1013
Rivron, N, and Fu, J (2021b) Snapshot: embryo models. *Stem Cell Reports* 16: 1142e1
Robinson, B, Gonnerman, C, and O'Rourke, M (2019) Experimental philosophy of science and philosophical differences across the sciences. *Philosophy of Science* 86: 551 576
Rogers, Y, Scaife, M, Rizzo, A (2005) Interdisciplinarity: an emergent or engineered process? In: Derry, SJ, Schunn, CD, and Gernsbacher, MA (eds.) *Interdisciplinary Collaboration: An Emerging Cognitive Science*. Mahwah, NJ: Lawrence Erlbaum Associates, 265–285
Rolin, K (2008) Science as collective knowledge. *Cognitive Systems Research* 9: 115–124
Rolin, K (2015) Values in science: the case of scientific collaboration. *Philosophy of Science* 82: 157–177
Ross, L (2015) Dynamical models and explanation in neuroscience. *Philosophy of Science* 82: 32–54
Ross, L (2021) Distinguishing topological and causal explanation. *Synthese* 198: 9803–9820
Roth, Abraham Sesshu, "Shared Agency," *Stanford Encyclopedia of Philosophy* (Summer 2017 Edition), Edward N Zalta (ed.), <https://plato.stanford.edu/archives/sum2017/entries/shared-agency/>
Roy, O, and Schwenkenbecher, A (2021) Shared intentions, loose groups, and pooled knowledge. *Synthese* 198: 4523–4541
Salmon, W (1971) *Statistical Explanation and Statistical Relevance*. University of Pittsburgh Press

Salmon, W (1978) Why ask 'Why?'? An inquiry concerning scientific explanation. *Proceedings and Addresses of the American Philosophical Association* 51: 683–705

Salmon, W (1984) *Scientific Explanation and the Causal Structure of the World.* Princeton, NJ: Princeton University Press

Salmon, W (1989) *Four Decades of Scientific Explanation.* University of Minnesota Press

Satne, G (2021) Understanding others by doing things together: an enactive account. *Synthese* 198 (Suppl 1): S507–S528

Schaffer, LV, and Ideker, T (2021) Mapping the multiscale structure of biological systems. *Cell Systems* 12: 622–635

Schaffner, K (1969) Chemical systems and chemical evolution: the philosophy of molecular biology. *American Scientist* 57: 410–420

Schaffner, K (1974) The peripherality of reductionism in the development of molecular biology. *Journal of the History of Biology* 7: 111–139

Schaffner, K (2016) *Behaving: What's Genetic, What's Not, and Why Should We Care*? Oxford: Oxford University Press

Schmid, HB (2014) Plural self-awareness. *Phenomenology and the Cognitive Sciences* 13: 7–24

Schmid, HB (2016) On knowing what we're doing together. In: Brady, M and Fricker, M (eds.) *The Epistemic Life of Groups.* Oxford: Oxford University Press, 51–72

Schmitz, M (2018) Co-subjective consciousness constitutes collectives. *Journal of Social Philosophy* 49: 137–160

Schweikard, David P, and Schmid, Hans Bernhard (2013) Collective Intentionality. *Stanford Encyclopedia of Philosophy* (Summer 2013 Edition), Edward N Zalta (ed.), <https://plato.stanford.edu/archives/sum2013/entries/collective-intentionality/>

Scientific American (1993) Special Issue: Life, death and the immune system. *Scientific American* 269(3)

Searle, J (1990) *The Construction of Social Reality.* New York: Free Press

Sette, A, and Crotty, S (2021) Adaptive immunity to SARS-CoV-2 and COVID-19. *Cell* 184: 861–880

Shrum, W, Genuth, J, Carlson, WB, and Chompalov, I (2007) *Structures of Scientific Collaboration.* Cambridge, MA: MIT Press

Silverstein, AM (1989) *A History of Immunology.* San Diego: Academic Press

Skow, B (2015) Are there genuine physical explanations of mathematical phenomena? *British Journal for the Philosophy of Science* 66: 1–25

Solomon, M (2001) *Social Empiricism.* Cambridge, MA: MIT Press

Stadtfeld, M, and Hochedlinger, K (2010) Induced pluripotency: history, mechanisms, and applications. *Genes & Development* 24: 2239–2263

Steinberg, MH, Forget, BG, Higgs, DR, and Weatherall, DJ (eds.) (2009) *Disorders of Hemoglobin: Genetics, Pathophysiology, and Clinical Management.* Cambridge: Cambridge University Press

Strevens, M (2008) *Depth.* Cambridge, MA: Harvard University Press

Strevens, M (2013) No understanding without explanation. *Studies in History and Philosophy of Science* 44: 510–515

Strogatz, SH (2000) *Nonlinear Dynamics and Chaos: with Applications to Physics, Biology, Chemistry and Engineering.* New York: Perseus Books

Stryer, L (1988) *Biochemistry*, 3rd edition. WH Freeman

Suárez, M (2004) An inferential conception of scientific representation. *Philosophy of Science* 71: 767–779

Susser, M (1973) *Causal Thinking in the Health Sciences: Concepts and Strategies of Epidemiology.* Oxford: Oxford University Press

Swiatczak, B, and Tauber, AI (2020) Philosophy of Immunology. *Stanford Encyclopedia of Philosophy* (Summer 2020 Edition), Edward N Zalta (ed.), <https://plato.stanford.edu/archives/sum2020/entries/immunology/>.

Takahashi, K, and Yamanaka, S (2006) Induction of pluripotent stem cells from mouse embryonic and adult fibroblast cultures by defined factors. *Cell* 126: 663–676

Tahko, TE (2020) Where do you get your protein? Or, biochemical realization. *British Journal for the Philosophy of Science* 71: 799–825

Tauber AI (1994) *The Immune Self: Theory or Metaphor?* Cambridge: Cambridge University Press

Tauber, AI (2017) *Immunity: The Evolution of an Idea*. Oxford: Oxford University Press

Teller, P (2001) Twilight of the perfect model model. *Erkenntnis* 55: 393–415

Thagard, P (1997) Collaborative knowledge. *Noûs* 31: 242–61

Theiner, G (2018) Groups as distributed cognitive systems. In: Jankovic, M, and Ludwig, K (eds.) *The Routledge Handbook of Collective Intentionality*. Routledge, 233–248

Thibodeau, GA, and Patton, KT (2007) *Anatomy and Physiology*, 6th edition. Elsevier

Thomson, J, Itskovitz-Eldor, J, Shapiro, S, Waknitz, M, Swiergel, J, Marshall, V, and Jones, J (1998) Embryonic stem cell lines derived from human blastocysts. *Science* 282: 1145–1147

Tollefsen, D (2002) Collective intentionality and the social sciences. *Philosophy of the Social Sciences* 32: 25–50

Tollefsen, D (2015) *Groups as Agents*. John Wiley & Sons

Tollefsen, D, and Gallagher, S (2017) We-narratives and the stability and depth of shared agency. *Philosophy of the Social Sciences* 47: 95–110

Tomasello, M (2009) *Why We Cooperate*. Cambridge, MA: MIT Press

Toon, A (2015) Where is the understanding? *Synthese* 192: 3859–3875

Tulodziecki, D (2021) How (not) to think about theory-change in epidemiology. *Synthese* 198: 2569–2588

Tuomela, R (2005) We-intentions revisited. *Philosophical Studies* 125: 327–269

University of Bielefeld, Centre for Interdisciplinary Research (ZiF) <https://www.uni-bielefeld.de/einrichtungen/zif/index.xml>

Valles, S (2018) *Philosophy of Population Health*. New York: Routledge/Taylor & Francis

van Fraassen, B (1977) The pragmatics of explanation. *American Philosophical Quarterly* 14: 143–150

van Fraassen, B (1980) *The Scientific Image*. Oxford: Oxford University Press

van Fraassen, B (2008) *Scientific Representation: Paradoxes of Perspective*. Oxford: Oxford University Press

van Noorden, R (2015) Interdisciplinary research by the numbers. *Nature* (Special Issue: Interdisciplinarity) 525: 306–307

Waddington, CH (1957) *The Strategy of the Genes*. London: Taylor & Francis

Warmflash, A, Sorre, B, Etoc, F, Siggia, ED, Brivanlou, AH (2014) A method to recapitulate early embryonic spatial patterning in human embryonic stem cells. *Nature Methods* 11: 847–854

Waters, CK (2007) Causes that make a difference. *Journal of Philosophy* 104: 551–579

Watson and Crick (1953) Molecular structure of nucleic acids. *Nature* 171: 737–738

Weingart, P (2010) A short history of knowledge formations. In: Frodeman, R (ed.) *The Oxford Handbook of Interdisciplinarity*. Oxford: Oxford University Press, 3–14

Weisberg, M (2013) *Simulation and Similarity*. Oxford: Oxford University Press

Westerhoff, HV, and Palsson, BO (2004) The evolution of molecular biology into systems biology. *Nature Biotechnology* 22: 1249–1252

Whewell, W (1834) Review of Mrs. Somerville's *On the Connexion of the Physical Sciences*. *Quarterly Review* 51: 58–61

Wilkenfeld, DA (2013) Understanding as representation manipulability. *Synthese* 190: 997–1016

Wilkenfeld, DA (2017) MUDdy understanding. *Synthese* 194: 1273–1293

Woodward, J (2002) What is a mechanism? A counterfactual account. *Philosophy of Science* 69: S366–S377

Woodward, J (2003) *Making Things Happen: A Theory of Causal Explanation*. New York: Oxford University Press

Woodward, J (2010) Causation in biology: stability, specificity, and the choice of levels of explanation. *Biology and Philosophy* 25: 287–318

Woodward, J (2013) Mechanistic explanation: its scope and limits. *Proceedings of the Aristotelian Society Supplementary Volume* 87: 39–65

Woodward, J (2019) "Scientific Explanation," *Stanford Encyclopedia of Philosophy* (Winter 2019 Edition), Edward N Zalta (ed.), <https://plato.stanford.edu/archives/win2019/entries/scientific-explanation/>

Woodward, James and Ross, Lauren (2021), "Scientific Explanation," *Stanford Encyclopedia of Philosophy* (Summer 2021 Edition), Edward N Zalta (ed.), <https://plato.stanford.edu/archives/sum2021/entries/scientific-explanation/>

Woody, A (2015) Re-orienting discussions of scientific explanation: a functional perspective. *Studies in History and Philosophy of Science Part A* 52: 79–87

Wray, B (2002) The epistemic significance of collaborative research. *Philosophy of Science* 69: 150–168

Wray, B (2003) Is science really a young man's game? *Social Studies of Science* 33: 137–149

Wray, B (2007) Who has scientific knowledge? *Social Epistemology* 21: 337–347

Wylie, A (1999) Rethinking unity as a "working hypothesis" for philosophy. *Perspectives on Science* 7: 293–317

Wylie, A (2003) Why standpoint theory matters: feminist standpoint theory. In: Figueroa, RM, and Harding, S (eds.) *Philosophical Explorations of Science, Technology, and Diversity.* Routledge, 26–48

Wylie, A (2015) A plurality of pluralisms: collaborative practice in archaeology. In: Padovani, F, Richardson, A, Tsou, JY (eds.) *Objectivity in Science.* Dordrecht: Springer, 189–210

Yilmaz, Özlem (2017) Causation and explanation in phenotype research. *Balkan Journal of Philosophy* 9: 112–129

Yin, X, Mead, BE, Safaee, H, Langer, R, Karp, JM, Levy, O (2016) Engineering stem cell organoids. *Cell Stem Cell* 18: 25–38

Yong, E (2020) Immunology is where intuition goes to die. *The Atlantic*, August 5

Zach, M, and Greslehner, G (2023) Understanding immunity: an alternative framework beyond defense and strength. *Biology and Philosophy* 38: 7 <https://doi.org/10.1007/s10539-023-09893-2>

Zagzebski, L (2001) Recovering understanding. In: Steup, M (ed.) *Knowledge, Truth, and Duty.* Oxford: Oxford University Press, 235–251

Zollman, K (2007) The communication structure of epistemic communities. *Philosophy of Science* 74: 574–587

Index

For the benefit of digital users, indexed terms that span two pages (e.g., 52–53) may, on occasion, appear on only one of those pages.